METHODS IN MOLECULAR BIOLOGY

For further volumes:
http://www.springer.com/series/7651

For over 35 years, biological scientists have come to rely on the research protocols and methodologies in the critically acclaimed *Methods in Molecular Biology* series. The series was the first to introduce the step-by-step protocols approach that has become the standard in all biomedical protocol publishing. Each protocol is provided in readily-reproducible step-by-step fashion, opening with an introductory overview, a list of the materials and reagents needed to complete the experiment, and followed by a detailed procedure that is supported with a helpful notes section offering tips and tricks of the trade as well as troubleshooting advice. These hallmark features were introduced by series editor Dr. John Walker and constitute the key ingredient in each and every volume of the *Methods in Molecular Biology* series. Tested and trusted, comprehensive and reliable, all protocols from the series are indexed in PubMed.

Purinergic Signaling

Methods and Protocols

Edited by

Pablo Pelegrín

BioMedical Research Institute of Murcia (IMIB), Clinical University Hospital Virgen de la Arrixaca, Murcia, Spain

Humana Press

Editor
Pablo Pelegrín
BioMedical Research Institute
of Murcia (IMIB)
Clinical University Hospital Virgen de la Arrixaca
Murcia, Spain

ISSN 1064-3745 ISSN 1940-6029 (electronic)
Methods in Molecular Biology
ISBN 978-1-4939-9716-9 ISBN 978-1-4939-9717-6 (eBook)
https://doi.org/10.1007/978-1-4939-9717-6

This Humana imprint is published by the registered company Springer Science+Business Media, LLC, part of Springer Nature.
The registered company address is: 233 Spring Street, New York, NY 10013, U.S.A.

Preface

The biological actions of purine nucleotides and nucleosides signaling in the extracellular milieu communicating different cells of the organism have been recognized since 1929. However, it was not until the 1970s that the term "purinergic signaling" was proposed and specific receptors for extracellular nucleotides were characterized. The receptors for nucleosides have been referred as P1 purinoceptors, whereas those receptors for nucleotides have been termed P2 purinoceptors. P2 receptors can be further sub-classified into two major families, P2X and P2Y. The P2X receptors are ligand-gated ion channels selective for cations, whereas the P2Y as well as the P1 receptors are G protein-coupled receptors. Extracellular nucleotides and nucleosides are also degraded by the action of several ectonucleotidases, which rapidly break down ATP to ADP, AMP, and adenosine. The specific expression patterns of purinoceptors and ectonucleotidases in different tissues and infiltrating immune cells exquisitely shape the response to extracellular nucleotides and nucleosides in different physiological and pathophysiological situations. Currently, the functions of purinergic signaling are recognized in almost every organ and system of the organism and drugs targeting purinergic receptors are already used in clinics. Developing new drugs for specific purinergic receptors acting as potent antagonists or agonists is a priority for different pharma companies.

This volume aims to cover all major methodological aspects of research into purinergic signaling and to provide a foundation for studying them at molecular, biochemical, pharmacological, and physiological levels. This is the need that this book aims to satisfy. In the first chapter, Geoffrey Burnstock provides a succinct albeit authoritative appraisal of the historic perspective and current knowledge of purinergic signaling. In the second chapter, Dariusz C. Górecki reports on the current knock-out and knock-in mouse models used to study purinergic signaling, which is followed by a comprehensive review prepared by Christa E. Müller on the current pharmacological weapons (agonists and antagonists) for purinergic receptors. Then, Ralf Schmid presents a chapter covering the methods for in silico modeling of the P2X receptor structure. Maria Jose da Silva Fernandes describes how to knock down purinoceptor expression, and the next five chapters, by Karine Massé, Gennady G. Yegutkin, Friedrich Koch-Nolte, Mark T. Young, and Vincent Compan, cover classic biochemical, microscopy and cellular techniques for detecting purinoceptor and ectonucleotidase expression at cellular or tissue level. Vincent Compan, François Rassendren, and Leigh A. Stoddart present two chapters on novel methods that use bioluminescence resonance energy transfer to study multimeric purinoceptors and antagonist binding. Then, the following five chapters by Francesco Di Virgilio, Tobias Engel, Kirstan A. Vessey, Friedrich Haag, and Miguel Díaz-Hernández cover the detection of the prototypic agonist of P2 receptors, ATP, in the extracellular milieu by reporting different techniques that include bioluminescence, Förster resonance energy transfer, and enzyme-based biosensors. Eric Boué-Grabot, Lin-Hua Jiang, Sébastien Roger, Leanne Stokes, and Carol J. Milligan detail the protocols for recording P2X receptor electrophysiology by applying the patch-clamp technique. Then, Liam E. Browne describes the procedure for controlling P2X receptors by optogenetics, inducing the rapid and reversible opening and closing of the receptor with light. Maria Teresa Miras-Portugal, with Felipe Ortega, Javier Gualix, Raquel Perez-Sen, Esmerilda G. Delicado, and Rosa Gomez-Villafuertes describe the protocol for studying one of the most common consequences of purinoceptor activation, namely the increase in intracellular calcium, and James S. Wiley presents a protocol to measure purinoceptor pore dilation. Finally, three chapters by Helios Martínez-Banaclocha, Michael R. Elliott, and Pablo Pelegrín describe protocols for studying cellular events that occur after purinoceptor activation, such as inflammasome activation, leukocyte migration, and cell adhesion.

The hope is that this book will provide a sound basis for molecular, cellular, and physiological research into purinergic signaling in health and disease and will spark interest in this fascinating signaling process among researchers in many other different and unrelated disciplines.

Murcia, Spain ***Pablo Pelegrín***

Contents

Preface .... *v*
Contributors .... *ix*

1 Introduction to Purinergic Signaling 1
Geoffrey Burnstock

2 Knockout and Knock-in Mouse Models to Study Purinergic Signaling 17
Robin M. H. Rumney and Dariusz C. Górecki

3 Agonists and Antagonists for Purinergic Receptors 45
Christa E. Müller, Younis Baqi, and Vigneshwaran Namasivayam

4 Homology Modeling of P2X Receptors 65
Anastasios Stavrou, Sudad Dayl, and Ralf Schmid

5 Using RNA Interference for Purinoceptor Knockdown In Vivo 77
Rebeca Padrão Amorim, Iscia Teresinha Lopes Cendes, and Maria Jose da Silva Fernandes

6 Developmental Expression of Ectonucleotidase and Purinergic Receptors Detection by Whole-Mount In Situ Hybridization in *Xenopus* Embryos 87
Camille Blanchard and Karine Massé

7 Histochemical Approach for Simultaneous Detection of Ectonucleotidase and Alkaline Phosphatase Activities in Tissues 107
Karolina Losenkova, Marius Paul, Heikki Irjala, Sirpa Jalkanen, and Gennady G. Yegutkin

8 Flow Cytometry of Membrane Purinoreceptors 117
Nicole Schwarz, Marten Junge, Friedrich Haag, and Friedrich Koch-Nolte

9 Studying Purinoceptor Cell-Surface Expression by Protein Biotinylation 137
Mark T. Young

10 Multimeric Ionotropic Purinoceptor Detection by Protein Cross-Linking 147
Vincent Compan and François Rassendren

11 Multimeric Purinoceptor Detection by Bioluminescence Resonance Energy Transfer 155
Vincent Compan and François Rassendren

12 Application of Fluorescent Purinoceptor Antagonists for Bioluminescence Resonance Energy Transfer Assays and Fluorescent Microscopy 163
Mark Soave, Joëlle Goulding, Robert Markus, Stephen J. Hill, and Leigh A. Stoddart

13 Detection of Extracellular ATP in the Tumor Microenvironment, Using the pmeLUC Biosensor 183
Elena De Marchi, Elisa Orioli, Anna Pegoraro, Elena Adinolfi, and Francesco Di Virgilio

14 Using Amperometric, Enzyme-Based Biosensors for Performing Longitudinal Measurements of Extracellular Adenosine 5-Triphosphate in the Mouse 197
Edward Beamer and Tobias Engel

15 Fluorescent Labeling and Quantification of Vesicular ATP Release Using Live Cell Imaging 209
Kirstan A. Vessey, Tracy Ho, Andrew I. Jobling, Anna Y. Wang, and Erica L. Fletcher

16 Using FRET-Based Fluorescent Sensors to Monitor Cytosolic and Membrane-Proximal Extracellular ATP Levels 223
Klaus E. Kaschubowski, Axel E. Kraft, Viacheslav O. Nikolaev, and Friedrich Haag

17 ATP Measurement in Cerebrospinal Fluid Using a Microplate Reader 233
Laura de Diego-García, Álvaro Sebastián-Serrano, Carolina Bianchi, Caterina Di Lauro, and Miguel Díaz-Hernández

18 P2X Electrophysiology and Surface Trafficking in *Xenopus* Oocytes 243
Eléonore Bertin, Audrey Martinez, and Eric Boué-Grabot

19 Heterologous Expression and Patch-Clamp Recording of P2X Receptors in HEK293 Cells 261
Lin-Hua Jiang and Sébastien Roger

20 Recording P2X Receptors Using Whole-Cell Patch Clamp from Native Monocytes and Macrophages 275
Leanne Stokes

21 Automated Planar Patch-Clamp Recording of P2X Receptors 285
Carol J. Milligan and Lin-Hua Jiang

22 Controlling Engineered P2X Receptors with Light 301
Benjamin N. Atkinson, Vijay Chudasama, and Liam E. Browne

23 Intracellular Calcium Recording After Purinoceptor Activation Using a Video-Microscopy Equipment 311
Maria Teresa Miras-Portugal, Felipe Ortega, Javier Gualix, Raquel Perez-Sen, Esmerilda G. Delicado, and Rosa Gomez-Villafuertes

24 Assays to Measure Purinoceptor Pore Dilation 323
Ben J. Gu, Pavan Avula, and James S. Wiley

25 Detection of Inflammasome Activation by P2X7 Purinoceptor Activation by Determining ASC Oligomerization 335
Helios Martínez-Banaclocha and Pablo Pelegrín

26 Measuring Leukocyte Migration to Nucleotides 345
Taylor J. Moon and Michael R. Elliott

27 Assessment of Cell Adhesion After Purinoceptor Activation 351
Juan Jose Martínez-García and Pablo Pelegrín

Index *359*

Contributors

ELENA ADINOLFI • *Department of Morphology, Surgery and Experimental Medicine, University of Ferrara, Ferrara, Italy*

REBECA PADRÃO AMORIM • *Departamento de Neurologia e Neurocirurgia, Disciplina de Neurociência, Universidade Federal de São Paulo—UNIFESP, São Paulo, Brazil*

BENJAMIN N. ATKINSON • *Department of Chemistry, University College London, London, UK; Wolfson Institute for Biomedical Research, University College London, London, UK; Department of Neuroscience, Physiology and Pharmacology, University College London, London, UK*

PAVAN AVULA • *The Florey Institute of Neuroscience and Mental Health, University of Melbourne, Parkville, VIC, Australia*

YOUNIS BAQI • *Department of Chemistry, Sultan Qaboos University, Muscat, Oman*

EDWARD BEAMER • *Department of Physiology and Medical Physics, Royal College of Surgeons in Ireland, Dublin, Ireland*

ELÉONORE BERTIN • *Institut des Maladies Neurodégénératives, CNRS UMR 5293, Université de Bordeaux, Bordeaux, France*

CAROLINA BIANCHI • *Department of Biochemistry and Molecular Biology, Veterinary School, Complutense University of Madrid, Madrid, Spain; Instituto de Investigación Sanitaria del Hospital Clínico San Carlos (IdISSC), Madrid, Spain*

CAMILLE BLANCHARD • *Institut des Maladies Neurodégénératives, UMR 5293, Univ. de Bordeaux, Bordeaux, France; CNRS, Institut des Maladies Neurodégénératives, UMR 5293, Bordeaux, France; INSERM, U1215, Neurocentre Magendie, Univ. de Bordeaux, Bordeaux, France*

ERIC BOUÉ-GRABOT • *Institut des Maladies Neurodégénératives, CNRS UMR 5293, Université de Bordeaux, Bordeaux, France*

LIAM E. BROWNE • *Wolfson Institute for Biomedical Research, University College London, London, UK; Department of Neuroscience, Physiology and Pharmacology, University College London, London, UK*

GEOFFREY BURNSTOCK • *Department of Pharmacology and Therapeutics, The University of Melbourne, Parkville, VIC, Australia*

VIJAY CHUDASAMA • *Department of Chemistry, University College London, London, UK*

VINCENT COMPAN • *IGF, University of Montpellier, CNRS, INSERM, Montpellier, France; Labex ICST, Montpellier, France*

MARIA JOSE DA SILVA FERNANDES • *Departamento de Neurologia e Neurocirurgia, Disciplina de Neurociência, Universidade Federal de São Paulo—UNIFESP, São Paulo, Brazil; Escola Paulista de Medicina, Universidade Federal de SĐo Paulo (EPM/UNIFESP), São Paulo, Brazil*

SUDAD DAYL • *Department of Chemistry, College of Science, University of Baghdad, Baghdad, Iraq*

LAURA DE DIEGO-GARCÍA • *Department of Biochemistry and Molecular Biology, Veterinary School, Complutense University of Madrid, Madrid, Spain; Instituto de Investigación Sanitaria del Hospital Clínico San Carlos (IdISSC), Madrid, Spain; Department of Physiology and Medical Physics, Royal College of Surgeons in Ireland, Dublin, Ireland; FutureNeuro Research Center, Dublin, Ireland*

Elena De Marchi • *Department of Morphology, Surgery and Experimental Medicine, University of Ferrara, Ferrara, Italy*

Esmerilda G. Delicado • *Departamento de Bioquímica y Biología Molecular, Facultad de Veterinaria, Universidad Complutense de Madrid, Madrid, Spain; Instituto Universitario de Investigación en Neuroquímica (IUIN), Madrid, Spain; Instituto de Investigación Sanitaria del Hospital Clínico San Carlos (IdISSC), Madrid, Spain*

Caterina Di Lauro • *Department of Biochemistry and Molecular Biology, Veterinary School, Complutense University of Madrid, Madrid, Spain; Instituto de Investigación Sanitaria del Hospital Clínico San Carlos (IdISSC), Madrid, Spain*

Francesco Di Virgilio • *Department of Morphology, Surgery and Experimental Medicine, University of Ferrara, Ferrara, Italy*

Miguel Díaz-Hernández • *Department of Biochemistry and Molecular Biology, Veterinary School, Complutense University of Madrid, Madrid, Spain; Instituto de Investigación Sanitaria del Hospital Clínico San Carlos (IdISSC), Madrid, Spain*

Michael R. Elliott • *David H. Smith Center for Vaccine Biology and Immunology, University of Rochester, Rochester, NY, USA; Department of Microbiology and Immunology, University of Rochester, Rochester, NY, USA*

Tobias Engel • *Department of Physiology and Medical Physics, Royal College of Surgeons in Ireland, Dublin, Ireland*

Erica L. Fletcher • *Visual Neuroscience Laboratory, Department of Anatomy and Neuroscience, The University of Melbourne, Parkville, VIC, Australia*

Rosa Gomez-Villafuertes • *Departamento de Bioquímica y Biología Molecular, Facultad de Veterinaria, Universidad Complutense de Madrid, Madrid, Spain; Instituto Universitario de Investigación en Neuroquímica (IUIN), Madrid, Spain; Instituto de Investigación Sanitaria del Hospital Clínico San Carlos (IdISSC), Madrid, Spain*

Dariusz C. Górecki • *School of Pharmacy and Biomedical Sciences, University of Portsmouth, Portsmouth, UK; Military Institute of Hygiene and Epidemiology, Warsaw, Poland*

Joëlle Goulding • *Cell Signalling and Pharmacology Research Group, Division of Physiology, Pharmacology and Neuroscience, School of Life Sciences, University of Nottingham, Nottingham, UK; Centre of Membrane Proteins and Receptors (COMPARE), University of Birmingham and University of Nottingham, Midlands, UK*

Ben J. Gu • *The Florey Institute of Neuroscience and Mental Health, University of Melbourne, Parkville, VIC, Australia; National Clinical Research Center for Aging and Medicine, Huashan Hospital, Fudan University, Shanghai, China*

Javier Gualix • *Departamento de Bioquímica y Biología Molecular, Facultad de Veterinaria, Universidad Complutense de Madrid, Madrid, Spain; Instituto Universitario de Investigación en Neuroquímica (IUIN), Madrid, Spain; Instituto de Investigación Sanitaria del Hospital Clínico San Carlos (IdISSC), Madrid, Spain*

Friedrich Haag • *Institute of Immunology, University Medical Center Hamburg-Eppendorf, Hamburg, Germany*

Stephen J. Hill • *Cell Signalling and Pharmacology Research Group, Division of Physiology, Pharmacology and Neuroscience, School of Life Sciences, University of Nottingham, Nottingham, UK; Centre of Membrane Proteins and Receptors (COMPARE), University of Birmingham and University of Nottingham, Midlands, UK*

Tracy Ho • *Visual Neuroscience Laboratory, Department of Anatomy and Neuroscience, The University of Melbourne, Parkville, VIC, Australia*

Heikki Irjala • *Department of Otorhinolaryngology, Head and Neck Surgery, Turku University Hospital and Turku University, Turku, Finland*

Sirpa Jalkanen • *MediCity Research Laboratory, University of Turku, Turku, Finland*

Lin-Hua Jiang • *School of Biomedical Sciences, Faculty of Biological Sciences, University of Leeds, Leeds, UK*

Andrew I. Jobling • *Visual Neuroscience Laboratory, Department of Anatomy and Neuroscience, The University of Melbourne, Parkville, VIC, Australia*

Marten Junge • *Institute of Immunology, University Medical Center Hamburg-Eppendorf, Hamburg, Germany*

Klaus E. Kaschubowski • *Institute of Immunology, University Medical Center Hamburg-Eppendorf, Hamburg, Germany*

Friedrich Koch-Nolte • *Institute of Immunology, University Medical Center Hamburg-Eppendorf, Hamburg, Germany*

Axel E. Kraft • *Institute of Experimental Cardiovascular Research, University Medical Center Hamburg-Eppendorf, Hamburg, Germany*

Iscia Teresinha Lopes Cendes • *Departamento de Genetica Médica, Escola de Ciências Médicas da Universidade de Campinas—UNICAMP, Campinas, São Paulo, Brazil*

Karolina Losenkova • *MediCity Research Laboratory, University of Turku, Turku, Finland*

Robert Markus • *School of Life Sciences Imaging (SLIM), School of Life Sciences, University of Nottingham, Nottingham, UK*

Audrey Martínez • *Institut des Maladies Neurodégénératives, CNRS UMR 5293, Université de Bordeaux, Bordeaux, France*

Helios Martínez-Banaclocha • *Biomedical Research Institute of Murcia (IMIB), Clinical University Hospital Virgen de la Arrixaca, Murcia, Spain*

Juan Jose Martínez-García • *Biomedical Research Institute of Murcia (IMIB), Clinical University Hospital Virgen de la Arrixaca, Murcia, Spain*

Karine Massé • *Institut des Maladies Neurodégénératives, UMR 5293, Univ. de Bordeaux, Bordeaux, France; CNRS, Institut des Maladies Neurodégénératives, UMR 5293, Bordeaux, France*

Carol J. Milligan • *The Florey Institute of Neuroscience and Mental Health, Melbourne Brain Centre, Parkville, VIC, Australia*

Maria Teresa Miras-Portugal • *Departamento de Bioquímica y Biología Molecular, Facultad de Veterinaria, Universidad Complutense de Madrid, Madrid, Spain; Instituto Universitario de Investigación en Neuroquímica (IUIN), Madrid, Spain; Instituto de Investigación Sanitaria del Hospital Clínico San Carlos (IdISSC), Madrid, Spain*

Taylor J. Moon • *David H. Smith Center for Vaccine Biology and Immunology, University of Rochester, Rochester, NY, USA; Department of Microbiology and Immunology, University of Rochester, Rochester, NY, USA*

Christa E. Müller • *PharmaCenter Bonn, Pharmaceutical Institute, Pharmaceutical & Medicinal Chemistry, University of Bonn, Bonn, Germany*

Vigneshwaran Namasivayam • *PharmaCenter Bonn, Pharmaceutical Institute, Pharmaceutical & Medicinal Chemistry, University of Bonn, Bonn, Germany*

Viacheslav O. Nikolaev • *Institute of Experimental Cardiovascular Research, University Medical Center Hamburg-Eppendorf, Hamburg, Germany*

Elisa Orioli • *Department of Morphology, Surgery and Experimental Medicine, University of Ferrara, Ferrara, Italy*

Felipe Ortega • *Departamento de Bioquímica y Biología Molecular, Facultad de Veterinaria, Universidad Complutense de Madrid, Madrid, Spain; Instituto Universitario de Investigación en Neuroquímica (IUIN), Madrid, Spain; Instituto de Investigación Sanitaria del Hospital Clínico San Carlos (IdISSC), Madrid, Spain*

MARIUS PAUL • *MediCity Research Laboratory, University of Turku, Turku, Finland*

ANNA PEGORARO • *Department of Morphology, Surgery and Experimental Medicine, University of Ferrara, Ferrara, Italy*

PABLO PELEGRÍN • *Biomedical Research Institute of Murcia (IMIB), Clinical University Hospital Virgen de la Arrixaca, Murcia, Spain*

RAQUEL PEREZ-SEN • *Departamento de Bioquímica y Biología Molecular, Facultad de Veterinaria, Universidad Complutense de Madrid, Madrid, Spain; Instituto Universitario de Investigación en Neuroquímica (IUIN), Madrid, Spain; Instituto de Investigación Sanitaria del Hospital Clínico San Carlos (IdISSC), Madrid, Spain*

FRANÇOIS RASSENDREN • *IGF, University of Montpellier, CNRS, INSERM, Montpellier, France; Labex ICST, Montpellier, France*

SÉBASTIEN ROGER • *EA4245, Transplantation, Immunology, Inflammation, Faculty of Medicine, University of Tours, Tours, France*

ROBIN M. H. RUMNEY • *School of Pharmacy and Biomedical Sciences, University of Portsmouth, Portsmouth, UK*

RALF SCHMID • *Department of Molecular and Cell Biology, University of Leicester, Leicester, UK; Leicester Institute of Structural and Chemical Biology (LISCB), University of Leicester, Leicester, UK*

NICOLE SCHWARZ • *Institute of Immunology, University Medical Center Hamburg-Eppendorf, Hamburg, Germany*

ÁLVARO SEBASTIÁN-SERRANO • *Department of Biochemistry and Molecular Biology, Veterinary School, Complutense University of Madrid, Madrid, Spain; Instituto de Investigación Sanitaria del Hospital Clínico San Carlos (IdISSC), Madrid, Spain; Instituto de Investigaciones Biomédicas "Alberto Sols", Consejo Superior de Investigaciones Científicas-Universidad Autónoma de Madrid (CSIC-UAM), Madrid, Spain*

MARK SOAVE • *Cell Signalling and Pharmacology Research Group, Division of Physiology, Pharmacology and Neuroscience, School of Life Sciences, University of Nottingham, Nottingham, UK; Centre of Membrane Proteins and Receptors (COMPARE), University of Birmingham and University of Nottingham, Midlands, UK*

ANASTASIOS STAVROU • *Department of Molecular and Cell Biology, University of Leicester, Leicester, UK*

LEIGH A. STODDART • *Cell Signalling and Pharmacology Research Group, Division of Physiology, Pharmacology and Neuroscience, School of Life Sciences, University of Nottingham, Nottingham, UK; Centre of Membrane Proteins and Receptors (COMPARE), University of Birmingham and University of Nottingham, Midlands, UK*

LEANNE STOKES • *School of Pharmacy, University of East Anglia, Norwich, UK*

KIRSTAN A. VESSEY • *Visual Neuroscience Laboratory, Department of Anatomy and Neuroscience, The University of Melbourne, Parkville, VIC, Australia*

ANNA Y. WANG • *Visual Neuroscience Laboratory, Department of Anatomy and Neuroscience, The University of Melbourne, Parkville, VIC, Australia*

JAMES S. WILEY • *The Florey Institute of Neuroscience and Mental Health, University of Melbourne, Parkville, VIC, Australia*

GENNADY G. YEGUTKIN • *MediCity Research Laboratory, University of Turku, Turku, Finland*

MARK T. YOUNG • *School of Biosciences, Cardiff University, Cardiff, UK*

Chapter 1

Introduction to Purinergic Signaling

Geoffrey Burnstock

Abstract

Purinergic signaling was proposed in 1972, after it was demonstrated that adenosine 5′-triphosphate (ATP) was a transmitter in nonadrenergic, noncholinergic inhibitory nerves supplying the guinea-pig taenia coli. Later, ATP was identified as an excitatory cotransmitter in sympathetic and parasympathetic nerves, and it is now apparent that ATP acts as a cotransmitter in most, if not all, nerves in both the peripheral nervous system and central nervous system (CNS). ATP acts as a short-term signaling molecule in neurotransmission, neuromodulation, and neurosecretion. It also has potent, long-term (trophic) roles in cell proliferation, differentiation, and death in development and regeneration. Receptors to purines and pyrimidines have been cloned and characterized: P1 adenosine receptors (with four subtypes), P2X ionotropic nucleotide receptors (seven subtypes) and P2Y metabotropic nucleotide receptors (eight subtypes). ATP is released from different cell types by mechanical deformation, and after release, it is rapidly broken down by ectonucleotidases. Purinergic receptors were expressed early in evolution and are widely distributed on many different nonneuronal cell types as well as neurons. Purinergic signaling is involved in embryonic development and in the activities of stem cells. There is a growing understanding about the pathophysiology of purinergic signaling and there are therapeutic developments for a variety of diseases, including stroke and thrombosis, osteoporosis, pain, chronic cough, kidney failure, bladder incontinence, cystic fibrosis, dry eye, cancer, and disorders of the CNS, including Alzheimer's, Parkinson's. and Huntington's disease, multiple sclerosis, epilepsy, migraine, and neuropsychiatric and mood disorders.

Key words ATP, Adenosine, Cotransmission, Purinoceptor, Development, Stem cells, Thrombosis, Neurodegenerative diseases, Pain, Cough

1 ATP as a Transmitter in Nonadrenergic, Noncholinergic Nerves

Transmitters released from autonomic nerves that were neither of the classical neurotransmitters, acetylcholine (ACh) or noradrenaline (NA), was discovered in 1963 when inhibitory junction potentials were shown in the smooth muscle of the guinea-pig taenia coli in response to nerve stimulation in the presence of atropine and guanethidine ([1] and *see* [2]). These inhibitory nonadrenergic, noncholinergic (NANC) responses were shown to be mediated by intrinsic enteric neurons controlled by vagal and sacral parasympathetic nerves [3], NANC excitatory transmission

Pablo Pelegrín (ed.), *Purinergic Signaling: Methods and Protocols*, Methods in Molecular Biology, vol. 2041, https://doi.org/10.1007/978-1-4939-9717-6_1, © Springer Science+Business Media, LLC, part of Springer Nature 2020

was later shown in the urinary bladder, vas deferens, and vascular system ([4] and *see* [5]).

The next step was to identify the transmitter released during NANC inhibitory transmission in the gut and by NANC excitatory transmission in the urinary bladder. To establish a neurotransmitter, several criteria need to be satisfied: storage and synthesis in nerve terminals; release usually by a Ca^{2+}-dependent mechanism; nerve-mediated responses need to be mimicked by the exogenously applied transmitter; inactivation by neuronal uptake and/or ectoenzymes; and block or potentiation of responses to stimulation of nerves mimicked by exogenously applied transmitter. Amino acids, monoamines, and neuropeptides were examined in the late 1960s, but none satisfied the criteria. Drury and Szent-Györgyi [6] showed extracellular actions of purines on heart and blood vessels, papers by Feldberg showed extracellular actions of ATP on autonomic ganglia [7] and a paper by Holton showed release of ATP during antidromic stimulation of sensory nerves supplying the rabbit ear artery [8]. ATP was therefore tested and, to the surprise of the experimenters, it perfectly satisfied all the criteria needed to establish it as a transmitter involved in NANC neurotransmission [9]. In 1972, an article in *Pharmacological Reviews* [10] that formulated the purinergic neurotransmission hypothesis was published. Unfortunately, most scientists rejected this hypothesis over the next 20 years and it was often ridiculed at meetings and workshops. Resistance to this concept was to some extent understandable because ATP was well established as an intracellular energy source involved in biochemical pathways, including the Krebs cycle, and it seemed unlikely that such a ubiquitous molecule would also act as an extracellular messenger. However, it seems that ATP as an early biological molecule evolved both as an intracellular energy source and an extracellular signaling molecule. When receptors for ATP and adenosine were cloned and characterized in the early 1990s (*see* [11]) and neuron–neuron synaptic transmission identified in sympathetic ganglia and in the brain in 1992 [12–14], the purinergic hypothesis began to be widely accepted (*see* [15]).

2 Purinergic Cotransmission

Neurotransmission was dominated for many years by the concept that one neuron releases only a single transmitter, known as "Dale's Principle." This concept arose from a misinterpretation of Dale's suggestion in 1935 that the same neurotransmitter was stored in and released from all terminals of a single neuron, which did not specifically preclude the possibility that more than one transmitter may be associated with the same nerve cell.

Experiments showed release of ATP with NA from sympathetic nerves [16] and there were many hints in the literature, so the

Table 1
ATP as a ubiquitous cotransmitter

	Cotransmitters	References
Peripheral nervous system		
Sympathetic nerves	ATP + NA + NPY	[21]
Parasympathetic nerves	ATP + ACh + VIP	[22]
Sensory-motor	ATP + CGRP + SP	[23]
NANC enteric nerves	ATP + NO + VIP	[24]
Motor nerves (in early development)	ATP + ACh	[25]
Central nervous system		
Cortex, caudate nucleus	ATP + ACh	[26]
Hypothalamus, locus coeruleus	ATP + NA	[27]
Hypothalamus, dorsal horn, retina	ATP + GABA	[28]
Mesolimbic system	ATP + DA	[29]
Hippocampus, dorsal horn, cortex	ATP + glutamate	[30–32]

ACh acetylcholine, *ATP* adenosine 5′-triphosphate, *CGRP* calcitonin gene-related peptide, *DA* dopamine, *GABA* γ-aminobutyric acid, *NA* noradrenaline, *NANC* nonadrenergic, noncholinergic, *NO* nitric oxide, *NPY* neuropeptide Y, *SP* substance P, *VIP* vasoactive polypeptide. Modified from [33] with permission

cotransmission hypothesis was formulated in 1976 [17]. Purinergic cotransmission is now well established, in sympathetic nerves and also in parasympathetic, sensory-motor, enteric nerves and developing motor nerves to skeletal muscle (*see* [18]). More recently ATP has been shown to be coreleased with glutamate, γ-aminobutyric acid, dopamine, NA, 5-hydroxytryptamine, and ACh in different populations of nerves in the central nervous system (CNS) (*see* [19, 20]). ATP is now established as a cotransmitter in most, if not all, nerves in the peripheral nervous system (PNS) and CNS (*see* Table 1). Purines and/or pyrimidines act as signaling molecules in virtually all nonneuronal tissues (*see* Tables 1, 2, and 3).

3 Receptors to Purines and Pyrimidines

Receptors for nucleotides and nucleosides needed to be identified. A basis for distinguishing two families of purinergic receptors was proposed [95]. One family was selective for adenosine (called P1 receptors), which were antagonized by methylxanthines. The other family was selective for ATP/ADP (called P2 receptors). This explained some of the early confusion in the literature resulting

Table 2
Tissue presence of principal P1 and P2 receptors

Neurons	
Sympathetic neurons	P2X1–7, $P2Y_1$, $P2Y_2$, $P2Y_4$, $P2Y_6$, A_1
Parasympathetic neurons	P2X2, P2X3, P2X4, $P2Y_1$, $P2Y_2$, $P2Y_4$, $P2Y_{11}$, A_1
Sensory neurons	P2X1–7, $P2Y_1$, $P2Y_2$, $P2Y_4$, A_{2A}
Enteric neurons	P2X3, P2X4, P2X7, $P2Y_1$, $P2Y_6$, $P2Y_{12}$, A_1, A_{2A}, A_{2B}
CNS neurons	P2X4, P2X6, $P2Y_1$, $P2Y_6$, $P2Y_{12}$, A_1, A_{2A}, A_{2B}, A_3
Glia	
Astrocytes	P2X1/5, P2X7 (reactive astroglia), P2Y, A_1
Oligodendrocytes	P2X7, $P2Y_1$, $P2Y_{11}$
Microglia	$P2Y_4$, $P2Y_7$, $P2Y_6$, $P2Y_{11}$, $P2Y_{12}$, $P2Y_{13}$, A_1, A_2
Special senses	
Inner ear	P2X1, P2X2, P2X3, P2X7, $P2Y_2$, $P2Y_4$, A_1
Eye	P2X2, P2X7, $P2Y_2$, A_1, A_2, A_3
Tongue	P2X2, P2X3, $P2Y_1$, A_1
Olfactory organ	P2X2, P2X4, $P2Y_1$, $P2Y_2$, A_{2A}, A_3
Muscle cells	
Smooth muscle	P2X1–7, $P2Y_1$, $P2Y_2$, $P2Y_4$, $P2Y_6$, A_1, A_{2A}, A_{2B}, A_3
Skeletal muscle	
– Developing	P2X2, P2X5, P2X6, $P2Y_1$, $P2Y_2$
– Adult	P2X1–7, $P2Y_2$, A_{2A}
Cardiac muscle	P2X1–6, $P2Y_1$, $P2Y_2$, $P2Y_4$, $P2Y_6$, A_1, A_3
Nonneuronal cells	
Osteoblasts	$P2X_7$, $P2Y_1$, $P2Y_2$
Cartilage	$P2X_2$, $P2Y_1$, $P2Y_2$, A_{2A}, A_{2B}
Keratinocytes	$P2X_5$, $P2X_2$, $P2X_3$, $P2X_7$, $P2Y_1$, $P2Y_2$, $P2Y_4$, A_{2B}
Fibroblasts	$P2X_7$, $P2Y_1$, $P2Y_2$, A_{2A}
Adipocytes	$P2X_1$, $P2Y_1$, $P2Y_2$, $P2Y_4$, A_1
Epithelial cells	P2X4–7, $P2Y_1$, $P2Y_2$, $P2Y_{11}$, A_1, A_{2A}, A_3
Hepatocytes	$P2Y_1$, $P2Y_2$, $P2Y_4$, $P2Y_6$, $P2Y_{13}$, A_{2A}, A_3
Sperm	P2X2, P2X7, $P2Y_2$, A_1
Endothelial cells	P2X1, $P2Y_1$, $P2Y_2$, $P2Y_4$, $P2Y_6$, A_1, A_{2A}
Erythrocytes	P2X2, P2X4, $P2X_7$, $P2Y_1$

(continued)

Table 2 (continued)

Platelets	P2X1, $P2Y_1$, $P2Y_{12}$, A_{2A}
Immune cells	P2X4, P2X7, $P2Y_1$, $P2Y_2$, A_{2A}, A_3
Exocrine secretary cells	P2X1, P2X4, P2X7, $P2Y_1$, $P2Y_2$, $P2Y_4$, A_1, A_{2A}
Endocrine secretory cells	P2X1-7, $P2Y_2$, $P2Y_4$, A_1, A_{2A}, A_{2B}, A_3

Modified from [34], with permission from Elsevier

from the rapid extracellular breakdown of ATP to adenosine. During purinergic neurotransmission, postjunctional receptors were shown to be P2, while prejunctional P1 receptors mediated negative feedback neuromodulatory responses or autoregulation of transmitter release. Two types of P2 receptors were distinguished on a pharmacological basis, defined as P2X and P2Y, in 1985 [96]. P2 receptors were cloned in the early 1990s [97–100] and second messenger mechanisms examined [101] identifying P2X ion channel receptors and P2Y G protein-coupled receptors. Four subtypes of P1 receptors are recognized, seven subtypes of P2X receptors and eight subtypes of P2Y receptors, including some responsive to the pyrimidines uridine 5′-triphosphate and uridine diphosphate [11, 102]. Three of the P2X receptor subtypes were shown to combine to form cation pores [103] either as homomultimers and heteromultimers, and more recently heterodimerization has been shown between P2Y receptor subtypes (*see* [104]). Most nonneural as well as neuronal cells express multiple purinoceptors [45] and this poses problems about how they mediate interacting physiological and pathophysiological events. Purinergic signaling has an early evolutionary basis (*see* [94]) with fascinating studies showing cloned receptors in two primitive invertebrates, *Dictyostelium* and *Schistosoma*, that resemble mammalian P2X receptors [105, 106], and ATP signaling in plants has also been described ([107–109] and *see* [110]).

4 Physiology of Purinergic Signaling

Early studies were largely focused on short-term signaling in neurotransmission, neuromodulation, secretion, chemoattraction, and acute inflammation, but long-term (trophic) purinergic signaling involving cell proliferation, differentiation, motility and death in development, regeneration, wound healing, restenosis, epithelial cell turnover, cancer, and ageing also occurs [111–113]. Dual short-term control of vascular tone is evoked by ATP released as an excitatory cotransmitter from perivascular sympathetic nerves to act on P2X receptors on smooth muscle, while ATP released

Table 3
Physiological role of purinergic signaling in living tissues

Tissue	Functional role	Reference
CNS	Fast excitatory cotransmission in CNS, modulation of synaptic plasticity, metabotropic transmission, regulation of growth and development, chemical transmission in neuronal–astroglial networks; signaling between axons and oligodendrocytes, CO_2 chemosensitivity, control of microglial motility and activation	[15, 35–44]
PNS	Nociception, thermal sensitivity, mechanosensitivity, chemosensitivity, neuronal-effector transmission	[45–51]
Cardiovascular system: Heart	Negative chronotropic and ionotropic effects in atria, positive chronotropic and ionotropic effect in ventricles, regulation of cardiomyocytes Ca^{2+} signaling, control of excitation of intrinsic cardiac neurons	[45, 52]
Cardiovascular system: Blood vessels	Vasodilation (P2Y-mediated) and vasoconstriction (P2X-mediated)	[45, 52–54]
Exocrine glands	Regulation of ionic permeability and Ca^{2+} signaling in salivary and lachrymal gland cells, induction of sweat production by sweat gland epithelial cells	[45]
Endocrine glands	Regulation of Ca^{2+} signaling in pituitary and thyroid cells, regulation of catecholamine release from adrenal chromaffin cells, stimulation of insulin, glucagon, and somatostatin secretion from endocrine pancreas	[45]
Immune system	Regulation of mitogenesis and DNA synthesis in thymocytes, regulation of activation and death of macrophages, aggregation of neutrophils, regulation of secretory response in basophles, and chemotactic response in eosinophils, modulation of proliferative response in lymphocytes, release of histamine and degranulation of mast cells, mediation of intercellular Ca^{2+} waves in mast cells, regulation of release of proinflammatory factors	[45, 55–60]
Lung	Bronchodilation, stimulation of surfactant release from airway epithelial cells; stimulation of mucin secretion from goblet cells; increase in ciliary beat frequency of ciliated epithelial cells, activation of lung myeloid dendritic cells; modulation of O_2 chemotransmission in cells of neuroepithelial bodies; contraction/relaxation of tracheal ring	[45, 61–64]
Gastrointestinal tract	Control of mucociliary activity of esophageal epithelial cells, regulation of acid secretion in gastric mucosa, regulation of contraction/relaxation of small intestine, inhibition of ACh release from enteric neurons, regulation of peristaltic activity of ileum and duodenum, inhibition of amino acid, sugar and ion transport in epithelial cells of small intestine, relaxation of taenia coli, control of contraction/relaxation of colon and rectum, relaxation of internal anal sphincter	[19, 45, 65–68]

(continued)

Table 3 (continued)

Tissue	Functional role	Reference
Liver	Stimulation of glycogenolysis, inhibition of glycolysis, regulation of bile formation and secretion via stimulation of Cl^- efflux, mediate chemosensitivity of cholangiocyte cilia	[45, 69–72]
Kidney	Regulation of renal blood flow, microvascular function and glomerular filtration rate, generation of prostanoids, regulation of Cl^- secretion, regulation of renal Na^+, glucose and water transport, possible involvement in biosensing activity of kidney macula densa cells	[73–78]
Bladder and urethra	Control of contraction/relaxation of mammalian bladder, relaxation of mammalian urethra	[45, 79–82]
Male genital system	Regulation of penile erection, contraction of prostate smooth muscle and seminal vesicles, micturition, peristalsis of the male excurrent duct system and thus sperm transport and ejaculation; control of steroid production by testis Leydig cells, inhibition of sperm motility, initiation of acrosome reaction	[83–86]
Female genital system	Regulation of myometrium contraction, modulation of ovarian function, control of blood flow in placenta, relaxation of vaginal smooth muscle, stimulation of vaginal moisture production, stimulation of Cl^- and mucus secretion from endocervical epithelial cells	[45, 87–92]
Bone and cartilage	Regulation of osteoclast/bone formation and resorption, formation of multinucleated osteoclasts, stimulation of resorption in cartilage, regulation of chondrocalcinosis	[45, 93]
Skeletal muscle	Regulation of proliferation and differentiation of developing myoblasts, modulation of contractile response of myocytes	[45]

Reproduced from [94] with Permission

from endothelial cells during changes in blood flow (causing shear stress) and hypoxia acts on P2X and P2Y receptors on endothelial cells leading to production of nitric oxide and relaxation [114, 115]. Long-term control by ATP and adenosine of cell proliferation and differentiation, migration, and death occurs in neovascularization, restenosis following angioplasty, and atherosclerosis [20, 116].

The source of ATP acting on receptors was considered for many years to be damaged or dying cells, apart from exocytotic vesicular release from nerves. However, many cell types release ATP physiologically in response to mechanical distortion, hypoxia and to some agents [117, 118]. The mechanism of ATP transport includes, in addition to vesicular release via the vesicular nucleotide transporter, VNUT [119], connexin or pannexin hemichannels, maxi-ion channels, and P2X7 receptors [15, 120, 121].

Extracellular breakdown of released ATP is by various ectonucleotidases, including E-NTPDases, E-NPPS, alkaline phosphatase, and ecto-5′-nucleotidose (*see* [122, 123]).

There is current interest in the roles of purinergic signaling in neuron–glial interactions in the CNS (*see* [104, 124, 125]).

5 Purinergic Signaling in Development

The autonomic nervous system shows high plasticity compared to the CNS. Substantial changes in cotransmitter and receptor expression occur during development and ageing, in the nerves that remain following trauma or surgery and in disease situations [126, 127]. A P2Y-like receptor was transiently expressed in the neural plate of *Xenopus* and again later in secondary neurulation in the tail bud, implicating involvement of purinergic signaling in the development of the nervous system [128]. Transient expression of P2X5 and P2X6 receptors occurs during development of myotubules and of P2X2 receptors during neuromuscular junction development [129]. P2X3 receptors are expressed first at embryological day (E)11; P2X2 and P2X7 receptors at E14; P2X4, P2X5, and P2X6 receptors at postnatal day (P)1; and P2X1 receptors at P16 in the rat brain [130].

Sprouting of central neurons occurred when the enteric nervous system was transplanted in the striatum of the brain [131]. A growth factor released from enteric glial cells acting synergistically with ATP, adenosine and nitric oxide was shown to be involved [132]. It is suggested that such synergistic activity of purines and growth factors might be involved in stem cell activity [133, 134].

6 Purinergic Pathophysiology and Therapeutic Potential

ATP is a major cotransmitter with ACh in parasympathetic nerves mediating contraction of the urinary bladder of rodents [135–137], but the role of ATP as a cotransmitter is minor in healthy human bladder. However, in pathological conditions, including interstitial cystitis, outflow obstruction and neurogenic bladder, the purinergic component is increased to about 40% [126, 138]. There is also a significantly greater cotransmitter role for ATP in sympathetic nerves in spontaneously hypertensive rats [139].

Clopidogrel and ticagrelor are $P2Y_{12}$ receptor antagonists that inhibit platelet aggregation and are very successful commercial drugs against thrombosis and stroke [140–142]. Purinergic compounds are also being developed for the treatment of hypertension and atherosclerosis (*see* [116, 141]), inflammatory bowel disease

[143–146], dry eye and cystic fibrosis [147], cancer [148–151], and a number of other diseases (*see* [15, 152]).

P2X3 receptors were cloned in 1995 and are located on small nociceptive sensory neurons that label with isolectin B4 in dorsal root ganglia [153, 154]. Central projections reach the inner lamina 2 of the dorsal horn of the spinal cord and peripheral extensions supplying skin, tongue and visceral organs. A purinergic hypothesis for the initiation of pain was proposed [155] and a hypothesis describing purinergic mechanosensory transduction in visceral organs, where ATP, released from epithelial cells during distension, acts on P2X3 and P2X2/3 receptors on subepithelial sensory nerve endings to send nociceptive messages to the conscious pain centres in the brain [156, 157]. Evidence includes epithelial release of ATP, localization of P2X3 receptors on subepithelial nerves, and activity recorded in sensory nerves during distension that is mimicked by ATP and reduced by P2X3 receptor antagonists in the urinary bladder [158], ureter [159], gut [160], and tongue [161]. Purinergic mechanosensory transduction is also active in urine voiding as shown in P2X3 knockout mice [162]. For neuropathic and inflammatory pain P2X4, P2X7, and $P2Y_{12}$ receptors on microglia have been implicated and antagonists to these receptors are very effective in abolishing allodynia [163–167]. The potential roles of purinergic signaling in trauma and ischemia, neurodegenerative conditions including Alzheimer's, Parkinson's, and Huntington's diseases and in multiple sclerosis and amyotrophic lateral sclerosis are also being studied, as well as in neuropsychiatric diseases, including depression, anxiety, and schizophrenia and in epileptic seizures (*see* [168, 169]).

References

1. Burnstock G, Campbell G, Bennett M, Holman ME (1964) Innervation of the guinea-pig taenia coli: are there intrinsic inhibitory nerves which are distinct from sympathetic nerves? Int J Neuropharmacol 3:163–166
2. Burnstock G (2004) A moment of excitement. Living history series. The discovery of non-adrenergic, non-cholinergic neurotransmission. Physiol News 56:7–9
3. Burnstock G, Campbell G, Rand MJ (1966) The inhibitory innervation of the taenia of the guinea-pig caecum. J Physiol 182:504–526
4. Burnstock G, Dumsday B, Smythe A (1972) Atropine resistant excitation of the urinary bladder: the possibility of transmission via nerves releasing a purine nucleotide. Br J Pharmacol 44:451–461
5. Burnstock G (1986) The non-adrenergic non-cholinergic nervous system. Arch Int Pharmacodyn Ther 280(Suppl):1–15
6. Drury AN, Szent-Györgyi A (1929) The physiological activity of adenine compounds with special reference to their action upon the mammalian heart. J Physiol Lond 68:213–237
7. Feldberg W, Hebb C (1948) The stimulating action of phosphate compounds on the perfused superior cervical ganglion of the cat. J Physiol 107:210–221
8. Holton P (1959) The liberation of adenosine triphosphate on antidromic stimulation of sensory nerves. J Physiol 145:494–504
9. Burnstock G, Campbell G, Satchell D, Smythe A (1970) Evidence that adenosine triphosphate or a related nucleotide is the transmitter substance released by non-adrenergic inhibitory nerves in the gut. Br J Pharmacol 40:668–688
10. Burnstock G (1972) Purinergic nerves. Pharmacol Rev 24:509–581

11. Ralevic V, Burnstock G (1998) Receptors for purines and pyrimidines. Pharmacol Rev 50:413–492
12. Edwards FA, Gibb AJ, Colquhoun D (1992) ATP receptor-mediated synaptic currents in the central nervous system. Nature 359 (6391):144–147
13. Evans RJ, Derkach V, Surprenant A (1992) ATP mediates fast synaptic transmission in mammalian neurons. Nature 357 (6378):503–505
14. Lalo U, Verkhratsky A, Burnstock G, Pankratov Y (2012) P2X receptor-mediated synaptic transmission. Paper presented at the WIREs membrane transport and signaling
15. Burnstock G (2007) Physiology and pathophysiology of purinergic neurotransmission. Physiol Rev 87:659–797
16. Su C, Bevan JA, Burnstock G (1971) [^{3}H] adenosine triphosphate: release during stimulation of enteric nerves. Science 173:337–339
17. Burnstock G (1976) Do some nerve cells release more than one transmitter? Neuroscience 1:239–248
18. Burnstock G (1990) Co-transmission. The fifth Heymans memorial lecture – Ghent, February 17, 1990. Arch Int Pharmacodyn Ther 304:7–33
19. Burnstock G (2009) Purinergic cotransmission. Exp Physiol 94(1):20–24
20. Burnstock G (2014) The Erasmus lecture 2012, academia Europaea. The concept of cotransmission: focus on ATP as a cotransmitter and its significance in health and disease. Eur Rev 22:1–17
21. Westfall DP, Stitzel RE, Rowe JN (1978) The postjunctional effects and neural release of purine compounds in the guinea-pig vas deferens. Eur J Pharmacol 50:27–38
22. Hoyle CHV (1996) Purinergic cotransmission: parasympathetic and enteric nerves. Semin Neurosci 8:207–215
23. Burnstock G (1993) Introduction: changing face of autonomic and sensory nerves in the circulation. In: Edvinsson L, Uddman R (eds) Vascular innervation and receptor mechanisms: new perspectives. Academic Press, USA, pp 1–22
24. Belai A, Burnstock G (1994) Evidence for coexistence of ATP and nitric oxide in non-adrenergic, non-cholinergic (NANC) inhibitory neurones in the rat ileum, colon and anococcygeus muscle. Cell Tissue Res 278:197–200
25. Silinsky EM, Hubbard JI (1973) Release of ATP from rat motor nerve terminals. Nature 243(5407):404–405
26. Richardson PJ, Brown SJ (1987) ATP release from affinity-purified rat cholinergic nerve terminals. J Neurochem 48(2):622–630
27. Sperlágh B, Sershen H, Lajtha A, Vizi ES (1998) Co-release of endogenous ATP and [^{3}H]noradrenaline from rat hypothalamic slices: origin and modulation by α_2-adrenoceptors. Neuroscience 82(2):511–520
28. Jo YH, Role LW (2002) Coordinate release of ATP and GABA at *in vitro* synapses of lateral hypothalamic neurons. J Neurosci 22 (12):4794–4804
29. Krügel U, Schraft T, Kittner H, Kiess W, Illes P (2003) Basal and feeding-evoked dopamine release in the rat nucleus accumbens is depressed by leptin. Eur J Pharmacol 482 (1–3):185–187
30. Mori M, Heuss C, Gahwiler BH, Gerber U (2001) Fast synaptic transmission mediated by P2X receptors in CA3 pyramidal cells of rat hippocampal slice cultures. J Physiol 535 (Pt 1):115–123
31. Pankratov Y, Lalo U, Krishtal O, Verkhratsky A (2002) Ionotropic P2X purinoreceptors mediate synaptic transmission in rat pyramidal neurones of layer II/III of somato-sensory cortex. J Physiol 542(Pt 2):529–536
32. Pankratov Y, Lalo U, Krishtal O, Verkhratsky A (2003) P2X receptor-mediated excitatory synaptic currents in somatosensory cortex. Mol Cell Neurosci 24(3):842–849
33. Burnstock G (1996) Cotransmission with particular emphasis on the involvement of ATP. In: Fuxe K, Hökfelt T, Olson L, Ottoson D, Dahlström A, Björklund A (eds) Molecular mechanisms of neuronal communication. A tribute to Nils-Ake Hillarp, Wenner-Gren international series. Pergamon Press, Oxford, pp 67–87
34. Knight GE (2009) Purinergic receptors. In: Squire LR (ed) Encyclopedia of neuroscience, 4th edn. Academic Press, Oxford, pp 1245–1252
35. Fields RD, Burnstock G (2006) Purinergic signaling in neuron-glia interactions. Nat Rev Neurosci 7(6):423–436
36. Farber K, Kettenmann H (2006) Purinergic signaling and microglia. Pflugers Arch 452 (5):615–621
37. North RA, Verkhratsky A (2006) Purinergic transmission in the central nervous system. Pflugers Arch 452(5):479–485
38. Kirischuk S, Moller T, Voitenko N, Kettenmann H, Verkhratsky A (1995) ATP-induced cytoplasmic calcium mobilization in Bergmann glial cells. J Neurosci 15 (12):7861–7871

39. Kirischuk S, Scherer J, Kettenmann H, Verkhratsky A (1995) Activation of P2-purinoreceptors triggered Ca^{2+} release from $InsP_3$-sensitive internal stores in mammalian oligodendrocytes. J Physiol 483 (Pt 1):41–57
40. Abbracchio MP, Burnstock G, Verkhratsky A, Zimmerman H (2009) Purinergic signaling in the nervous system: an overview. Trends Neurosci 32(1):19–29
41. Gourine AV, Llaudet E, Dale N, Spyer KM (2005) ATP is a mediator of chemosensory transduction in the central nervous system. Nature 436(7047):108–111
42. Lalo U, Pankratov Y, Wichert SP, Rossner MJ, North RA, Kirchhoff F, Verkhratsky A (2008) $P2X_1$ and $P2X_5$ subunits form the functional P2X receptor in mouse cortical astrocytes. J Neurosci 28(21):5473–5480
43. Pankratov Y, Lalo U, Krishtal OA, Verkhratsky A (2009) P2X receptors and synaptic plasticity. Neuroscience 158(1):137–148
44. Hamilton N, Vayro S, Kirchhoff F, Verkhratsky A, Robbins J, Gorecki DC, Butt AM (2008) Mechanisms of ATP- and glutamate-mediated calcium signaling in white matter astrocytes. Glia 56(7):734–749
45. Burnstock G, Knight GE (2004) Cellular distribution and functions of P2 receptor subtypes in different systems. Int Rev Cytol 240:31–304
46. Burnstock G, Wood JN (1996) Purinergic receptors: their role in nociception and primary afferent neurotransmission. Curr Opin Neurobiol 6(4):526–532
47. Khakh BS, North RA (2006) P2X receptors as cell-surface ATP sensors in health and disease. Nature 442(7102):527–532
48. Khmyz V, Maximyuk O, Teslenko V, Verkhratsky A, Krishtal O (2008) $P2X_3$ receptor gating near normal body temperature. Pflugers Arch 456(2):339–347
49. Souslova V, Cesare P, Ding Y, Akopian AN, Stanfa L, Suzuki R, Carpenter K, Dickenson A, Boyce S, Hill R, Nebenuis-Oosthuizen D, Smith AJ, Kidd EJ, Wood JN (2000) Warm-coding deficits and aberrant inflammatory pain in mice lacking P2X3 receptors. Nature 407(6807):1015–1017
50. Cook SP, Vulchanova L, Hargreaves KM, Elde R, McCleskey EW (1997) Distinct ATP receptors on pain-sensing and stretch-sensing neurons. Nature 387(6632):505–508
51. Rong W, Gourine AV, Cockayne DA, Xiang Z, Ford AP, Spyer KM, Burnstock G (2003) Pivotal role of nucleotide P2X2 receptor subunit of the ATP-gated ion channel mediating ventilatory responses to hypoxia. J Neurosci 23(36):11315–11321
52. Erlinge D, Burnstock G (2008) P2 receptors in cardiovascular regulation and disease. Purinergic Signal 4(1):1–20
53. Harrington LS, Evans RJ, Wray J, Norling L, Swales KE, Vial C, Ali F, Carrier MJ, Mitchell JA (2007) Purinergic $2X_1$ receptors mediate endothelial dependent vasodilation to ATP. Mol Pharmacol 72(5):1132–1136
54. Harrington LS, Mitchell JA (2004) Novel role for P2X receptor activation in endothelium-dependent vasodilation. Br J Pharmacol 143(5):611–617
55. Osipchuk Y, Cahalan M (1992) Cell-to-cell spread of calcium signals mediated by ATP receptors in mast cells. Nature 359 (6392):241–244
56. Chen L, Brosnan CF (2006) Regulation of immune response by $P2X_7$ receptor. Crit Rev Immunol 26(6):499–513
57. Coutinho-Silva R, Knight GE, Burnstock G (2005) Impairment of the splenic immune system in $P2X_2/P2X_3$ knockout mice. Immunobiology 209(9):661–668
58. Brough D, Le Feuvre RA, Wheeler RD, Solovyova N, Hilfiker S, Rothwell NJ, Verkhratsky A (2003) Ca^{2+} stores and Ca^{2+} entry differentially contribute to the release of IL-1 beta and IL-1 alpha from murine macrophages. J Immunol 170(6):3029–3036
59. Pelegrin P, Barroso-Gutierrez C, Surprenant A (2008) $P2X_7$ receptor differentially couples to distinct release pathways for IL-1β in mouse macrophage. J Immunol 180 (11):7147–7157
60. Vaughan KR, Stokes L, Prince LR, Marriott HM, Meis S, Kassack MU, Bingle CD, Sabroe I, Surprenant A, Whyte MK (2007) Inhibition of neutrophil apoptosis by ATP is mediated by the $P2Y_{11}$ receptor. J Immunol 179(12):8544–8553
61. Idzko M, Hammad H, van Nimwegen M, Kool M, Willart MA, Muskens F, Hoogsteden HC, Luttmann W, Ferrari D, Di Virgilio F, Virchow JC Jr, Lambrecht BN (2007) Extracellular ATP triggers and maintains asthmatic airway inflammation by activating dendritic cells. Nat Med 13(8):913–919
62. Hayashi T, Kawakami M, Sasaki S, Katsumata T, Mori H, Yoshida H, Nakahari T (2005) ATP regulation of ciliary beat frequency in rat tracheal and distal airway epithelium. Exp Physiol 90(4):535–544
63. Fu XW, Nurse CA, Cutz E (2004) Expression of functional purinergic receptors in pulmonary neuroepithelial bodies and their role in

hypoxia chemotransmission. Biol Chem 385 (3–4):275–284
64. Mounkaila B, Marthan R, Roux E (2005) Biphasic effect of extracellular ATP on human and rat airways is due to multiple P2 purinoceptor activation. Respir Res 6:143
65. Giaroni C, Knight GE, Zanetti E, Chiaravalli AM, Lecchini S, Frigo G, Burnstock G (2006) Postnatal development of P2 receptors in the murine gastrointestinal tract. Neuropharmacology 50(6):690–704
66. Cooke HJ, Wunderlich J, Christofi FL (2003) "The force be with you": ATP in gut mechanosensory transduction. News Physiol Sci 18:43–49
67. Van Crombruggen K, Van Nassauw L, Timmermans JP, Lefebvre RA (2007) Inhibitory purinergic P2 receptor characterisation in rat distal colon. Neuropharmacology 53 (2):257–271
68. Furuzono S, Nakayama S, Imaizumi Y (2005) Purinergic modulation of pacemaker Ca^{2+} activity in interstitial cells of Cajal. Neuropharmacology 48(2):264–273
69. Roman RM, Feranchak AP, Salter KD, Wang Y, Fitz JG (1999) Endogenous ATP release regulates Cl− secretion in cultured human and rat biliary epithelial cells. Am J Phys 276(6 Pt 1):G1391–G1400
70. Dutta AK, Woo K, Doctor RB, Fitz JG, Feranchak AP (2008) Extracellular nucleotides stimulate Cl^- currents in biliary epithelia through receptor-mediated IP_3 and Ca^{2+} release. Am J Physiol Gastrointest Liver Physiol 295(5):G1004–G1015
71. Masyuk AI, Gradilone SA, Banales JM, Huang BQ, Masyuk TV, Lee SO, Splinter PL, Stroope AJ, Larusso NF (2008) Cholangiocyte primary cilia are chemosensory organelles that detect biliary nucleotides via P2Y12 purinergic receptors. Am J Physiol Gastrointest Liver Physiol 295(4): G725–G734
72. Doctor RB, Matzakos T, McWilliams R, Johnson S, Feranchak AP, Fitz JG (2005) Purinergic regulation of cholangiocyte secretion: identification of a novel role for P2X receptors. Am J Physiol Gastrointest Liver Physiol 288(4):G779–G786
73. Liu R, Bell PD, Peti-Peterdi J, Kovacs G, Johansson A, Persson AE (2002) Purinergic receptor signaling at the basolateral membrane of macula densa cells. J Am Soc Nephrol 13(5):1145–1151
74. Bell PD, Lapointe JY, Sabirov R, Hayashi S, Peti-Peterdi J, Manabe K, Kovacs G, Okada Y (2003) Macula densa cell signaling involves ATP release through a maxi anion channel. Proc Natl Acad Sci U S A 100(7):4322–4327
75. Vallon V (2008) P2 receptors in the regulation of renal transport mechanisms. Am J Physiol Renal Physiol 294(1):F10–F27
76. Lee YJ, Park SH, Han HJ (2005) ATP stimulates Na+-glucose cotransporter activity via cAMP and p38 MAPK in renal proximal tubule cells. Am J Physiol Cell Physiol 289 (5):C1268–C1276
77. Wildman SS, King BF (2008) P2X receptors: epithelial ion channels and regulators of salt and water transport. Nephron Physiol 108 (3):60–67
78. Guan Z, Osmond DA, Inscho EW (2007) P2X receptors as regulators of the renal microvasculature. Trends Pharmacol Sci 28 (12):646–652
79. Werkstrom V, Andersson KE (2005) ATP- and adenosine-induced relaxation of the smooth muscle of the pig urethra. BJU Int 96(9):1386–1391
80. Ford AP, Gever JR, Nunn PA, Zhong Y, Cefalu JS, Dillon MP, Cockayne DA (2006) Purinoceptors as therapeutic targets for lower urinary tract dysfunction. Br J Pharmacol 147 (Suppl 2):S132–S143
81. Chopra B, Gever J, Barrick SR, Hanna-Mitchell AT, Beckel JM, Ford AP, Birder LA (2008) Expression and function of rat urothelial P2Y receptors. Am J Physiol Renal Physiol 294(4):F821–F829
82. Ruggieri MR Sr (2006) Mechanisms of disease: role of purinergic signaling in the pathophysiology of bladder dysfunction. Nat Clin Pract Urol 3(4):206–215
83. Gur S, Kadowitz PJ, Hellstrom WJ (2007) Purinergic (P_2) receptor control of lower genitourinary tract function and new avenues for drug action: an overview. Curr Pharm Des 13 (31):3236–3244
84. Banks FC, Knight GE, Calvert RC, Thompson CS, Morgan RJ, Burnstock G (2006) The purinergic component of human vas deferens contraction. Fertil Steril 85(4):932–939
85. Poletto Chaves LA, Pontelli EP, Varanda WA (2006) P2X receptors in mouse Leydig cells. Am J Physiol Cell Physiol 290(4): C1009–C1017
86. Lau DH, Metcalfe MJ, Mumtaz FH, Mikhailidis DP, Thompson CS (2009) Purinergic modulation of human corpus cavernosum relaxation. Int J Androl 32(2):149–155
87. Piper AS, Hollingsworth M (1996) P2-purinoceptors mediating spasm of the isolated uterus of the non-pregnant guinea-pig. Br J Pharmacol 117(8):1721–1729

88. Ziganshin AU, Zaitcev AP, Khasanov AA, Shamsutdinov AF, Burnstock G (2006) Term-dependency of P2 receptor-mediated contractile responses of isolated human pregnant uterus. Eur J Obstet Gynecol Reprod Biol 129(2):128–134
89. Papka RE, Hafemeister J, Storey-Workley M (2005) P2X receptors in the rat uterine cervix, lumbosacral dorsal root ganglia, and spinal cord during pregnancy. Cell Tissue Res 321 (1):35–44
90. Katugampola H, Burnstock G (2004) Purinergic signaling to rat ovarian smooth muscle: changes in P2X receptor expression during pregnancy. Cells Tissues Organs 178 (1):33–47
91. Min K, Munarriz R, Yerxa BR, Goldstein I, Shaver SR, Cowlen MS, Traish AM (2003) Selective P2Y2 receptor agonists stimulate vaginal moisture in ovariectomized rabbits. Fertil Steril 79(2):393–398
92. Bardini M, Lee HY, Burnstock G (2000) Distribution of P2X receptor subtypes in the rat female reproductive tract at late pro-oestrus/early oestrus. Cell Tissue Res 299 (1):105–113
93. Gallagher JA (2004) ATP P2 receptors and regulation of bone effector cells. J Musculoskelet Neuronal Interact 4(2):125–127
94. Burnstock G, Verkhratsky A (2009) Evolutionary origins of the purinergic signaling system. Acta Physiol 195:415–447
95. Burnstock G (1978) A basis for distinguishing two types of purinergic receptor. In: Straub RW, Bolis L (eds) Cell membrane receptors for drugs and hormones: a multidisciplinary approach. Raven Press, New York, pp 107–118
96. Burnstock G, Kennedy C (1985) Is there a basis for distinguishing two types of P_2-purinoceptor? Gen Pharmacol 16:433–440
97. Brake AJ, Wagenbach MJ, Julius D (1994) New structural motif for ligand-gated ion channels defined by an ionotropic ATP receptor. Nature 371(6497):519–523
98. Lustig KD, Shiau AK, Brake AJ, Julius D (1993) Expression cloning of an ATP receptor from mouse neuroblastoma cells. Proc Natl Acad Sci U S A 90:5113–5117
99. Valera S, Hussy N, Evans RJ, Adani N, North RA, Surprenant A, Buell G (1994) A new class of ligand-gated ion channel defined by P_{2X} receptor for extra-cellular ATP. Nature 371:516–519
100. Webb TE, Simon J, Krishek BJ, Bateson AN, Smart TG, King BF, Burnstock G, Barnard EA (1993) Cloning and functional expression of a brain G-protein-coupled ATP receptor. FEBS Lett 324:219–225
101. Dubyak GR (1991) Signal transduction by P_2-purinergic receptors for extracellular ATP. Am J Respir Cell Mol Biol 4:295–300
102. Burnstock G (2007) Purine and pyrimidine receptors. Cell Mol Life Sci 64 (12):1471–1483
103. Nicke A, Baumert HG, Rettinger J, Eichele A, Lambrecht G, Mutschler E, Schmalzing G (1998) $P2X_1$ and $P2X_3$ receptors form stable trimers: a novel structural motif of ligand-gated ion channels. EMBO J 17 (11):3016–3028
104. Burnstock G, Verkhratsky A (2012) Purinergic signaling and the nervous system. Springer, Heidelberg/Berlin, pp 1–715
105. Agboh KC, Webb TE, Evans RJ, Ennion SJ (2004) Functional characterization of a P2X receptor from *Schistosoma mansoni*. J Biol Chem 279(40):41650–41657
106. Fountain SJ, Parkinson K, Young MT, Cao L, Thompson CR, North RA (2007) An intracellular P2X receptor required for osmoregulation in *Dictyostelium discoideum*. Nature 448(7150):200–203
107. Demidchik V, Nichols C, Oliynyk M, Dark A, Glover BJ, Davies JM (2003) Is ATP a signaling agent in plants? Plant Physiol 133 (2):456–461
108. Jeter CR, Roux SJ (2006) Plant responses to extracellular nucleotides: cellular processes and biological effects. Purinergic Signal 2 (3):443–449
109. Kim SY, Sivaguru M, Stacey G (2006) Extracellular ATP in plants. Visualization, localization, and analysis of physiological significance in growth and signaling. Plant Physiol 142 (3):984–992
110. Verkhratsky A, Burnstock G (2014) Biology of purinergic signaling: its ancient evolutionary roots, its omnipresence and its multiple functional significance. Bioessays 36 (7):697–705
111. Abbracchio MP, Burnstock G (1998) Purinergic signaling: pathophysiological roles. Jpn J Pharmacol 78:113–145
112. Burnstock G (2016) Short- and long-term (trophic) purinergic signaling. Philos Trans R Soc Lond B Biol Sci 371(1700):20150422
113. Burnstock G, Verkhratsky A (2010) Long-term (trophic) purinergic signaling: purinoceptors control cell proliferation, differentiation and death. Cell Death Dis 1:e9

114. Burnstock G (2002) Purinergic signaling and vascular cell proliferation and death. Arterioscler Thromb Vasc Biol 22(3):364–373
115. Burnstock G (2008) Dual control of vascular tone and remodelling by ATP released from nerves and endothelial cells. Pharmacol Rep 60(1):12–20
116. Erlinge D, Burnstock G (2008) P2 receptors in cardiovascular physiology and disease. Purinergic Signal 4(1):1–20
117. Bodin P, Burnstock G (2001) Purinergic signaling: ATP release. Neurochem Res 26 (8–9):959–969
118. Burnstock G, Knight GE (2017) Cell culture: complications due to mechanical release of ATP and activation of purinoceptors. Cell Tissue Res 370:1–11
119. Moriyama Y, Hiasa M, Sakamoto S, Omote H, Nomura M (2017) Vesicular nucleotide transporter (VNUT): appearance of an actress on the stage of purinergic signaling. Purinergic Signal 13(3):387–404
120. Dahl G (2015) ATP release through pannexon channels. Philos Trans R Soc Lond Ser B Biol Sci 370:20140191
121. Lazarowski ER, Sesma JI, Seminario-Vidal L, Kreda SM (2011) Molecular mechanisms of purine and pyrimidine nucleotide release. Adv Pharmacol 61:221–261
122. Yegutkin GG (2014) Enzymes involved in metabolism of extracellular nucleotides and nucleosides: functional implications and measurement of activities. Crit Rev Biochem Mol Biol 49(6):473–497
123. Zimmermann H, Mishra SK, Shukla V, Langer D, Gampe K, Grimm I, Delic J, Braun N (2007) Ecto-nucleotidases, molecular properties and functional impact. An R Acad Nac Farm 73:537–566
124. Fields D, Burnstock G (2006) Purinergic signaling in neuron-glial interactions. Nat Rev Neurosci 7(6):423–436
125. Verkhratsky A, Krishtal OA, Burnstock G (2009) Purinoceptors in neuroglia. Mol Neurobiol 39(3):190–208
126. Burnstock G (2006) Pathophysiology and therapeutic potential of purinergic signaling. Pharmacol Rev 58(1):58–86
127. Burnstock G, Dale N (2015) Purinergic signaling in development and ageing. Purinergic Signal 11(3):277–305
128. Bogdanov YD, Dale L, King BF, Whittock N, Burnstock G (1997) Early expression of a novel nucleotide receptor in the neural plate of *Xenopus* embryos. J Biol Chem 272 (19):12583–12590
129. Ryten M, Hoebertz A, Burnstock G (2001) Sequential expression of three receptor subtypes for extracellular ATP in developing rat skeletal muscle. Dev Dyn 221:331–341
130. Cheung K-K, Chan WY, Burnstock G (2005) Expression of P2X receptors during rat brain development and their inhibitory role on motor axon outgrowth in neural tube explant cultures. Neuroscience 133(4):937–945
131. Tew EMM, Anderson PN, Burnstock G (1992) Implantation of the myenteric plexus into the corpus striatum of adult rats: survival of the neurones and glia and interactions with host brain. Restor Neurol Neurosci 4:311–321
132. Höpker VH, Saffrey MJ, Burnstock G (1996) Neurite outgrowth of striatal neurons *in vitro*: involvement of purines in the growth promoting effect of myenteric plexus explants. Int J Dev Neurosci 14(4):439–451
133. Burnstock G, Ulrich H (2011) Purinergic signaling in embryonic and stem cell development. Cell Mol Life Sci 68:1369–1394
134. Grimm I, Messemer N, Stanke M, Gachet C, Zimmermann H (2009) Coordinate pathways for nucleotide and EGF signaling in cultured adult neural progenitor cells. J Cell Sci 122(Pt 14):2524–2533
135. Burnstock G (2013) Purinergic signaling in the lower urinary tract. Acta Physiol 207 (1):40–52
136. Burnstock G (2014) Purinergic signaling in the urinary tract in health and disease. Purinergic Signal 10(1):103–155
137. Burnstock G, Cocks T, Kasakov L, Wong HK (1978) Direct evidence for ATP release from non-adrenergic, non-cholinergic ("purinergic") nerves in the guinea-pig taenia coli and bladder. Eur J Pharmacol 49:145–149
138. Burnstock G (2001) Purinergic signaling in lower urinary tract. In: Abbracchio MP, Williams M (eds) Handbook of experimental pharmacology, vol 151/I. Purinergic and pyrimidinergic signaling: I – Molecular, nervous and urinogenitary system function, vol 151/I. Handbook of experimental pharmacology. Springer-Verlag, Berlin, pp 423–515
139. Vidal M, Hicks PE, Langer SZ (1986) Differential effects of α,β-methylene ATP on responses to nerve stimulation in SHR and WKY tail arteries. Naunyn Schmiedebergs Arch Pharmacol 332:384–390
140. Boeynaems JM, Communi D, Gonzalez NS, Robaye B (2005) Overview of the P2 receptors. Semin Thromb Hemost 31(2):139–149

141. Burnstock G, Ralevic V (2014) Purinergic signaling and blood vessels in health and disease. Pharmacol Rev 66(1):102–192
142. Thachil J (2016) Antiplatelet therapy – a summary for the general physicians. Clin Med 16 (2):152–160
143. Burnstock G (2008) Commentary. Purinergic receptors as future targets for treatment of functional GI disorders. Gut 57 (9):1193–1194
144. Burnstock G, Jacobson KA, Christofi FL (2017) Purinergic drug targets for gastrointestinal disorders. Curr Opin Pharmacol 37:131–141
145. Dal Ben D, Antonioli L, Lambertucci C, Fornai M, Blandizzi C, Volpini R (2018) Purinergic ligands as potential therapeutic tools for the treatment of inflammation-related intestinal diseases. Front Pharmacol 9:212
146. Longhi MS, Moss A, Jiang ZG, Robson SC (2017) Purinergic signaling during intestinal inflammation. J Mol Med (Berl) 95 (9):915–925
147. Yerxa BR (2001) Therapeutic use of nucleotides in respiratory and ophthalmic diseases. Drug Dev Res 52:196–201
148. Burnstock G, Di Virgilio F (2013) Purinergic signaling in cancer. Purinergic Signal 9:491–540
149. Di Virgilio F, Sarti AC, Falzoni S, De ME, Adinolfi E (2018) Extracellular ATP and P2 purinergic signaling in the tumour microenvironment. Nat Rev Cancer 18(10):601–618
150. Shabbir M, Burnstock G (2009) Purinergic receptor-mediated effects of ATP in urogenital malignant diseases. Int J Urol 16 (2):143–150
151. White N, Burnstock G (2006) P2 receptors and cancer. Trends Pharmacol Sci 27 (4):211–217
152. Burnstock G (2017) Purinergic signaling: therapeutic developments. Front Pharmacol 8:661
153. Bradbury EJ, Burnstock G, McMahon SB (1998) The expression of $P2X_3$ purinoceptors in sensory neurons: effects of axotomy and glial-derived neurotrophic factor. Mol Cell Neurosci 12(4–5):256–268
154. Chen CC, Akopian AN, Sivilotti L, Colquhoun D, Burnstock G, Wood JN (1995) A P2X purinoceptor expressed by a subset of sensory neurons. Nature 377:428–431
155. Burnstock G (1996) A unifying purinergic hypothesis for the initiation of pain. Lancet 347:1604–1605
156. Burnstock G (1999) Release of vasoactive substances from endothelial cells by shear stress and purinergic mechanosensory transduction. J Anat 194(3):335–342
157. Burnstock G (2009) Purinergic mechanosensory transduction and visceral pain. Mol Pain 5:69
158. Vlaskovska M, Kasakov L, Rong W, Bodin P, Bardini M, Cockayne DA, Ford APDW, Burnstock G (2001) $P2X_3$ knockout mice reveal a major sensory role for urothelially released ATP. J Neurosci 21(15):5670–5677
159. Rong W, Burnstock G (2004) Activation of ureter nociceptors by exogenous and endogenous ATP in guinea pig. Neuropharmacology 47(7):1093–1101
160. Wynn G, Burnstock G (2006) Adenosine 5-′-triphosphate and it's relationship with other mediators that activate pelvic afferent neurons in the rat colorectum. Purinergic Signal 2:517–526
161. Rong W, Burnstock G, Spyer KM (2000) P2X purinoceptor-mediated excitation of trigeminal lingual nerve terminals in an *in vitro* intra-arterially perfused rat tongue preparation. J Physiol 524(Pt 3):891–902
162. Cockayne DA, Hamilton SG, Zhu Q-M, Dunn PM, Zhong Y, Novakovic S, Malmberg AB, Cain G, Berson A, Kassotakis L, Hedley L, Lachnit WG, Burnstock G, McMahon SB, Ford APDW (2000) Urinary bladder hyporeflexia and reduced pain-related behaviour in $P2X_3$-deficient mice. Nature 407 (6807):1011–1015
163. Bele T, Fabbretti E (2015) P2X receptors, sensory neurons and pain. Curr Med Chem 22(7):845–850
164. Burnstock G (2009) Purinergic receptors and pain. Curr Pharm Des 15(15):1717–1735
165. Burnstock G (2016) Purinergic receptors and pain – an update. In: Front Med Chem, vol 9, pp 3–55
166. Inoue K (2007) P2 receptors and chronic pain. Purinergic Signal 3:135–144
167. Tsuda M, Inoue K (2016) Neuron-microglia interaction by purinergic signaling in neuropathic pain following neurodegeneration. Neuropharmacology 104:76–81
168. Burnstock G (2008) Purinergic signaling and disorders of the central nervous system. Nat Rev Drug Discov 7:575–590
169. Cheffer A, Castillo ARG, Corrêa-Velloso J, Gonçalves MCB, Naaldijk Y, Nascimento I, Burnstock G, Ulrich H (2017) Purinergic system in psychiatric diseases. Mol Psychiatry 23(1):94–106

Chapter 2

Knockout and Knock-in Mouse Models to Study Purinergic Signaling

Robin M. H. Rumney and Dariusz C. Górecki

Abstract

Purinergic signaling involves extracellular purines and pyrimidines acting upon specific cell surface purinoceptors classified into the P1, P2X, and P2Y families for nucleosides and nucleotides. This widespread signaling mechanism is active in all major tissues and influences a range of functions in health and disease. Orthologs to all but one of the human purinoceptors have been found in mouse, making this laboratory animal a useful model to study their function. Indeed, analyses of purinoceptors via knock-in or knockout approaches to produce gain or loss of function phenotypes have revealed several important therapeutic targets. None of the homozygous purinoceptor knockouts proved to be developmentally lethal, which suggest that either these receptors are not involved in key developmental processes or that the large number of receptors in each family allowed for functional compensation. Different models for the same purinoceptor often show compatible phenotypes but there have been examples of significant discrepancies. These revealed unexpected differences in the structure of human and mouse genes and emphasized the importance of the genetic background of different mouse strains. In this chapter, we provide an overview of the current knowledge and new trends in the modifications of purinoceptor genes in vivo. We discuss the resulting phenotypes, their applications and relative merits and limitations of mouse models available to study purinoceptor subtypes.

Key words Knock-in, Knockout, Genetically modified animals, Purinergic signaling, Purinoceptor

1 Overview of the Purinergic Signaling

The first data to suggest that extracellular nucleotides could influence physiological activity came from a 1929 study by Drury and Szent-Györgyi. Adenosine monophosphate (then described as adenylic acid) and adenosine were isolated from heart muscle extract and yeast, respectively. Both were identified as active substances capable of transiently decreasing heart rate in a range of animal species [1]. Notwithstanding this early finding, it was not until the 1970s when Geoff Burnstock proposed ATP as a neurotransmitter within the autonomic nervous system and coined the term "purinergic signaling" [2]. Despite growing pharmacological and functional evidences, the idea that extracellular nucleotides could act as

Pablo Pelegrín (ed.), *Purinergic Signaling: Methods and Protocols*, Methods in Molecular Biology, vol. 2041,
https://doi.org/10.1007/978-1-4939-9717-6_2,

signaling molecules remained somewhat controversial, until the early 1990s when the first purinergic receptor was cloned by Eric A. Barnard's group [3]. This was followed by the identification of a series of other purinoceptors [4]. Purinoceptors have subsequently been classified according to their pharmacological properties into adenosine sensitive P1 receptors and ATP/ADP sensitive P2 receptors. The P2 receptors are further subdivided into P2X transmembrane ligand gated ion channels (formed by homomeric or heteromeric trimers of subunits) and P2Y nucleotide activated G-protein-coupled receptors. There are 19 human purinoceptors in total including four P1 adenosine receptors (A_1-AR, A_{2a}-AR, A_{2b}-AR, and A_3-AR), seven P2X (P2X1–7), and eight P2Y nucleotide receptors (P2Y1, P2Y2, P2Y4, P2Y6, and P2Y11–14) [5–7].

Purinergic signaling is regarded as the most widespread autocrine and paracrine extracellular signaling mechanism because purinoceptors have been identified across all major tissue types and are involved in processes as diverse as secretion, cell proliferation, cell death, immune responses, inflammation, pain, vasodilatation, and haemostasis. Given purinoceptors are functionally important in regulating so many physiological and pathological processes, they have been researched as 'druggable' targets, with some found to be clinically important indeed. For example, the antithrombotic drug clopidogrel effectively inhibits the P2Y12 receptor on platelets [8]. Because of the complex expression patterns and intricate, often opposing effects of purinoceptors' activation in health and disease, therapeutic advances have been possible thanks to in vivo studies involving a range of model organisms.

Indeed, purinoceptors are present not only in humans but across the phylogenetic range including in the key laboratory organisms: mouse, rat, chick, *Xenopus laevis*, zebrafish, and even in the single cell amoeba [9, 10]. Orthologs to all but the P2Y11 receptor have been found in mouse thus allowing this widely used model organism to be applied to the study of purinoceptor functions. However, not all therapeutic explorations have been successful and some drug trials ended in disappointment. For example, several P2X7 antagonists have been developed but these failed to meet expectations to be superior anti-inflammatory therapeutics. The reason for this appears to be the incomplete understanding of purinoceptor function in health and disease stemming from the properties of animal models used in preclinical studies. Therefore, this chapter aims to provide a succinct summary of various mouse models used to study purinergic receptors, their advantages and potential weaknesses.

2 Knockout and Knock-in Mice

The laboratory mouse has been the main animal model in biomedical studies for decades and therefore of great interest as a species for genetic modifications at the organismal level.

The initial transgenic mice were made by microinjection of a DNA construct containing a gene expression cassette under the control of a specific promoter–enhancer combination into the pronuclei of fertilized embryos at the one-cell stage. The random integration of one or more copies of a transgene into the mouse genome ensures that all daughter cells contain the transgene. However, the random integration carries risks of mutagenesis while the heterologous expression of a transgene offers limited opportunities to investigate the effects of its mutations.

Gene targeted knockout mice were first developed as a result of the combined work of Mario R. Capecchi, Oliver Smithies, and Martin J. Evans. In the 1980s, Evans's group isolated and cultured murine embryonic stem (ES) cells [11] and subsequently injected these into the blastocyst to generate the chimeric mouse so that a transgene introduced to ES cells would be carried down to the germ line [12]. Meanwhile, the groups of Capecchi and Smithies simultaneously carried out research into homologous recombination in mammalian cells. In 1988 Capecchi published the method of positive-negative antibiotic selection whereby the neomycin resistance gene (neo^r) is introduced to disrupt the target gene. The second element was the negative selection marker thymidine kinase (TK), which sensitizes cells to ganciclovir to engender cell death. If the targeting vector successfully undergoes homologous recombination with the target gene then only the neo^r is included, which allows the embryonic stem cells to survive positive selection with G418. However, if there is random integration then both neo^r and TK will be functional and any such modified ES cells will be excluded by negative TK selection [13]. Unsurprisingly, these significant advancements making it possible to modify chosen genes in embryonic stem cells and to generate mice that express the modified gene were recognized with the 2007 Nobel Prize in Physiology or Medicine awarded to Capecchi, Evans and Smithies. This methodology has proven invaluable in establishing gene functions, understanding cellular pathways and for manipulating these in order to model human diseases, understand developmental processes, and evaluate new therapeutic approaches.

A recent innovation is the clustered regularly interspaced short palindromic repeats (CRISPR)/Cas9 system wherein short RNA sequences are used to target a specific DNA sequence for inactivation with the DNA endonuclease CRISPR associated protein 9 (Cas9). This method has been successfully used to generate gene knockouts in both mouse and rat [14].

Gene knockout resulting in the elimination of a gene or part of a gene critically important for the expression of its encoded protein can provide key information on its function. However, the phenotypes of standard knockout mice can be complex because all tissues of the mouse may be affected in the absence of functional gene expression from the earliest stage of development. If a gene is involved in early development or viability, its knockout may results in early/embryonic lethality. On the other hand, it is not uncommon for knockout mice to show little or no phenotype due to early compensatory overexpression of other genes. Because of these complexities, more sophisticated models of controlled gene knockout have been developed. Conditional knockouts can be generated via *Cre-lox* recombination, whereby transgenic mice carrying *Cre* recombinase (an enzyme that excises DNA segments flanked by *loxP* sites) are crossed with mice carrying the gene of interest flanked by *loxP*. The resulting offspring have the target gene excised only in cells expressing the *Cre* enzyme [15]. A similar method utilizes flippase (*Flp*), a recombinase for removing sequences between two flippase recognition target sites (FRTs) [16]. Reversible conditional gene knockout has been made possible by the *loxP*-FRT Trap (LOFT) system, which utilizes a floxed allele (between two *loxP* sites) and a reversibly trapped null allele that can be conditionally switched to the wild type [17].

Recent advances have enabled the development of knockout rats. Knockout rats allow for useful comparison as they can be used to confirm results from mouse models, identify phenotypes specific to rodents or investigate genes not available in the mouse. Moreover, some behavioral and immune functions are closer to their human counterparts in rats than in mice. The first knockout rats were made using zinc-finger nucleases (ZFN), which edit the genome by creating double-strand breaks in DNA at targeted locations following injection into rat embryos [18]. In addition to the sequence-specific DNA modifying nucleases, such as ZFN and transcription activator-like effector nucleases (TALENS), the current methods involve transposon vector systems and CRISPR/Cas9 mentioned above [19, 20].

Although knockout organisms can be used to study the effect of the loss of function they cannot explain the impact of a specific gene mutation in vivo. This can be investigated using the 'knock-in' approach, where a specific DNA sequence is inserted to replace the endogenous sequence in its locus without any other gene disruptions. Given that the knocked-in sequence remains under the control of the endogenous promoter, its expression can be tissue- and temporal-specific. Such a knocked-in region could contain a particular mutation or the entire sequence from another species (ortholog), a terminal tag allowing for highly specific immunodetection and protein pulldown—with a plethora of resulting research applications. Moreover, this approach can also be used for conditional gene modifications using *Cre-lox* and Flp-FRT systems allowing the modified gene to be knocked out in a tightly controlled spatiotemporal manner. Another method involves the use of drug-controlled transcriptional activation where the administration of the drug (e.g., doxycycline, tamoxifen) allows for the reversible regulation of a target gene [21]. Because a gene is affected in a subset of cells and/or at a chosen time only, its function in the development, physiology, or pathology can be studied in a highly specific manner.

The detailed methods for generating knockouts have been reviewed extensively [22–24]. Continued development of this technology plays a key role in the rapid progress in life sciences and in the emergence of innovative medical therapies.

3 P1 Receptor Models

An overview of the different models for P1 adenosine receptors is summarized in Table 1.

Table 1
Adenosine receptor models

Receptor (*gene*)	Phenotype	Modification	Publication
A_1-AR (*Adora1*)	Protected against cardiac ischemia	Transgenic overexpression in the heart	[25]
	Impaired kidney function	Removal of exon 6	[26]
	Impaired kidney function; behavioral and neurological defects	Excision of entire coding region	[27]
	Cardiomyopathy	Cardiac overexpression that is reducible with Tet-OFF system	[28]
	Decreased depressive behavior	Tetracycline responsive overexpression in forebrain neurons	[29]
A_{2A}-AR (*Adora2A*)	Increased aggressiveness Increased blood pressure Hypoalgesia	*neo*^{r} insertion into exon 1	[30]
	Protected against ischemic injury	Replaced 3′ end of exon 12	[31]
	Neurological interaction with cannabinoid CB_1 receptor, increased resistance to synaptotoxin	Overexpression of human A_{2A}-AR sequence in rat	[32]
	More rapid recovery from ischemia Increased basal heart rate Increased systolic diameter	Cardiac overexpression	[33]
A_{2B}-AR (*Adora2B*)	Increased adipogenesis	Removal of exon 1	[34]
	Increased macrophage activity	Myeloid-specific knockout	[35]
A_3-AR (*Adora3*)	Impaired mast cell activation Increased tolerance to ischemia Protective against hypertension Fewer hematopoietic cells Increased sensitivity of nervous system	Disruption of A_3-AR coding region	[36]
	Embryonic lethal	Overexpression with SM22α vascular smooth muscle specific promoter	[37]

3.1 Adenosine A_1 Receptor

The adenosine A_1 receptor (A_1-AR) encoded by the *Adora1* gene has been investigated using two knockout mouse models, both described in 2001. Sun et al. used a targeting vector with lacZ and *neo*^{r} to replace the entire coding region [27], while Johansson et al. used a PGK-neo cassette to replace exon 6 only [26]. Both models are consistent with each other in showing a loss of tubulo-glomerular feedback in the absence of A_1-AR [27, 38]. The Sun et al. model has been used to show that the A_1-AR protects against renal injury following ischemia by decreasing necrosis and inflammation [39]. In an elegant experiment, A_1-AR expression in the inflamed kidneys of *Adora1*$^{-/-}$ mice was reconstituted by lentiviral transfection, which reduced tubular inflammation [40]. An additional kidney-related effect of *Adora1* knockout is reduced secretion of the enzyme renin required to regulate blood pressure [41]. Clinically, adenosine is used as an antiarrhythmic agent to treat supraventricular tachycardia and this therapeutic response to adenosine was found to be absent in the Sun et al. model, which lacks a functional A_1-AR [42].

The model by Johansson et al. is the more widely applied, having been used to demonstrate additional pathophysiological alterations caused by the lack of functional A_1-AR such as increased anxiety, hyperalgesia, and reduced neuronal response to hypoxia [26]. There is a neuroprotective aspect to the A_1-AR function as in the knockout, brain injury-induced epileptic seizures progress to the lethal *status epilepticus* [43].

Overexpression of *Adora1* in the heart tissue of mature mice has been engendered in the transgenic approach by subcloning a fragment of rat *Adora1* cDNA into a construct with the human growth hormone polyadenylation signal and the α-myosin heavy chain (MHC) promoter with a MEF-2 mutation. Although there were no noticeable differences in baseline heart function in A_1-AR overexpressing hearts, time to ischaemic contracture was increased and there was a higher rate of recovery in contractile function following reperfusion compared to controls [25]. Further study into this model revealed that overexpression of *Adora1* was associated with decreased calcium transport into the sarcoplasmic reticulum, which could protect against ischemia by delaying cytosolic calcium overload [44]. Paradoxically, overexpression of A_1-AR has also been associated with cardiomyopathy, so Funakoshi et al. developed an overexpression model by cloning human A_1-AR cDNA into an inducible cardiac-specific vector containing a mouse-MHC minimal promoter fused with the Tet-OFF system (rTA). Mice exhibited cardiomyopathy when A_1-AR was overexpressed in heart tissue unless treated with doxycycline, which inhibited rTA transactivation to attenuate A_1-AR expression [28].

Overexpression of the A_1-AR specifically in forebrain neurons has been engendered by a doxycycline activated tetracycline responsive bidirectional promoter to drive concurrent expression of

Adora1 and mCherry reporter with Tet-OFF. The resultant mice were less likely to exhibit depressive behavior, in contrast to *Adora1*$^{-/-}$ mice, which exhibited more depression-like symptoms [29].

3.2 Adenosine $A_{2}a$ Receptor

The adenosine A_{2A} receptor (A_{2A}-AR) encoded by the *Adora2A* gene was the first adenosine receptor for which a knockout was generated in 1997 by Ledent et al. [30]. In this model, *Adora2A* was inactivated with a *neo*^{r} containing cassette inserted into exon 1 [30]. This mouse was characterized by aggressiveness, increased blood pressure and a lack of response to the A_{2A} agonist CGS-21680, which in WT mice caused tachycardia [30]. However, in contrast to the *Adora1* knockouts, which exhibited hyperalgesia (an increased sensitivity to pain) [26], *Adora2A*$^{-/-}$ mice exhibited hypoalgesia (a decreased sensitivity to pain) [30]. A second *Adora2A*$^{-/-}$ model was generated by Chen et al. with a targeting vector that replaced the 3′ end of exon 12 and adjacent introns with a selection cassette to remove a region of protein located between the third and fourth transmembrane domains. This A_{2A}-AR knockout was neuroprotective as mice exhibited attenuated brain damage in response to transient focal ischemia, which suggests a role for the A_{2A}-AR in the development of ischemic injury [31].

A model of heart-specific overexpression of A_{2A}-AR was created with a cDNA construct containing an *Adora2A* cDNA fragment and the murine cardiac α-myosin heavy chain promoter. The results from this model were complex as although these mice exhibited speedier recovery from ischemia, they also had a higher basal heart rate and higher systolic diameter [33]. In animals overexpressing both A_{1}-AR and A_{2A}-AR in hearts, the overexpression of A_{2A}-AR appeared to mitigate the effect of A_{1}-AR overexpression upon calcium transport and suggested coordination of these two receptors to be required for healthy cardiac function [45].

The A_{2A}-AR has been further investigated in transgenic and knockout rats. Transgenic rats overexpressing the A_{2A}-AR were generated by injection of a DNA construct containing the full length human A_{2A}-AR sequence into the male pronucleus of rat zygotes. A_{2A}-AR overexpressing rats demonstrated lessened working memory in maze tests [32]. Subsequent studies into A_{2A}-AR overexpressing rats revealed interactions with the cannabinoid CB_{1} receptor in the striatum [46] and a reduction in the synaptotoxic effect of 3-nitropropionic acid, an inhibitor of the mitochondrial citric acid cycle, which would otherwise lead to symptoms of Huntington's disease [47].

3.3 Adenosine A2b Receptor

A knockout mouse model for the adenosine A_{2B} receptor (A_{2B}-AR) was generated with a construct containing β- galactosidase to replace exon 1 of the *Adora2B* gene. In this model, there was increased release of proinflammatory cytokines from macrophages

[34]. Subsequently, it was found that this model is more prone to obesity than WT controls placed on a high fat diet [48] because A_{2B}-AR is normally expressed on adipocyte progenitors, where its function is to prevent adipogenesis [49]. When crossed with the ApoE model of atherosclerosis, mice lacking A_{2B}-AR fared worse with increased plasma cholesterol, suggesting that functional A_{2B}-AR is protective against atherosclerosis [50].

A myeloid-specific knockout was generated by disruption of exon 2 of *Adora2B* with a floxed targeting vector containing *loxP* (LOFT), FRT, and PGK-*neo*^{r} cassettes, with subsequent crosses to create $A_2AR^{flox/flox}$ LysM-cre$^{+/-}$ mice. These *Adora2B* myeloid-specific knockouts showed increased IL-6 and TNF-α release from macrophages with increased phagocytosis and improved bacterial clearance [35].

3.4 Adenosine A3 Receptor

There is one very widely used *Adora3*$^{-/-}$ knockout for the adenosine A3a receptor (A_3-AR), generated by Salvatore et al. The targeting construct engineered to disrupt the A_3-AR coding region with the neo^r cassette was electroporated into ES cells and G418 resistant clones injected into C57BL/6 blastocysts. The resulting chimeras were bred onto B6D2 and C57BL/6 mice. This model was first used to identify the effects of the absence of this receptor upon mast cell degranulation and that it is required for in vivo generation of TNFα following LPS treatment [36]. Subsequently, it was demonstrated that the A_3-AR is required for mast cell activation [51].

In line with other adenosine receptors, the A_3-AR appears to be pathogenic in ischemia as *Adora3*$^{-/-}$ mice have an ischemia-tolerant phenotype with their hearts having enhanced postischemic recovery time [52], improved tolerance to ischemia-induced injury [53] and also their kidneys were protected against ischemia-driven failure [54]. Moreover, absence of the A_3-AR protects against hypertension [55] and affects hematopoietic cells: *Adora3*$^{-/-}$ mice have erythrocytes containing less haemoglobin [56], fewer platelets and immune cells [56, 57], with chemotaxis being impaired in both neutrophils [58] and macrophages [59]. The analysis of the nervous system of the *Adora3*$^{-/-}$ has revealed a role for this receptor in neuroprotection against hypoxia-induced damage [60, 61], excitotoxic cell death [62], responses to psychostimulants [63], and an anti-nociceptive pathway [64].

A tissue-specific A_3-AR overexpression model has been created with the mouse *Adora3* gene being driven by the SM22α vascular smooth muscle specific promoter. Interestingly, such targeted A_3-AR overexpression was embryonic lethal [37], while the *Adora3* knockout protected the cardiovascular system.

4 P2X Receptor Models

An overview of the different models for P2X receptors is summarized in Table 2.

4.1 P2X Purinoceptor 1

P2X purinoceptor 1 (P2X1) knockout has been generated using a targeting vector containing lacZ, herpes simplex virus (HSV)-TK and *neo*^{r} to delete a 350 bp region in exon 1 of the *P2rx1* gene,

Table 2
P2X receptor models

Receptor (*gene*)	Phenotype	Modification	Publication
P2X1 (*P2rx1*)	Decreased fertility, impaired neutrophil activation and chemotaxis	Removal of exon 1	[65]
P2X2 (*P2rx2*)	Dysregulated neuromuscular synaptogenesis, impaired carotid body function	Removal of exons 2–11	[66]
P2X3 (*P2rx3*)	Increased bladder capacity, decreased nociception, and impaired gastrointestinal function	1 kb removed from *P2X3* gene	[67]
P2X4 (*P2rx4*)	Impaired neuronal function, behavioral defects	Removal of exon 1	[68]
	Vascular defects, reduced nociception	Removal of exons 3–5	[69]
	Impaired electrophysiological response of macrophages to ATP	Cassette insertion into exon 2	[70]
	Severe heart failure phenotype	Insertion of *loxP* sites into introns 1 and 4	[71]
P2X5 (*P2rx5*)	Suppressed osteoclast differentiation	*P2rx5* excision	[72]
P2X6 (*P2rx6*)	No difference in renal electrolytes	Knockout cassette inserted within exon 2	[73]
P2X7 (*P2rx7*)	Decreased cytokine release from macrophages	Deletion of sequences encoding Cys506 to Pro532 region in the carboxyl-terminal domain on B6xDBA/2 background	[74]
	Lack of chronic inflammatory hypersensitivity Impaired cytokine release	Disruption of P2X7(a) variant only	[75]
	Suggests functional P2RX7 in neuroglia and glutamatergic pyramidal neurons	Humanized P2RX7 (*P2rx7*^{hP2RX7}) mice with conditional knockouts in neuronal populations	[76]
	Identified P2X7 expression on microglia and oligodendrocytes	BAC transgenic model with EGFP-tagged P2X7 receptor	[77]

corresponding to the first 45 amino acids of the P2X1 protein [65]. This model by Mulryan et al. was originally explored to investigate male infertility as ejaculate from male *P2rx1*$^{-/-}$ mice contained fewer sperm, which the authors suggested might make P2X1 a non-hormonal target for male contraception [65]. The same model of P2X1 ablation was subsequently used to demonstrate requirements for this receptor in neutrophil function, specifically as a promoter of chemotaxis and a negative regulator for activation [78, 79].

4.2 P2X Purinoceptors 2 and 3

The *P2rx2* and *P2rx3* genes both encode proteins that interact to form homomeric P2X purinoceptor 2 and 3, respectively or heterotrimeric receptors composed of both P2X2 and P2X3 subunits. The *P2rx2* knockout mouse was generated by Cockayne et al. by removing exons 2 through to 11 [66]. This model has been used to demonstrate roles for P2X2 in neural function, including regulation of the fast excitatory postsynaptic potentials in neurons [80], coordination of synaptogenesis within the neuromuscular junction (which is disorganized in *P2rx2*$^{-/-}$ mice) [81], and regulation of the carotid body response to hypoxia [82].

P2X3 deficient mice were developed with a *loxP*-flanked targeting vector containing *neo*^{r} and HSV-TK to delete 1 kb from the *P2rx3* sequence [67]. Initial studies with *P2rx3*$^{-/-}$ mice described an increase in bladder capacity with decreased frequency of urination and decreased pain sensitivity [67]. Dorsal root ganglion (DRG) neurons from *P2rx3*$^{-/-}$ mice lack the rapid desensitizing response to ATP that is present in WT controls [83]. This model has also demonstrated a requirement for P2X3 in peristalsis and mechanosensation along the gastrointestinal tract [84, 85].

While there are noticeable phenotypes of either *P2rx2*$^{-/-}$ or *P2rx3*$^{-/-}$ *mice*, there is at least some redundancy between these two receptors as there is a set of specific phenotypes in the *P2rx2*$^{-/-}$/ *P2rx3* $^{-/-}$ double knockout. Multiple studies have focused on an intriguingly unusual feature of *P2rx2*$^{-/-}$/ *P2rx3* $^{-/-}$ mice in that they are "taste blind", unable to detect salt or artificial sweetener and have decreased preference for fat or maltodextrin [86, 87].

4.3 P2X Purinoceptor 4

There are at least four available murine *P2rx4* knockouts. One model, first described by Sim et al., was generated by the insertion of a LacZ neomycin cassette in place of the first exon of the *P2rx4* gene [68]. This model is by far the most widely used to study P2X4 deficiency. In the nervous system, P2X4 depletion has been used to demonstrate a requirement for this receptor for synaptic potentiation in the hippocampus [68], synaptic strengthening [88], and activation of microglia following status epilepticus [89]. Moreover, the Sim et al. *P2rx4* knockout exhibits behavioral defects including increased ethanol intake [90, 91] and, in the male mice, phenotypic abnormalities consistent with autism-spectrum disorders [92]. The

Sim et al. model also demonstrated requirements for P2X4 in pain and inflammation. Peripheral nerve injury triggers de novo expression of P2X4 on microglia, while *P2rx4*$^{-/-}$ mice have decreased sensitivity to neuropathic pain following such injury, possibly due to impaired brain-derived neurotrophic factor signaling [93]. Inflammatory pain hypersensitivity can be triggered by P2X4-mediated release of PGE2 from macrophages, a mechanism that can be transferred to naive mice by an injection with ATP-primed macrophages from WT but not P2X4 deficient mice [94].

Another model of P2X4 depletion was generated by Yamamoto et al. via replacement of exons 3–5 with a floxed *neo*^{r} cassette susceptible to remove by Cre-recombinase [69]. This model was first used to identify the effects of a lack of P2X4 upon blood vessels, as it presented with lack of vessel dilation, increased blood pressure, and reduced vascular remodeling [69]. In common with the previous model by Sim et al., this model by Yamamoto et al. exhibits decreased pain sensitivity, in particular chronic neuropathic and inflammatory pain [95].

A third P2X4 loss-of-function model was generated by Brône et al. by insertion of a LacZ reporter-containing cassette into the second coding exon of the *P2rx4* gene [70]. Patch clamping on peritoneal macrophages from this model identified electrophysiological responses to ATP, which were absent in *P2rx4*$^{-/-}$ cells and thus suggested that P2X4 is required for macrophage activation, a finding that is consistent with the Sim et al. model.

In 2014 a fourth model was made by Yang et al. by insertion of *loxP* sites into introns 1 and 4. The cassette was removed and floxed pups were generated by crossing chimeric males with ROSA26-Flpe females to generate *P2rx4*$^{floxed/+}$ F1 offspring, which were backcrossed over ten generations onto C57BL/6 mice [71]. This model was used to demonstrate a protective role for P2X4 expression in cardiac myocytes in heart failure [71], an effect which may be consistent with the vascular role identified in the Yamamoto model.

4.4 P2X Purinoceptor 5

P2X5 has been less well studied than other members of the P2X subfamily of purinoceptors. To date, there is only one report we are aware of that studied *P2rx5*$^{-/-}$ mice generated with sperm obtained from the International Mouse Strain Resource (IMSR). The role of P2X5 was investigated in murine osteoclasts, as it is highly expressed during the differentiation of this cell type, and inhibition of P2X5 was shown to suppress osteoclast differentiation. Consistent with these observations, fewer osteoclasts formed from precursor cells derived from *P2rx5*$^{-/-}$ mice, and there was decreased resorption of dentine in vitro. Surprisingly, there were no differences in bone development between WT and knockout mice, however LPS-stimulated inflammatory bone loss associated with increased osteoclast numbers was absent in *P2rx5*$^{-/-}$ mice, thus suggesting a key role for this receptor [72].

4.5 P2X Purinoceptor 6

To our knowledge there has been only one *P2rx6* knockout analyzed to date. In this study by Baaij et al., mice were made by the insertion in exon 2 of a LacZ-neo containing knockout cassette and backcrossed onto C57BL/6 mice. Renal electrolytes were measured from serum and urine only to find no significant difference between samples from *P2rx6* knockout and WT controls, which may suggest a lack of role for P2X6 in electrolyte homeostasis or a redundancy with other purinoceptors [73].

4.6 P2X Purinoceptor 7

P2X7, perhaps the most comprehensively studied member of the family, has several features that make it unique among purinoceptors. P2X7 has uniquely lower affinity for ATP [96]. Its activation can evoke tonic, prosurvival effects while chronic stimulation can be damaging and these effects depend on agonist concentration and stimulus duration. The downstream responses involve caspase activation, inflammatory mediator processing and release, membrane blebbing and cell death, with roles in both the innate and adaptive immune responses [97, 98]. Importantly, the list of P2X7 functions in health and disease is still growing. Structurally, it has a uniquely long carboxyl tail, which appears responsible for P2X7 function as an ion channel, which can also form a nonselective transmembrane pore permeable to molecules of up to 900 Da [99].

Because of these unique features and widespread functions, P2X7 has been the subject of intense research including in knockout models. However, generation of these mice has proven more challenging than initially anticipated. In the first *P2rx7*$^{-/-}$ mouse model described by Solle et al. the targeting vector was generated to delete sequences encoding Cys506 to Pro532 region in the carboxyl-terminal domain. The first characterization of this model observed a lower cytokine release from *P2rx7*$^{-/-}$ macrophages in response to ATP stimulus [74].

In 2005, Chessell et al. disrupted exon 1 of *P2rx7*. Homozygotes were backcrossed six times with C57Bl/6J. In this *P2rx7*$^{-/-}$ model there was a lack of thermal hypersensitivity following partial nerve ligation and a lack of chronic inflammatory hypersensitivity following injection with Freund's complete adjuvant, both of which could be explained by impaired cytokine release [75]. Both these models were maintained on a C57Bl6 or B6xDBA/2 backgrounds, with these strains subsequently found to have a naturally occurring proline to leucine substitution at position 451 in the cytoplasmic tail, which impairs the function of P2X7 receptor. Thus, comparisons of the effects of the gene knockout against the already impaired wild type receptor in *P2rx7*$^{+/+}$ animals could be somewhat misleading [100]. Moreover, the Chessell et al. model was also found to be a hypomorph. It was revealed that the murine *P2rx7* gene has two promoters driving transcripts encoding two N-terminal isoforms. Although the P2X7(a) variant was inactivated in the Chessell et al. model, the P2X7(k) variant (which is less abundant but 8–10 times more sensitive to agonist) remained intact. Expression of these transcript variants also differs by cell type, so although P2X7(k) is predominant on T cells, P2X7(a) is predominant on macrophages and loss of this splice variant could still account for decreased cytokine release [101].

In bone, the P2X7 receptor is functionally important in both bone-forming osteoblasts and resorbing osteoclasts. As such, the bone phenotypes of both the Solle et al. and the Chessell et al. models have been studied and revealed apparently contradicting phenotypes.

The Solle et al. model showed decreases in femoral bone content together with increased bone resorption by osteoclasts [102] and a reduction in bone formation in response to mechanical loading [103]. In contrast, the Chessell et al. model exhibited increased cortical thickness [104]. These seemingly contradictory results may be connected to a confounding factor, as the Solle et al. model was maintained on a B6xDBA/2 background having inherently low bone mass [105]. To overcome these issues, Sydberg et al. bred female B6 *P2rx7*$^{-/-}$ from the Chessell et al. model with BALB/cJ for five generations to generate *P2rx7*$^{-/-}$ mice without the confounding factors of either the P451L mutation or pre-existing low bone mass. Bone mineral density was greatly increased in this *P2rx7* knockout model [106].

In 2017, Metzger et al. generated a mouse model with human *P2RX7* cDNA knocked in to the murine *P2rx7* locus. The targeting vector included a *loxP* site with the 3′-end of the first intron of mouse *P2rx7* but with the second murine *P2rx7* exon replaced with exons 2–13 of human *P2RX7* cDNA followed by selection markers including *neo*^{r}, a poly-A signals and a second *loxP* site. Chimeras were bred to generate *P2rx7*$^{hP2RX7\text{-}neo}$ mice, which were crossed with *Deleter-Flp* mice to remove the FRT-flanked region of the selection cassette to generate mice with humanized *P2RX7* (*P2rx7* ^{hP2RX7}) on a 129S2/Sv × C57BL/6N background [76]. The humanized *P2RX7* ^{hP2RX7} mice were used to generate a set of conditional *P2RX7* knockout models by breeding with mice carrying a set of Cre drivers that were either ubiquitously expressed or tailored to specific neuronal cell types to have controlled P2X7 knockout. This model was used to show the presence of P2X7 in neuroglia and glutamatergic pyramidal neurons of the hippocampus [76].

In 2018, Kaczmarek-Hájek et al. generated P2X7-bacterial artificial chromosome (BAC) transgenic mice to allow in vivo observations of EGFP-tagged P2X7 receptors. The BAC clone contained the full-length *P2rx7* sequence with an EGFP containing cassette upstream of the stop codon in exon 13. Specificity of GFP labeling and concomitance with endogenous P2X7 was confirmed by nanobody immunohistochemical labeling. Immunofluorescent colocalization with cell specific markers suggested that within the central nervous system P2X7 is expressed only in microglia and oligodendrocytes but not in neurons [77], in contrast to pyramidal hippocampal neuron expression identified in the previous study.

Moreover, recent data expanded the pathophysiological significance of P2RX7 even further but also increased the uncertainty of its cellular localization, by showing that P2X7 receptor located on pericytes of cerebral microvessels may influence the integrity of the blood-brain barrier in certain pathologies [107].

5 P2Y Receptor Models

An overview of the different models for P2Y receptors is summarized in Table 3.

5.1 P2Y Purinoceptor 1

Two *P2ry1* knockouts were both described in 1999. In one made by Leon et al., the entire coding sequence for the *P2ry1* gene was cloned into the pBluescript-KS plasmid and interrupted with a cassette containing PGK, *neo*^{r}, and poly(A) [109]. Fabre et al. developed a *P2ry1* knockout by replacing a 4.7 kb fragment with a construct containing the *neo*^{r} cassette [108]. Both of these models were remarkably similar in that they both exhibited decreased thrombosis and increased bleeding time [108, 109]. A couple of recently published reports described *P2ry1* knockout models from The Jackson Laboratory. The B6.129P2-*P2ry1*^{tm1Bhk}/J model has been used to demonstrate a requirement for functional P2Y1 receptors in motor control of the colon [110], and the knockout showed lack of neuromuscular transmission in parts of the gastrointestinal tract [122].

5.2 P2Y Purinoceptor 2

The *P2y2* knockout was generated with a targeting vector corresponding to 597 bp of the *P2ry2* cDNA sequence that was replaced with *neo*^{r} via homologous recombination. The model was first used to demonstrate that P2Y2 is required for calcium signaling in lung fibroblasts [111], and since has been exploited to demonstrate gradient sensing in neutrophil chemotaxis (together with adenosine receptor A3) [58].

5.3 P2Y Purinoceptor 4

The *P2ry4*$^{-/-}$ knockout was developed by deleting the first 398 bp of its coding sequence [112]. The main roles for the P2Y4 receptor identified in this model have been related to ion transport and cardiac function. The first study exploiting this model found P2Y4R to be required for epithelial chloride transport in the jejunum [112] with a subsequent study finding it was needed for potassium secretion in the colon [123]. In addition to ion transport through the epithelium of the intestinal tract, P2Y4 appears to play an important role in the cardiovascular system. *P2ry4*$^{-/-}$ mice exhibit abnormal postnatal development of the heart with decreased

Table 3
P2Y receptor models

Receptor (*gene*)	Phenotype	Modification	Publication
P2RY1 (*P2ry1*)	Both models by Fabre et al. and Leon et al. exhibit decreased thrombosis and increased bleeding time	4.7 kb fragment removed and replaced	[108]
		Sequence interrupted with a PGK, *neo*^{r} and poly(A) containing cassette	[109]
	Impaired function of gastrointestinal tract	B6.129P2-*P2ry1*^{tm1Bhk}/J model from Jackson Labs	[110]
P2RY2 (*P2ry2*)	Impaired calcium signaling, impaired neutrophil chemotaxis, bone loss	597 bp replaced with *neo*^{r} containing cassette	[111]
P2RY4 (*P2ry4*)	Reduced ion transport in jejunum and colon. Decreased left ventricle in heart. Reduction in myocardial hypertrophy following exercise	Deletion of 398 bp of coding sequence	[112]
P2RY6 (*P2ry6*)	Ablated macrophage and smooth muscle responses to UDP/UTP nucleotides; cardiac defects, reduced colorectal cancer progression	One *loxP* site upstream of the start codon, another downstream of the polyadenylation signal	[113]
	Increased cytokine production following stimulation of lung with dust mite extract	*P2ry6* (flox/flox) cre/+ tamoxifen inducible knockout	[114]
	Decreased potentiation of macrophage inflammatory signals	Replacement of third coding exon	[115]
	Decreased ability to fight vesicular stomatitis virus infection	151 bp CRISPR deletion	[116]
P2RY12 (*P2ry12*)	Impaired inflammatory response	Subcloned fragments of *P2ry12* into pOSDupDel plasmid flanking *neo*^{r}	[117]
	Impaired function of cells from monocyte-macrophage lineage	Fragments of *P2ry12* subcloned into pLNL vector	[118]
P2RY13 (*P2ry13*)	Impaired reverse cholesterol transport, impaired bone formation	Replacement of exon 1 and the first 182 bp of exon 2 with neo^r cassette	[119]
P2RY14 (*P2ry14*)	Impaired gastrointestinal tract function	Replacement of 892 bp of coding sequence	[120]
	Impaired insulin release and smooth muscle function	Partial replacement of the *P2ry14* with an EMCV-IRES/β-galactosidase cassette	[121]

mass of the left ventricle [124]. Subsequent adaptations to exercise in adult mice, including myocardial hypertrophy, are dependent upon functional P2Y4 as these are reduced in *P2ry4*$^{-/-}$ mice [125].

5.4 P2Y Purinoceptor 6

To date, at least four distinct knockout models for this purinoceptor have been described. The first, described by Bar et al., was generated with a targeting construct containing a pair of *loxP* sites (one upstream of the start codon and another downstream of the polyadenylation signal) and a neo^r cassette flanked by FRT sites in the pPKOEZ plasmid. G418 resistant R1 ES clones were aggregated with CD1 morulae for implantation into CD1 mice. Resulting male progeny were crossed with PGK-Cre female mice to generate the $P2ry6^{-/-}$ mouse. Functionally, the UDP and UTP responses in macrophages from these animals were ablated compared to WT controls, in which these nucleotides increased inositol phosphate production. Moreover, there was a decrease in cytokine release by LPS-stimulated $P2ry6^{-/-}$ macrophages. Furthermore, there was an ablated response to UDP and UTP evoked vascular smooth muscle contraction [113]. Further cardiovascular characteristics of this $P2ry6^{-/-}$ mouse model include macrocardia, cardiac hypertrophy, and loss of myogenic tone [126, 127]. Intriguingly, colorectal cancer progression was found to be reduced in $P2ry6^{-/-}$ mice and it has been suggested that the functional receptor facilitates cancer progression by reducing apoptosis [128].

An inducible P2Y6 ablation model was described by Giannattasio et al. *LoxP* sites were inserted either side of exon 3 with an FRT-flanked PGK-Neo cassette at the 3′ end. 129/Sv ES cells surviving G418 selection were injected into C57BL/6 blastocysts and chimeras were crossed with C56BL/6 mice to generate $P2ry6^{-/-}$ (flox/+) mice. With sequential generations intracrossed, crossed with homozygous cre/cre mice crossed with flox/flox mice, *P2ry6* (flox/flox) cre/+ mice were generated in which injection of tamoxifen could induce Cre-recombinase-driven *P2ry6* ablation. This model was used to investigate the effect of dust mite extract upon lung inflammation where, in contrast to the model by Bar et al., cytokine production was increased in *P2ry6* depleted mice [114].

Two more models were described in 2014. In one model by Garcia et al., the targeting vector was made with the lambda KOS vector system to replace the third coding exon of the *P2ry6* gene with a LacZ neo^r cassette. 129/SvEvBrd ES cells containing the modification were injected into C57BL/6 albino blastocysts and backcrossed four times onto the C57BL/6 strain [115]. This model was used to further demonstrate a role for functional P2Y6 in the potentiation of macrophage inflammatory signals consistent with Bar et al. and in contrast to Giannattasio et al. Finally, Li et al. used CRISPR/Cas9 to delete a random 151 bp sequence causing the receptor loss of function. Infection with vesicular stomatitis virus was more severe in mice lacking functional P2Y6, whereas viral quantity could be decreased by UDP treatment in cells expressing P2Y6 [116].

5.5 P2Y Purinoceptor 12

The ADP sensitive P2Y12 receptor is perhaps best known as the target of the blockbuster drug clopidogrel used to prevent platelet aggregation [8]. Two *P2ry12* knockouts were established in the early 2000s. The first, described by Foster et al. was generated with a targeting vector made by subcloning genomic fragments of *P2ry12* into the pOSDupDel plasmid flanking *neo*^{r}. The linearized vector was electroporated into 129/SV ES cells, which were selected with G418 and ganciclovir prior to injection into C57BL/6 blastocysts [117]. In the second model by André et al. fragments of the *P2ry12* gene were subcloned into the pLNL vector, linearized and electroporated into the E14Tg2a ES cell line. Subsequently generated chimeric mice were bred onto the C57BL/6 background [118]. The Foster et al. model has been used to demonstrate further complex roles for the P2Y12 receptor as it is protective against LPS-stimulated inflammation [129] and augments inflammation in atherosclerosis [130]. A similar role for the P2Y12 receptor has been identified in the model by André et al. where this receptor has been found promoting the development of atherosclerosis by driving leukocyte activity [131]. The model by André et al. has been further used to identify roles for the P2Y12 receptor in specific immune cell populations: In dendritic cells, P2Y12 appears to be required for endocytosis and antigen presentation [132]. Meanwhile, absence of the P2Y12 receptor abolishes nucleotide driven chemotaxis in microglia [133] and its presence is required for microglia involvement in neuropathic pain [134]. Another cell from the monocyte–macrophage lineage is the bone resorbing osteoclast, whose activity is increased by extracellular ADP. Mice lacking the P2Y12 receptor were protected against age, arthritis, osteoporosis, and tumor-associated bone loss [135]. Lastly, one intriguing role for P2Y12 is in lung carcinoma model where the spread of Lewis cancer cells depended upon platelet function. In *P2ry12*$^{-/-}$ mice, primary tumor burden was unaltered compared to *P2ry12*$^{+/+}$ controls, but metastasis was decreased as cancer-induced changes in platelet activity are mediated by the P2Y12 receptor [136].

5.6 P2Y Purinoceptor 13

The *P2ry13* knockout was generated by replacing exon 1 and the first 182 bp of exon 2 of the *P2ry13* gene with a *neo*^{r} knockout cassette [119]. This model was first used to demonstrate a role for this receptor in reverse cholesterol transport [119] but its major roles have since been identified in bone. *P2ry13*$^{-/-}$ mice have a lower bone formation rate that WT controls [137], although interestingly the knockout exhibits a greater increase in bone formation following mechanical loading [138]. In the *P2ry13*$^{-/-}$ model, serum FGF-23 is higher while ALP is lower in juvenile animals suggesting a role for this receptor in FGF-23 secretion and phosphate metabolism, both of which influence bone formation [139].

5.7 P2Y Purinoceptor 14

The P2Y14 receptor is a relatively recently discovered UDP-sensitive member of the purinoceptor family, formerly known as the orphan GPCR KIAA0001. To date at least two knockout mouse models have been described. In the first of these by Bassil et al., an 892-bp sequence encoding six transmembrane domains was replaced with IRES-lacZ expression cassette [120]. The positive selection cassette contained the neomycin phosphotransferase gene driven by the PGK promoter. The construct was injected into E14.1 ES cells with successfully targeted clones injected into C57BL/6J blastocysts. Chimeras were crossed with C57BL/6J and backcrossed with C57BL/6J for nine generations. LacZ staining revealed expression in the stomach and functional P2Y14 was found to be required for stomach circular muscle tension and gastric emptying [120].

More recently, another P2RY14$^{-/-}$ mouse was developed by partial replacement of the *P2ry14* gene with an EMCV-IRES/β-galactosidase floxed *neo^r* cassette. The *neo^r* was removed by breeding with EIIa-Cre mice and the resultant *P2ry14*$^{-/-}$ progeny were backcrossed onto the 129S6 line. X-Gal staining revealed widespread expression of P2Y14 in mouse tissues. Physiological roles of the P2Y14 receptor determined using this model included influence over insulin release and smooth muscle function in both airways and gastrointestinal tract [121], consistent with reports from Bassil et al.

6 Discussion

The methodology to develop knock-in and knockout models has been a vital tool contributing to our understanding of the in vivo function of purinergic receptors. To date, there is at least one knockout mouse model per receptor that has the mouse ortholog. Where there are more than one, these often show matching phenotypes (e.g., both *P2ry1*$^{-/-}$ models exhibit decreased thrombosis and increased bleeding time [108, 109]) or phenotypes that are complementary (e.g., the André et al. [118] *P2ry12*$^{-/-}$ model has impaired immune cell function while the Foster et al. model [117] has an impaired inflammatory response). In instances such as these, where independently developed models that target the same gene in different ways deliver matching or complementary results, there is a strong indication that results obtained are a genuine representation of a given purinoceptor function.

It is notable that none of the homozygous purinoceptor knockouts proved to be developmentally lethal while the overall embryonic lethality of mouse gene knockouts is estimated at 15%. This might suggest that none of these receptors is involved in critically important developmental processes. The alternative explanation could be that the large number of receptors in each family affords a level of genetic redundancy that can overcome the loss of expression of a single one. Limited developmental expression data for most purinoceptors makes answering this question difficult. Several of the P2X purinoceptors can function as hetero-oligomers. Therefore, it is not impossible that the functional receptor *trimer* in knockouts can contain different subunits that the wild type counterpart. However, few multiple knockouts have been studied [86, 87].

It is also important to note that gene knockout may result in a phenotype significantly different to that produced by mutations of the human ortholog or may fail to produce an observable phenotype in a mouse. This might be the result of divergence in gene functions between the two species or the insufficiently subtle testing tools (what Lewis Wolpert summed up as an inability to take the knockout mouse for a test night at the opera). An example of the significance of genetic differences is the presence of two promoters in the murine but not human *P2X7* gene, which had confused the knockout analyses for some time.

Another difficulty with the interpretation of gene knockouts is the apparent phenotypic variability. In some studies, knockout mice were screened to identify a broad spectrum of phenotypes while in others the analyses have been focused on the specific alteration or organ. Most laboratories would have expertise focused around specific scientific interests and therefore examine knockout mice for certain phenotypic traits and emphasis on specific outcomes. Moreover, the broad and the focused approaches often require different screening pipelines, which may result in perceived differences: a broad screen is likely to miss subtle organotypic or temporal alterations that would be the main focus of targeted phenotypic analyses.

Some problems with reproducibility in knockout mouse models have been tracked to the presence of environmental factors in different colonies but also environment-genetic interactions [140, 141]. For example, pathogen-free vs. standard conditions would have profound effect on the immune status of animals, which would be particularly important in analyses of purinoceptors involved in inflammation and immunity. On the other hand, mouse immune responses are one of the functions that is least aligned with humans.

Furthermore, if data are collected from one colony, where the use of small batches of animals is common, various batches of mice can show a degree of phenotypic variation, which can, however, be alleviated by appropriate methodology [142].

An example of specific models being used to describe particular changes but without reassessment of phenotypes described in another model is the P2X4 purinoceptor. It has been studied in several different *P2rx4*$^{-/-}$ mice, which have demonstrated overlapping results in the areas of nociception [68, 93–95], immune cell function [70, 94], and cardiovascular health [69, 71]. However, no direct comparison between these mouse strains has been attempted. It appears that a concerted effort of the purinergic community to share and, where necessary, re-evaluate the models and the available body of phenotyping data could cut through this considerable variability and result in confirmation of key phenotypes and perhaps identification of new ones.

Importantly, there are examples where knockouts that target the same receptor do not exhibit fully complementary phenotypes or appear inconsistent. Moreover, some knockouts displayed altered physiological parameters that were not anticipated, or abnormal phenotypes manifested under certain conditions only. Although the *P2ry6* knockout models that have been described so far all suggest an important role for this receptor in the immune system, the precise roles described in the literature appear to be somewhat contradictory. Total knockout of *P2ry6* results in decreased macrophage function suggesting that P2Y6 functions to increase macrophage activity [113, 115, 116]. In contrast, conditional *P2ry6* knockout actually increases cytokine production following an inflammatory stimulus, which would suggest that P2Y6 function is to downregulate macrophage activity [114]. In fact, the effects of a global receptor ablation via a standard knockout approach can differ from the effects of its ablation in a conditional manner. These might be due to compensatory overexpression of homologous genes occurring during the development of the knockout mouse and the lack of such compensation in an adult tissue.

Moreover, while it is expected that genetic modifications disrupting the open reading frame should result in the lack of a functional protein, it is not always the case due to alternative promoters' usage and alternative splicing events. For example, in the *P2rx7*$^{-/-}$ model by Chessell et al. [75] only the P2X7 (a) variant was inactivated [101]. In cases where more complex genetic modifications have been used, the effects of such modification on the chromatin structure and functioning of the neighboring genes needs to be considered and excluded as a source of phenotypic differences.

The genetic background of knockout mice is another important factor for consideration when comparing the phenotypes of different models. Nowhere the difference between knockouts of the same purinoceptor has been more pertinent than in P2X7. The first *P2rx7*$^{-/-}$ models [74] were on a background carrying a debilitating mutation in the *P2rx7* gene [100] thus weakening the functional impact of gene ablation. Furthermore, the contrasting results of bone function analyses in P2X7 knockouts provide a convincing lesson in the effects of differences in genetic backgrounds. On the other hand, knockout mice are inbred and therefore do not reproduce the impact of the genetic variation existing in human populations.

In some cases the conclusive way to determine if a gene disruption is responsible for a specific phenotype is to rescue this phenotype by in vivo complementation or, particularly suitable for purinoceptors, by comparing the gene knockout against the effects of specific receptor antagonist. Moreover, the recent developments in the generation of more refined models allow for the function of individual purinoceptors to be elucidated in a time and tissue specific manner. The time, cost, and effort required to generate such new models are high, but the results are often significantly less ambiguous. On the other hand, the existing knockouts of several purinoceptors have been investigated for specific functions only. P2X5 ablation produced significant effect upon osteoclast functions [72], but the effect on other cell types remains to be investigated. The P2X6 knockout study did not find the expected renal electrolytes phenotype [73], yet that does not mean that other tissues or other physiological process are unaffected by the lack of this purinoceptor. These examples indicate that despite decades of purinoceptor research there might be significant findings to be made in the existing models by extending investigations into other cell types or physiological processes. The aforementioned heteromerization of functional P2X purinoceptors is also worth consideration.

Although the mouse is by far and wide the most widely used mammalian model for the study of purinoceptor function, other animal models have also been developed. Already, as described in this chapter, there are rat models for some receptors. This model animal is especially suitable for studies of receptor functions in immune responses and in the brain, where the increased similarity to humans makes rat a better model for the study of cognition and neurodegenerative diseases [143].

Gene knockouts in other species (rat, rabbit, pig, sheep) are also emerging, and it might only be a matter of time before such animal models exist for specific purinoceptors. This will allow for a greater array of research questions to be answered.

Overall, the past two decades of research into purinoceptor function in knock-in/knockout animals has provided invaluable data toward our understanding of this diverse family of receptors and this technology will continue to deliver with new tools to develop refined, well-controlled spatiotemporal on- and off-tuning of gene expression across more species.

Acknowledgments

The authors would like to acknowledge the Polish Ministry of National Defence project "Kościuszko" no: 523/2017/DA and the EU COST Program (BM1406).

References

1. Drury AN, Szent-Györgyi A (1929) The physiological activity of adenine compounds with especial reference to their action upon the mammalian heart. J Physiol 68:213–237
2. Burnstock G (1972) Purinergic nerves. Pharmacol Rev 24:509–581
3. Webb TE, Simon J, Krishek BJ et al (1993) Cloning and functional expression of a brain G-protein-coupled ATP receptor. FEBS Lett 324:219–225
4. Ralevic V, Burnstock G (1998) Receptors for purines and pyrimidines. Pharmacol Rev 50:413–492
5. Abbracchio MP (2006) International Union of Pharmacology. LVIII: Update on the P2Y G protein-coupled nucleotide receptors: from molecular mechanisms and pathophysiology to therapy. Pharmacol Rev 58:281–341
6. Khakh BS, Burnstock G, Kennedy CL et al (2001) International union of pharmacology. XXIV. Current status of the nomenclature and properties of P2X receptors and their subunits. Pharmacol Rev 53:107–118
7. Fredholm BB, IJzerman AP, Jacobson KA et al (2001) International Union of Pharmacology. XXV. Nomenclature and classification of adenosine receptors. Pharmacol Rev 53:527–552
8. Herbert JM, Savi P (2003) P2Y12, a new platelet ADP receptor, target of clopidogrel. Semin Vasc Med 3:113–122
9. Fountain SJ, Parkinson K, Young MT et al (2007) An intracellular P2X receptor required for osmoregulation in Dictyostelium discoideum. Nature 448:200–203
10. Burnstock G (2007) Purine and pyrimidine receptors. Cell Mol Life Sci 64:1471–1483
11. Evans MJ, Kaufman MH (1981) Establishment in culture of pluripotential cells from mouse embryos. Nature 292:154–156
12. Bradley A, Evans M, Kaufman MH et al (1984) Formation of germ-line chimaeras from embryo-derived teratocarcinoma cell lines. Nature 309:255–256
13. Mansour SL, Thomas KR, Capecchi MR (1988) Disruption of the proto-oncogene int-2 in mouse embryo-derived stem cells: a general strategy for targeting mutations to non-selectable genes. Nature 31:681–683
14. Li D, Qiu Z, Shao Y et al (2013) Heritable gene targeting in the mouse and rat using a CRISPR-Cas system. Nat Biotechnol 31:681–683
15. Metzger D, Clifford J, Chiba H et al (1995) Conditional site-specific recombination in mammalian cells using a ligand-dependent chimeric Cre recombinase. Proc Natl Acad Sci U S A 92:6991–6995
16. Dymecki SM (1996) A modular set of Flp, FRT and lacZ fusion vectors for manipulating genes by site-specific recombination. Gene 171:197–201
17. Chaiyachati BH, Kaundal R, Zhao J et al (2012) LoxP-FRT Trap (LOFT): a simple and flexible system for conventional and reversible gene targeting. BMC Biol 10:96
18. Geurts AM, Cost GJ, Freyvert Y et al (2009) Knockout rats via embryo microinjection of zinc-finger nucleases. Science 325:433
19. Kitada K, Keng VW, Takeda J et al (2009) Generating mutant rats using the Sleeping Beauty transposon system. Methods 49:236–242
20. Furushima K, Jang CW, Chen DW et al (2012) Insertional mutagenesis by a hybrid piggyBac and Sleeping Beauty transposon in the rat. Genetics 192:1235–1248
21. Carter M, Shieh J, Carter M et al (2010) Making and using transgenic organisms. In: Guide to research techniques in neuroscience. Elsevier B.V., Amsterdam, pp 243–262
22. Bouabe H, Okkenhaug K (2013) Gene targeting in mice: a review. Methods Mol Biol 1064:315–336
23. Eisener-Dorman AF, Lawrence DA, Bolivar VJ (2009) Cautionary insights on knockout mouse studies: the gene or not the gene? Brain Behav Immun 23:318–324
24. Guan C, Ye C, Yang X et al (2010) A review of current large-scale mouse knockout efforts. Genesis 48:73–85
25. Matherne GP, Linden J, Byford AM et al (1997) Transgenic A1 adenosine receptor overexpression increases myocardial resistance to ischemia. Proc Natl Acad Sci U S A 94:6541–6546
26. Johansson B, Halldner L, Dunwiddie TV et al (2001) Hyperalgesia, anxiety, and decreased hypoxic neuroprotection in mice lacking the adenosine A1 receptor. Proc Natl Acad Sci U S A 98:9407–9412
27. Sun D, Samuelson LC, Yang T et al (2001) Mediation of tubuloglomerular feedback by adenosine: evidence from mice lacking adenosine 1 receptors. Proc Natl Acad Sci U S A 98:9983–9988

28. Funakoshi H, Chan TO, Good JC et al (2006) Regulated overexpression of the A1-adenosine receptor in mice results in adverse but reversible changes in cardiac morphology and function. Circulation 114:2240–2250
29. Serchov T, Clement HW, Schwarz MK et al (2015) Increased signaling via adenosine A1 receptors, sleep deprivation, imipramine, and ketamine inhibit depressive-like behavior via induction of homer1a. Neuron 87:549–562
30. Ledent C, Vaugeois JM, Schiffmann SN et al (1997) Aggressiveness, hypoalgesia and high blood pressure in mice lacking the adenosine A2a receptor. Nature 388:674–678
31. Chen JF, Huang Z, Ma J et al (1999) A2A adenosine receptor deficiency attenuates brain injury induced by transient focal ischemia in mice. J Neurosci 19:9192–9200
32. Giménez-Llort L, Schiffmann SN, Shmidt T et al (2007) Working memory deficits in transgenic rats overexpressing human adenosine A2A receptors in the brain. Neurobiol Learn Mem 87:42–56
33. Boknik P, Drzewiecki K, Eskandar J et al (2018) Phenotyping of mice with heart specific overexpression of A2A-adenosine receptors: evidence for cardioprotective effects of A2A-adenosine receptors. Front Pharmacol 9:13
34. Yang D, Zhang Y, Nguyen HG et al (2006) The A2B adenosine receptor protects against inflammation and excessive vascular adhesion. J Clin Invest 116:1913–1923
35. Belikoff BG, Hatfield S, Georgiev P et al (2011) A2B adenosine receptor blockade enhances macrophage-mediated bacterial phagocytosis and improves polymicrobial sepsis survival in mice. J Immunol 275:4429–4434
36. Salvatore CA, Tilley SL, Latour AM et al (2000) Disruption of the A(3) adenosine receptor gene in mice and its effect on stimulated inflammatory cells. J Biol Chem 275:4429–4434
37. Zhao Z, Yaar R, Ladd D et al (2002) Overexpression of A3 adenosine receptors in smooth, cardiac, and skeletal muscle is lethal to embryos. Microvasc Res 63:61–69
38. Brown R, Ollerstam A, Johansson B et al (2001) Abolished tubuloglomerular feedback and increased plasma renin in adenosine A1 receptor-deficient mice. Am J Physiol Regul Integr Comp Physiol 281:R1362–R1367
39. Lee HT, Xu H, Nasr SH et al (2004) A1 adenosine receptor knockout mice exhibit increased renal injury following ischemia and reperfusion. Am J Physiol Ren Physiol 286: F298–F306
40. Kim M, Chen SWC, Park SW et al (2009) Kidney-specific reconstitution of the A1 adenosine receptor in A1 adenosine receptor knockout mice reduces renal ischemia-reperfusion injury. Kidney Int 75:809–823
41. Schweda F, Segerer F, Castrop H et al (2005) Blood pressure-dependent inhibition of renin secretion requires A1 adenosine receptors. Hypertension 46:780–786
42. Koeppen M, Eckle T, Eltzschig HK (2009) Selective deletion of the A1 adenosine receptor abolishes heart-rate slowing effects of intravascular adenosine in vivo. PLoS One 4: e6784
43. Kochanek PM, Vagni VA, Janesko KL et al (2006) Adenosine A1 receptor knockout mice develop lethal status epilepticus after experimental traumatic brain injury. J Cereb Blood Flow Metab 26:565–575
44. Zucchi R, Cerniway RJ, Ronca-Testoni S et al (2002) Effect of cardiac A1 adenosine receptor overexpression on sarcoplasmic reticulum function. Cardiovasc Res 53:326–333
45. Chan TO, Funakoshi H, Song J et al (2008) Cardiac-restricted overexpression of the A2A-adenosine receptor in FVB mice transiently increases contractile performance and rescues the heart failure phenotype in mice overexpressing the A1-adenosine receptor. Clin Transl Sci 1:126–133
46. Chiodi V, Ferrante A, Ferraro L et al (2016) Striatal adenosine-cannabinoid receptor interactions in rats over-expressing adenosine A2A receptors. J Neurochem 136:907–917
47. Domenici MR, Chiodi V, Averna M et al (2018) Neuronal adenosine A2Areceptor overexpression is neuroprotective towards 3-nitropropionic acid-induced striatal toxicity: a rat model of Huntington's disease. Purinergic Signal 14:235–243
48. Johnston-Cox H, Koupenova M, Yang D et al (2012) The A2b adenosine receptor modulates glucose homeostasis and obesity. PLoS One 7:e40584
49. Eisenstein A, Carroll SH, Johnston-Cox H et al (2014) An adenosine receptor-Krüppel-like factor 4 protein axis inhibits adipogenesis. J Biol Chem 289:21071–21081
50. Koupenova M, Johnston-Cox H, Vezeridis A et al (2012) A2b adenosine receptor regulates hyperlipidemia and atherosclerosis. Circulation 125:354–363
51. Zhong H, Shlykov SG, Molina JG et al (2003) Activation of murine lung mast cells by the

adenosine A3 receptor. J Immunol 53:147–155
52. Harrison G (2002) Effects of A3 adenosine receptor activation and gene knock-out in ischemic-reperfused mouse heart. Cardiovasc Res 53:147–155
53. Cerniway RJ, Yang Z, Jacobson MA et al (2001) Targeted deletion of A(3) adenosine receptors improves tolerance to ischemia-reperfusion injury in mouse myocardium. Am J Physiol Heart Circ Physiol 281: H1751–H1758
54. Lee HT, Ota-Setlik A, Xu H et al (2003) A3 adenosine receptor knockout mice are protected against ischemia- and myoglobinuria-induced renal failure. Am J Physiol Renal Physiol 284:F267–F273
55. Yang T, Zollbrecht C, Winerdal ME et al (2016) Genetic abrogation of adenosine A3 receptor prevents uninephrectomy and high salt-induced hypertension. J Am Heart Assoc 5:e003868
56. Hofer M, Pospíšil M, Dušek L et al (2014) Lack of adenosine A3 receptors causes defects in mouse peripheral blood parameters. Purinergic Signal 10:509–514
57. Hofer M, Pospíšil M, Dušek L et al (2013) Erythropoiesis-and thrombopoiesis-characterizing parameters in adenosine A3 receptor knock-out mice. Physiol Res 62:305–311
58. Chen Y, Corriden R, Inoue Y et al (2006) ATP release guides neutrophil chemotaxis via P2Y2 and A3 receptors. Science 314:1792–1795
59. Joós G, Jákim J, Kiss B et al (2017) Involvement of adenosine A3 receptors in the chemotactic navigation of macrophages towards apoptotic cells. Immunol Lett 183:62–72
60. Fedorova IM, Jacobson MA, Basile A et al (2003) Behavioral characterization of mice lacking the A3 adenosine receptor: sensitivity to hypoxic neurodegeneration. Cell Mol Neurobiol 23:431–447
61. Björklund O, Shang M, Tonazzini I et al (2008) Adenosine A1and A3receptors protect astrocytes from hypoxic damage. Eur J Pharmacol 596:6–13
62. Rosito M, Deflorio C, Limatola C et al (2012) CXCL16 orchestrates adenosine A3 receptor and MCP-1/CCL2 activity to protect neurons from excitotoxic cell death in the CNS. J Neurosci 32:3154–3163
63. Björklund O, Halldner-Henriksson L, Yang J et al (2008) Decreased behavioral activation following caffeine, amphetamine and darkness in A3 adenosine receptor knock-out mice. Physiol Behav 95:668–676
64. Little JW, Ford A, Symons-Liguori AM et al (2015) Endogenous adenosine A3receptor activation selectively alleviates persistent pain states. Brain 138:28–35
65. Mulryan K, Gitterman DP, Lewis CJ et al (2000) Reduced vas deferens contraction and male infertility in mice lacking P2X1 receptors. Nature 183:2801–2809
66. Cockayne DA, Dunn PM, Zhong Y et al (2005) P2X2 knockout mice and P2X2/P2X3 double knockout mice reveal a role for the P2X2 receptor subunit in mediating multiple sensory effects of ATP. J Physiol 567:621–639
67. Cockayne DA, Hamilton SG, Zhu QM et al (2000) Urinary bladder hyporeflexia and reduced pain-related behaviour in P2X3-deficient mice. Nature 407:1011–1015
68. Sim JA (2006) Altered hippocampal synaptic potentiation in P2X4 knock-out mice. J Neurosci 26:9006–9009
69. Yamamoto K, Sokabe T, Matsumoto T et al (2006) Impaired flow-dependent control of vascular tone and remodeling in P2X4-deficient mice. Nat Med 12:133–137
70. Brône B, Moechars D, Marrannes R et al (2007) P2X currents in peritoneal macrophages of wild type and P2X4−/− mice. Immunol Lett 113:83–89
71. Yang T, Shen JB, Yang R et al (2014) Novel protective role of endogenous cardiac myocyte P2X4 receptors in heart failure. Circ Heart Fail 7:510–518
72. Kim H, Walsh MC, Takegahara N et al (2017) The purinergic receptor P2X5 regulates inflammasome activity and hyper-multinucleation of murine osteoclasts. Sci Rep 7:196
73. de Baaij JHF, Kompatscher A, Viering DHHM et al (2016) P2X6 knockout mice exhibit normal electrolyte homeostasis. PLoS One 11:e0156803
74. Solle M, Labasi J, Perregaux DG et al (2001) Altered cytokine production in mice lacking P2X(7) receptors. J Biol Chem 276:125–132
75. Chessell IP, Hatcher JP, Bountra C et al (2005) Disruption of the P2X7 purinoceptor gene abolishes chronic inflammatory and neuropathic pain. Pain 114:386–396
76. Metzger MW, Walser SM, Aprile-Garcia F et al (2017) Genetically dissecting P2rx7 expression within the central nervous system using conditional humanized mice. Purinergic Signal 13:153–170
77. Kaczmarek-Hajek K, Zhang J, Kopp R et al (2018) Re-evaluation of neuronal P2X7

expression using novel mouse models and a P2X7-specific nanobody. Elife 7:e36217

78. Lecut C, Frederix K, Johnson DM et al (2009) P2X1 ion channels promote neutrophil chemotaxis through rho kinase activation. J Immunol 183:2801–2809
79. Lecut C, Faccinetto C, Delierneux C et al (2012) ATP-gated P2X1 ion channels protect against endotoxemia by dampening neutrophil activation. J Thromb Haemost 10:453–465
80. Ren J, Bian X, DeVries M et al (2003) P2X2 subunits contribute to fast synaptic excitation in myenteric neurons of the mouse small intestine. J Physiol 552:809–821
81. Ryten M, Koshi R, Knight GE et al (2007) Abnormalities in neuromuscular junction structure and skeletal muscle function in mice lacking the P2X2 nucleotide receptor. Neuroscience 148:700–711
82. Rong W, Gourine AV, Cockayne DA et al (2003) Pivotal role of nucleotide P2X2 receptor subunit of the ATP-gated ion channel mediating ventilatory responses to hypoxia. J Neurosci 23:11315–11321
83. Zhong Y, Dunn PM, Bardini M et al (2001) Changes in P2X receptor responses of sensory neurons from P2X3-deficient mice. Eur J Neurosci 14:1784–1792
84. Bian X, Ren J, DeVries M et al (2003) Peristalsis is impaired in the small intestine of mice lacking the P2X3 subunit. J Physiol 551:309–322
85. McIlwrath SL, Davis BM, Bielefeldt K (2009) Deletion of P2X3 receptors blunts gastro-oesophageal sensation in mice. Neurogastroenterol Motil 21:890–e66
86. Eddy MC, Eschle BK, Barrows J et al (2009) Double P2X2/P2X3 purinergic receptor knockout mice do not taste NaCl or the artificial sweetener SC45647. Chem Senses 34:789–797
87. Sclafani A, Ackroff K (2014) Maltodextrin and fat preference deficits in "taste-blind" P2X2/P2X3 knockout mice. Chem Senses 39:507–514
88. Baxter AW, Choi SJ, Sim JA et al (2011) Role of P2X4 receptors in synaptic strengthening in mouse CA1 hippocampal neurons. Eur J Neurosci 34:213–220
89. Ulmann L, Levavasseur F, Avignone E et al (2013) Involvement of P2X4 receptors in hippocampal microglial activation after status epilepticus. Glia 61:1306–1319
90. Khoja S, Huynh N, Asatryan L et al (2018) Reduced expression of purinergic P2X4 receptors increases voluntary ethanol intake in C57BL/6J mice. Alcohol 68:63–70
91. Wyatt LR, Finn DA, Khoja S et al (2014) Contribution of P2X4 receptors to ethanol intake in male C57BL/6 mice. Neurochem Res 39:1127–1139
92. Wyatt LR, Godar SC, Khoja S et al (2013) Sociocommunicative and sensorimotor impairments in male P2X4-deficient mice. Neuropsychopharmacology 38:1993–2002
93. Ulmann L, Hatcher JP, Hughes JP et al (2008) Up-regulation of P2X4 receptors in spinal microglia after peripheral nerve injury mediates BDNF release and neuropathic pain. J Neurosci 28:11263–11268
94. Ulmann L, Hirbec H, Rassendren F (2010) P2X4 receptors mediate PGE2 release by tissue-resident macrophages and initiate inflammatory pain. EMBO J 29:2290–2300
95. Tsuda M, Kuboyama K, Inoue T et al (2009) Behavioral phenotypes of mice lacking purinergic P2X4 receptors in acute and chronic pain assays. Mol Pain 5:28
96. Yan Z, Khadra A, Li S et al (2010) Experimental characterization and mathematical modeling of P2X7 receptor channel gating. J Neurosci 30:14213–14224
97. Di Virgilio F, Dal Ben D, Sarti AC et al (2017) The P2X7 receptor in infection and inflammation. Immunity 47(1):15–31
98. Young CNJ, Gorecki DC (2017) P2RX7 purinoceptor as a therapeutic target – the second coming? Front Chem 6:248
99. Di Virgilio F, Schmalzing G, Markwardt F (2018) The elusive P2X7 macropore. Trends Cell Biol 28:392–404
100. Adriouch S, Dox C, Welge V et al (2002) Cutting edge: a natural P451L mutation in the cytoplasmic domain impairs the function of the mouse P2X7 receptor. J Immunol 169:4108–4112
101. Nicke A, Kuan YH, Masin M et al (2009) A functional P2X7 splice variant with an alternative transmembrane domain 1 escapes gene inactivation in P2X7 knock-out mice. J Biol Chem 284:25813–25822
102. Ke HZ, Qi H, Weidema AF et al (2003) Deletion of the P2X 7 nucleotide receptor reveals its regulatory roles in bone formation and resorption. Mol Endocrinol 17:1356–1367
103. Li J, Liu D, Ke HZ et al (2005) The P2X7 nucleotide receptor mediates skeletal mechanotransduction. J Biol Chem 13:243–253
104. Gartland A, Buckley KA, Hipskind RA et al (2003) Multinucleated osteoclast formation

in vivo and in vitro by P2X7 receptor-deficient mice. Crit Rev Eukaryot Gene Expr 13:243–253

105. Syberg S, Schwarz P, Petersen S et al (2012) Association between P2X7 receptor polymorphisms and bone status in mice. J Osteoporos 2012:637986

106. Syberg S, Petersen S, Beck Jensen JE et al (2012) Genetic background strongly influences the bone phenotype of P2X7 receptor knockout mice. J Osteoporos 2012:391097

107. Grygorowicz T, Dąbrowska-Bouta B, Strużyńska L (2018) Administration of an antagonist of P2X7 receptor to EAE rats prevents a decrease of expression of claudin-5 in cerebral capillaries. Purinergic Signal 14(4):385–393

108. Fabre JE, Nguyen M, Latour A et al (1999) Decreased platelet aggregation, increased bleeding time and resistance to thromboembolism in P2Y1-deficient mice. Nat Med 5:1199–1202

109. Léon C, Hechler B, Freund M et al (1999) Defective platelet aggregation and increased resistance to thrombosis in purinergic P2Y1receptor-null mice. J Clin Invest 104:1731–1737

110. Hwang SJ, Blair PJ, Durnin L et al (2012) P2Y1 purinoreceptors are fundamental to inhibitory motor control of murine colonic excitability and transit. J Physiol 590:1957–1972

111. Homolya L, Watt WC, Lazarowski ER et al (1999) Nucleotide-regulated calcium signaling in lung fibroblasts and epithelial cells from normal and P2Y2 receptor (−/−) mice. J Biol Chem 274:26454–26460

112. Robaye B, Ghanem E, Wilkin F et al (2003) Loss of nucleotide regulation of epithelial chloride transport in the jejunum of P2Y4-null mice. Mol Pharmacol 63:777–783

113. Bar I, Guns P-J, Metallo J et al (2008) Knockout mice reveal a role for P2Y6 receptor in macrophages, endothelial cells, and vascular smooth muscle cells. Mol Pharmacol 74:777–784

114. Giannattasio G, Ohta S, Boyce JR et al (2011) The P2Y6 receptor inhibits effector T cell activation in allergic pulmonary inflammation. J Immunol 187:1486–1495

115. Garcia RA, Yan M, Search D et al (2014) P2Y6receptor potentiates pro-inflammatory responses in macrophages and exhibits differential roles in atherosclerotic lesion development. PLoS One 9:e111385

116. Li R, Tan B, Yan Y et al (2014) Extracellular UDP and P2Y6 function as a danger signal to protect mice from vesicular stomatitis virus infection through an increase in IFN-β production. J Immunol 193:4515–4526

117. Foster CJ, Prosser DM, Agans JM et al (2001) Molecular identification and characterization of the platelet ADP receptor targeted by thienopyridine antithrombotic drugs. J Clin Invest 107:1591–1598

118. André P, Delaney SM, LaRocca T et al (2003) P2Y12regulates platelet adhesion/activation, thrombus growth, and thrombus stability in injured arteries. J Clin Invest 112:398–406

119. Fabre AC, Malaval C, Ben Addi A et al (2010) P2Y13 receptor is critical for reverse cholesterol transport. Hepatology 52:1477–1483

120. Bassil AK, Bourdu S, Townson KA et al (2009) UDP-glucose modulates gastric function through P2Y14 receptor-dependent and -independent mechanisms. Am J Physiol Gastrointest Liver Physiol 296:G923–G930

121. Meister J, Le Duc D, Ricken A et al (2014) The G protein-coupled receptor P2Y14 influences insulin release and smooth muscle function in mice. J Biol Chem 289:23353–23366

122. Gil V, Martínez-Cutillas M, Mañé N et al (2013) P2Y1 knockout mice lack purinergic neuromuscular transmission in the antrum and cecum. Neurogastroenterol Motil 25: e170–e182

123. Matos JE, Robaye B, Boeynaems JM et al (2005) K+ secretion activated by luminal P2Y2 and P2Y4 receptors in mouse colon. J Physiol 564:269–279

124. Horckmans M, Robaye B, Léon-Gómez E et al (2012) P2Y4 nucleotide receptor: a novel actor in post-natal cardiac development. Angiogenesis 15:349–360

125. Horckmans M, Leon-Gomez E, Robaye B et al (2012) Gene deletion of P2Y4 receptor lowers exercise capacity and reduces myocardial hypertrophy with swimming exercise. AJP Heart Circ Physiol 303:H835–H843

126. Kauffenstein G, Tamareille S, Prunier F et al (2016) Central role of P2Y6 UDP receptor in arteriolar myogenic tone. Arterioscler Thromb Vasc Biol 36:1598–1606

127. Clouet S, Di Pietrantonio L, Daskalopoulos EP et al (2016) Loss of mouse P2Y6nucleotide receptor is associated with physiological macrocardia and amplified pathological cardiac hypertrophy. J Biol Chem 291:15841–15852

128. Placet M, Arguin G, Molle CM et al (2018) The G protein-coupled P2Y6 receptor promotes colorectal cancer tumorigenesis by inhibiting apoptosis. Biochim Biophys Acta Mol basis Dis 1864:1539–1551

129. Liverani E, Rico MC, Yaratha L et al (2014) LPS-induced systemic inflammation is more severe in P2Y12 null mice. J Leukoc Biol 95:313–323
130. Li D, Wang Y, Zhang L et al (2012) Roles of purinergic receptor P2Y, G protein-coupled 12 in the development of atherosclerosis in apolipoprotein E-deficient mice. Arterioscler Thromb Vasc Biol 32:e81–e89
131. Harada K, Matsumoto Y, Umemura K (2011) Adenosine diphosphate receptor P2Y12-mediated migration of host smooth muscle-like cells and leukocytes in the development of transplant arteriosclerosis. Transplantation 92:148–154
132. Ben Addi A, Cammarata D, Conley PB et al (2010) Role of the P2Y12 receptor in the modulation of murine dendritic cell function by ADP. J Immunol 185:5900–5906
133. Haynes SE, Hollopeter G, Yang G et al (2006) The P2Y12 receptor regulates microglial activation by extracellular nucleotides. Nat Neurosci 9:1512–1519
134. Gu N, Eyo UB, Murugan M et al (2016) Microglial P2Y12 receptors regulate microglial activation and surveillance during neuropathic pain. Brain Behav Immun 55:82–92
135. Su X, Floyd DH, Hughes A et al (2012) The ADP receptor P2RY12 regulates osteoclast function and pathologic bone remodeling. J Clin Invest 122:3579–3592
136. Wang Y, Sun Y, Li D et al (2013) Platelet P2Y12 is involved in murine pulmonary metastasis. PLoS One 8:e80780
137. Wang N, Robaye B, Agrawal A et al (2012) Reduced bone turnover in mice lacking the P2Y13 receptor of ADP. Mol Endocrinol 26:142–152
138. Wang N, Rumney RMH, Yang L et al (2013) The P2Y13 receptor regulates extracellular ATP metabolism and the osteogenic response to mechanical loading. J Bone Miner Res 28:1446–1456
139. Wang N, Robaye B, Gossiel F et al (2014) The P2Y13 receptor regulates phosphate metabolism and FGF-23 secretion with effects on skeletal development. FASEB J 28:2249–2259
140. Kafkafi N, Benjamini Y, Sakov A et al (2005) Genotype-environment interactions in mouse behavior: a way out of the problem. Proc Natl Acad Sci U S A 102:4619–4624
141. Richter SH, Garner JP, Würbel H (2009) Environmental standardization: cure or cause of poor reproducibility in animal experiments? Nat Methods 6:257–261
142. Karp NA, Melvin D, Mott RF (2012) Robust and sensitive analysis of mouse knockout phenotypes. PLoS One 7:e52410
143. Ellenbroek B, Youn J (2016) Rodent models in neuroscience research: is it a rat race? Dis Model Mech 9:1079–1087

Chapter 3

Agonists and Antagonists for Purinergic Receptors

Christa E. Müller, Younis Baqi, and Vigneshwaran Namasivayam

Abstract

Membrane receptors that are activated by the purine nucleoside adenosine (adenosine receptors) or by purine or pyrimidine nucleotides (P2Y and P2X receptors) transduce extracellular signals to the cytosol. They play important roles in physiology and disease. The G protein-coupled adenosine receptors comprise four subtypes: A_1, A_{2A}, A_{2B}, and A_3. The G-protein-coupled P2Y receptors are subdivided into eight subtypes: $P2Y_1$, $P2Y_2$, $P2Y_4$, $P2Y_6$, $P2Y_{11}$, $P2Y_{12}$, $P2Y_{13}$, and $P2Y_{14}$, while the P2X receptors represent ATP-gated homomeric or heteromeric ion channels consisting of three subunits; the most important subunits are P2X1, P2X2, P2X3, P2X4, and P2X7. This chapter provides guidance for selecting suitable tool compounds for studying these large and important purine receptor families.

Key words Adenosine receptors, Agonists, Allosteric modulators, Antagonists, Binding site, Ligands, P2Y receptors, P2X receptors, Purine receptors, Structure, Tool compounds

1 Introduction

Receptors involved in purinergic signaling are subdivided into two families termed P1 and P2 receptors (Fig. 1) [1]. P1 or adenosine receptors, of which four subtypes exist (A_1, A_{2A}, A_{2B}, and A_3) are G protein-coupled receptors (GPCRs) that are activated by the nucleoside adenosine (Fig. 2) [2]. P2 receptors or nucleotide receptors are further subdivided into P2Y and P2X receptors (Fig. 1) [3–5]. The P2Y receptor family comprises eight different GPCRs each of which has a specific agonist profile, termed $P2Y_1$ (activated by ADP), $P2Y_2$ (ATP, UTP), $P2Y_4$ (UTP), $P2Y_6$ (UDP), $P2Y_{11}$ (ATP), $P2Y_{12}$ (ADP), $P2Y_{13}$ (ADP), and $P2Y_{14}$ (UDP, UDPglucose, UDPgalactose) [6]. The P2X receptors are homotrimeric or heterotrimeric ATP-gated ion channel receptors. Seven different subunits are known, P2X1–P2X7 [4, 7]. There is a metabolic link between P1 and P2 receptor agonists since the nucleotides ATP and ADP (P2 receptor agonists) are hydrolyzed by various ectonucleotidases producing the P1 receptor agonist adenosine (Fig. 1). While ATP is a danger signal mediating proinflammatory effects,

Pablo Pelegrín (ed.), *Purinergic Signaling: Methods and Protocols*, Methods in Molecular Biology, vol. 2041,
https://doi.org/10.1007/978-1-4939-9717-6_3, © Springer Science+Business Media, LLC, part of Springer Nature 2020

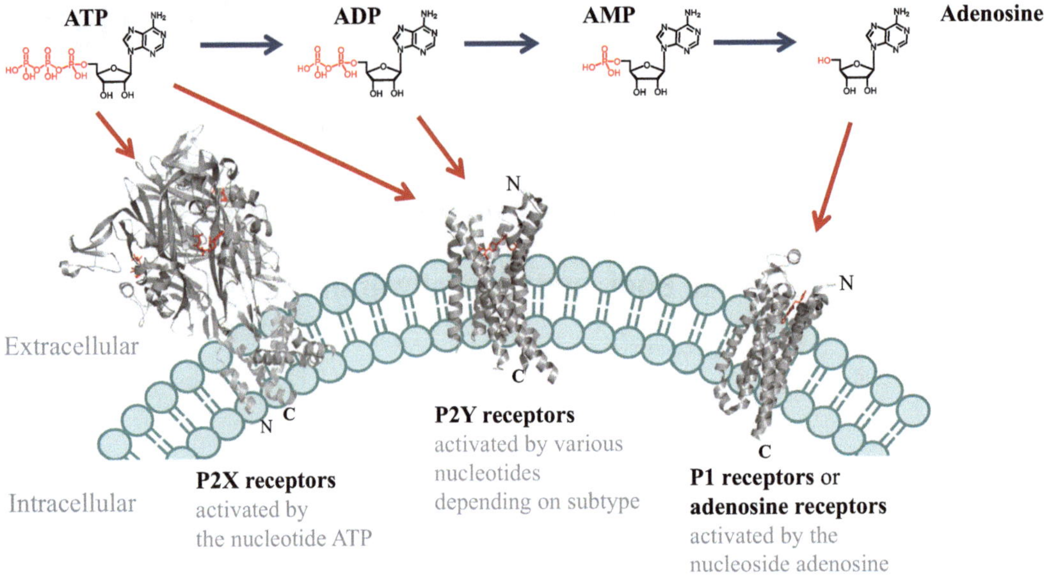

Fig. 1 Purine receptor subfamilies

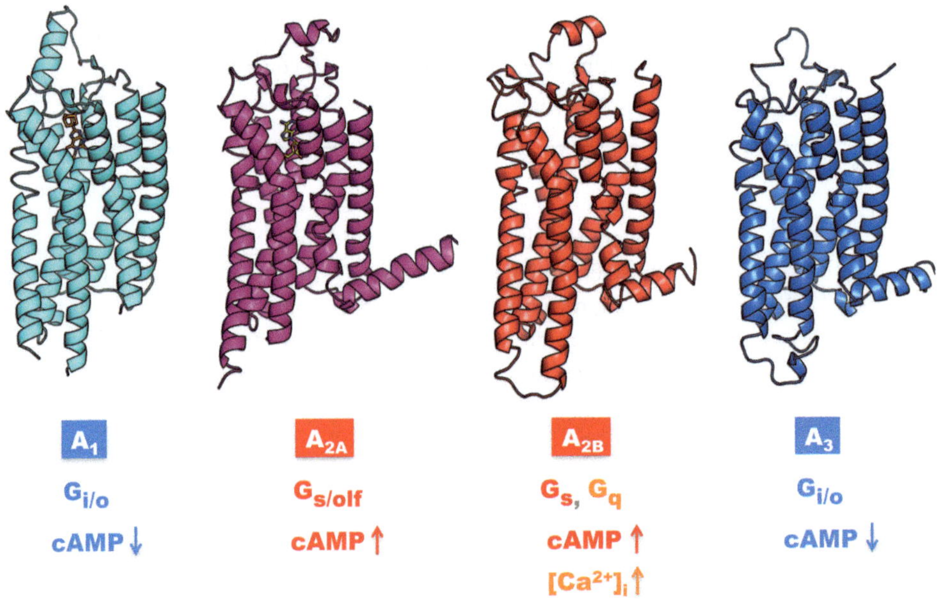

Fig. 2 Structures, G protein coupling, and second messengers of AR subtypes. Shown are the crystal structures of the AR subtypes A_1 (cyan) in complex with the antagonist PSB-36 (5N2S.pdb), and A_{2A} (magenta) in complex with the cognate agonist adenosine (2YDO.pdb), and homology models of the AR subtypes A_{2B} (red) and A_3 (blue) generated based on the crystal structures of the A_{2A}AR (for A_{2B}) and the A_1AR (for A_3). The receptors are represented in cartoon models, and the cocrystallized ligands PSB-36 (carbon atoms colored orange) and adenosine (carbon atoms colored yellow) are shown as stick models. Oxygen atoms are colored red, nitrogen atoms blue, and phosphorus atoms orange

adenosine acts as a stop signal inducing anti-inflammatory and immunosuppressive activities.

In addition to P1 and P2 receptors, G-protein-coupled receptors activated by the purine nucleobase adenine have been cloned from in mouse, rat and hamster, and designated P0 receptors [8, 9]. Direct human orthologs of the rodent P0 receptors have not been detected, but there is evidence that adenine-activated G protein-coupled P0 receptor subtypes may also exist in humans.

Despite decades of research, only few drugs have been approved so far that interact with purine receptors, most prominently the $P2Y_{12}$ receptor antagonists (clopidogrel, prasugrel, cangrelor, and ticagrelor) which have become an important class of antithrombotic drugs [10–12]. Recently, new hypes and hopes have been created in the field, due to (1) successful clinical trials for the P2X3 receptor antagonist gefapixant in chronic cough and other inflammatory conditions [10, 13], (2) the advancement of the partial A_1 adenosine receptor agonist neladenoson bialanate into phase III clinical trials for heart failure [14], and, most importantly, (3) the gold rush fever in immuno-oncology [15–17]. Blockade of A_{2A} and A_{2B} adenosine receptors and/or inhibition of adenosine formation by blocking ectonucleotidases, such as CD39 or CD73, are being pursued as novel principles leading to an activation of the immune system to defeat cancer, and possibly infections as well. In this context, our group has recently shown that the A_{2B} adenosine receptor, which is typically upregulated under hypoxic conditions, that is, in inflammation and cancer, forms stable heteromeric complexes with the A_{2A} receptor subtype and thereby completely blocks A_{2A} receptor signaling [18]. This finding will likely have implications for the development of drugs for those targets.

The following paragraphs will introduce a selection of ligands for the individual receptor subtypes that are expected to be suitable for in vitro and/or in vivo studies. Potency and selectivity in human, rat, and mouse and physicochemical properties are discussed as far as the respective data are available. The (partly subjective) selection of compounds is based on potency and selectivity and takes into account the compounds' accessibility and the degree of their characterization.

2 Adenosine Receptors

The G protein-coupled adenosine receptors (ARs) are divided into four subtypes, A_1, A_{2A}, A_{2B}, and A_3 (*see* Fig. 2) [2]. The A_1- and A_3AR are $G_{i/o}$ protein-coupled leading to an inhibition of adenylate cyclase, while the A_{2A}- and A_{2B}AR are G_s protein-coupled mediating adenylate cyclase activation. The A_{2B}AR is additionally coupled to G_q proteins in many cellular systems leading to phospholipase C activation and subsequent calcium mobilization (*see* Fig. 2) [19]. The A_{2A}AR is already an established drug target for cardiac imaging (utilizing the nonselective short-acting physiological agonist adenosine or the A_{2A}-selective agonist regadenoson) and for Parkinson's disease (with the A_{2A}-selective antagonist istradefylline which was approved in Japan) [5, 10]. In recent years, several X-ray structures of the human A_{2A}AR and very recently also of the human A_1AR were published. These provided important knowledge on the orthosteric ligand binding site and on receptor activation, and have crucially supported drug development efforts [20, 21].

3 Adenosine Receptor Agonists

The physiological agonist adenosine (**1**, *see* Fig. 3) is significantly more potent at A_1-, A_{2A}-, and A_3ARs than at A_{2B}ARs (*see* Table 3). Reliable radioligand binding data, however, are not available since adenosine is present in tissues, cells, and cell membranes and is constantly produced; therefore, it has to be removed by the addition of adenosine deaminase (ADA). The enzyme is typically present during the incubation with radioligand and test compound. In contrast to radioligand binding data, potencies determined in functional, G protein-dependent assays depend on receptor expression levels. Therefore, data for agonists obtained in different cellular systems are not comparable. In addition, adenosine shows a short half-life and is degraded by ADA or adenosine kinase (AdoK) or removed by cellular uptake, which additionally influences the test results. Therefore, metabolically (more) stable adenosine analogs have been developed. The closely related analog NECA (**2**) is not degraded by ADA or phosphorylated by AdoK. Like adenosine, it is much more potent at A_1-, A_{2A}-, and A_3ARs than at A_{2B}ARs. Nevertheless, NECA is still one of the most potent full A_{2B}AR agonists and represents a useful tool to study A_{2B}ARs in combination with selective agonists for the other AR subtypes, since truly selective, and at the same time fully efficacious A_{2B}AR agonists are still lacking [19].

Highly selective A_1AR agonists have been developed by N^6-substitution of adenosine (*see* Fig. 3 and Table 1). CCPA (**3**) is suitable for rat and mouse studies, where it shows >100-fold selectivity versus all other AR subtypes, however it is somewhat less selective in humans versus the A_3AR subtype (46-fold). For studies at the human A_1AR, its 2′-methyl-substituted derivative, 2′-MeCCPA (**4**) is more selective (>300-fold) [22]. Neladenoson bialanate (**5**), an ester prodrug containing an L-Ala-L-Ala dipeptide connected via the C-terminal carboxylate to the hydroxy group of neladenoson, is quickly cleaved by esterases in the body releasing the partial A_1-selective AR agonists neladenoson [14]. Formation

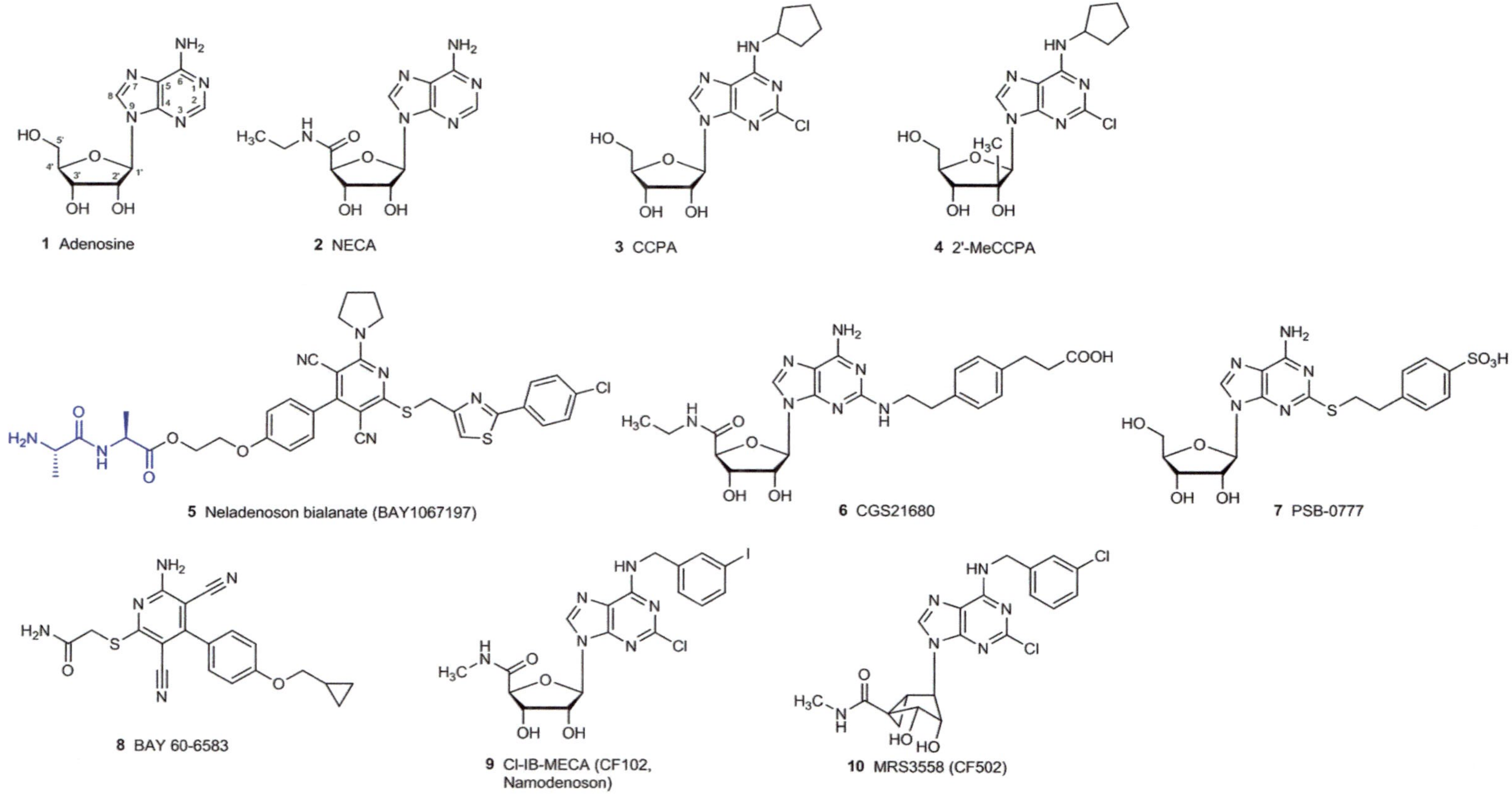

Fig. 3 Structures of adenosine receptor agonists

Table 1
Affinities of selected adenosine receptor agonists

		K_i (nM)[a]			
		A_1	A_{2A}	A_{2B}[b]	A_3
Nonselective agonists					
1	Adenosine[c]	ca. 100 (h) 73 (r)	310 (h) 150 (r)	15,000 (h) 5,100 (r)	290 (h) 6,500 (r)
2	NECA	14 (h) 5.1 (r) 2.49 (m)	20 (h) 9.7 (r) 43.4 (m)	1,890 (h) 1,110 (r) 656 (m)	25 (h) 113 (r) 13.2 (m)
A_1-selective agonists					
3	CCPA	0.83 (h) 1.3 (r) 0.269 (m)	2270 (h) 950 (r) 988 (m)	18,800 (h) 6,160 (r) 25,300 (m)	38 (h) 237 (r) 15.6 (m)
4	2′-MeCCPA	3.3 (h)	9,580 (h)	37,600 (h)	1.150 (h)
5	Neladenoson bialanate (prodrug) - data for neladenoson	0.1 (h)[c]	670 (h)	80 (h)	>3000 (h)
A_{2A}-selective agonists					
6	CGS21680	289 (h) 1,800 (r) 961 (m)	27 (h) 19 (r) 13.7 (m)	>10,000 (h) >10,000 (r) >10,000 (m)	67 (h) 584 (r) 93.0 (m)
7	PSB-0777	541 (h) ≥10,000 (r)	360 (h) 44.4 (r)	>10,000 (h)	>10,000 (h)
A_{2B}-selective (partial) agonist					
8	BAY 60-6583	387 (h) 514 (r) 351 (m)	>10,000 (h) >10,000 (r) >10,000 (m)	114 (h) 100 (r) 136 (m)	223 (h) 2,750 (r) 3,920 (m)
A_3-selective agonists					
9	Cl-IB-MECA (CF102)	220 (h) 280 (r) 35 (m)	5360 (h) 470 (r) 290 (m)	>10,000 (h) 1,210 (r) 44,300 (m)	1.4 (h) 0.33 (r) 0.18 (m)
10	MRS3558 (CF502)	260 (h) 105 (r) 15.8 (m)	2330 (h) 1080 (r) 10,400 (m)	>10,000 (h)	0.29 (h) 1.0 (r) 1.49 (m)

[a] *h* human, *m* mouse, *r* rat; data are taken from literature [14, 22–25]
[b] Most data are from functional studies
[c] Data from functional studies (functional data are strongly dependent on receptor expression levels and assay system)

of a hydrochloride salt of the terminal amino function results in high water solubility, while neladenoson itself displays only very low solubility. Compound **5** has been in phase 3 clinical trials as a peroral drug for heart failure. It is a partial A_1 agonist (ca. 67%

efficacy compared to CCPA), and, in contrast to full A_1AR agonists, it did not cause atrioventricular (AV) block. Although it penetrated into the brain to some extent, it was devoid of sedative side effects, probably due to its partial agonistic character [14].

AR agonists with selectivity for the $A_{2A}AR$ subtype have been obtained by introducing large, bulky substituents into the 2-position of adenosine or NECA. Most of the developed compounds are not highly selective in humans versus the A_1- and/or A_3AR subtype. CGS21680 (**6**) is potent and highly selective in rat, but displays moderate selectivity in humans (vs. A_1 and A_3). In mice, it is more selective than in humans (*see* Table 1). The water-soluble $A_{2A}AR$ agonist PSB-0777 (**7**) bearing a phenylsulfonate group is very useful for injection, or for local application in the gut since it is not perorally absorbed. It shows high selectivity in rat, but not in humans.

Potent and selective full agonists for the $A_{2B}AR$ are lacking. BAY 60–6583 (**8**), structurally related to the partial A_1AR agonist neladenoson, is a partial $A_{2B}AR$ agonist [26] (*see* Fig. 3 and Table 1). It was shown to act as an antagonist at other AR subtypes [23], and data obtained with **8** should therefore be carefully interpreted.

For the A_3AR, very potent and selective agonists are available, such as Cl-IB-MECA (**9**, CF102, namodenoson), which is being evaluated in clinical trials for treating hepatocellular carcinoma and nonalcoholic steatohepatitis (NASH). A related potent A_3 agonist is MRS3558 (**10**, CF502), which is more potent in humans, but less potent in rat, and only moderately potent in mice (*see* Table 1). For pharmacological studies, the doses of **9** and **10** have to be carefully chosen in order not to activate the A_1AR as well.

4 Adenosine Receptor Antagonists

The natural alkaloid caffeine (trimethylxanthine, **11**) represents the prototypic AR antagonist (*see* Fig. 4). In humans, it is about equally potent at all four AR subtypes (*see* Table 2). However, it is inactive at rat and mouse A_3ARs and only blocks A_1-, A_{2A}-, and A_{2B}-, but not A_3ARs in rodents [27]. Potent, A_1-selective AR antagonists related to caffeine are the xanthine derivatives DPCPX (CPX, **12**) and PSB-36 (**13**) [23, 28]. While DPCPX shows only moderate selectivity in humans, PSB-36 is highly selective in all three species, human, rat, and mouse. SLV320 (**14**) is an A_1AR antagonists with a non-xanthine structure [29]; it is derived from adenine, a 7-deaza-adenine derivative bearing a cyclohexyl group at the exocyclic amino function. The compound is potent and selective in humans, but complete data in other species are lacking.

Xanthine derivatives have also been optimized for the $A_{2A}AR$ (*see* Fig. 4). Istradefylline (**15**) was the first A_{2A} antagonist that was

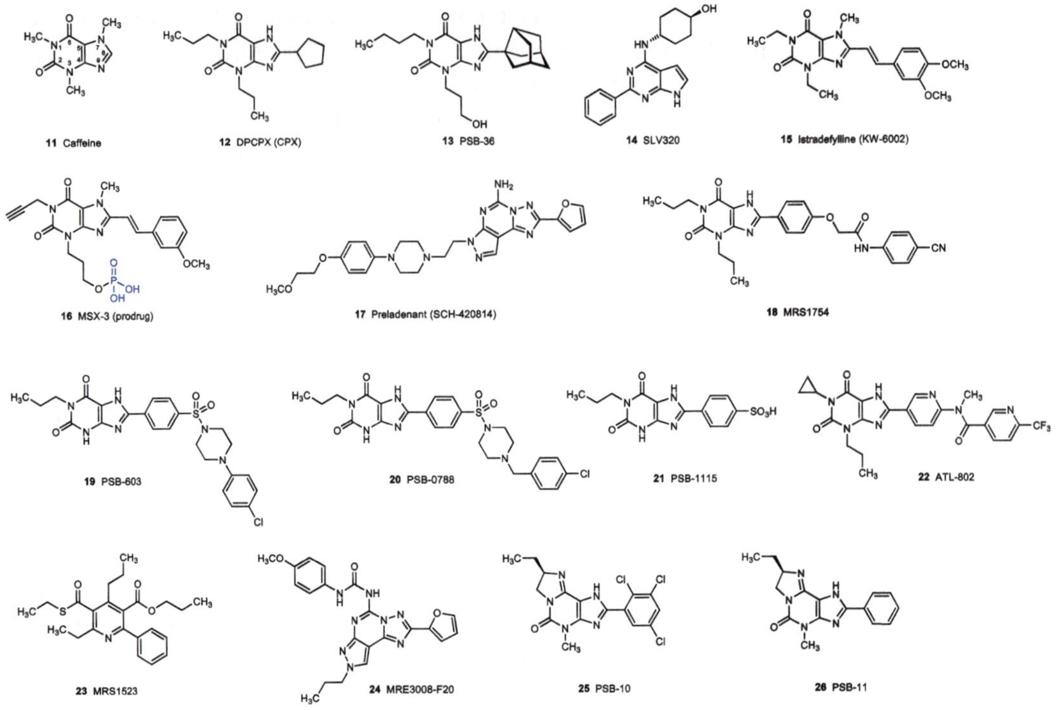

Fig. 4 Structures of adenosine receptor antagonists

Table 2
Affinities of selected adenosine receptor antagonists

		K_i (nM)[a]			
		A_1	A_{2A}	A_{2B}	A_3
Nonselective antagonists					
11	Caffeine	44,900 (h) 41,000 (r) 50,700 (m)	23,400 (h) 43,000 (r) 11,100 (m)	33,800 (h) 30,000 (r) 23,000 (m)	13,300 (h) >100,000 (r) >100,000 (m)
A_1-selective antagonists					
12	DPCPX (CPX)	3.0 (h) 0.50 (r) 0.413 (m)	129 (h) 157 (r) 263 (m)	51 (h) 186 (r) 86.2 (m)	243 (h) >10,000 (r) >10,000 (m)
13	PSB-36	0.7 (h) 0.124 (r) 1.58 (m)	980 (h) 552 (r) 697 (m)	187 (h) 350 (r) 704 (m)	2,300 (h) 6,500 (r) >10,000 (m)
14	SLV320	1.00 (h) 2.51 (r)	398 (h)	3,981 (h) 501 (r)	200 (h)

(continued)

Table 2 (continued)

		K_i (nM)[a]			
		A_1	A_{2A}	A_{2B}	A_3
A_{2A}-selective antagonists					
15	Istradefylline (KW6002)	841 (h) 230 (r) 438 (m)	12 (h) 4.46 (r) 6.83 (m)	>10,000 (h) 5,940 (r) 3,590 (m)	4,470 (h) >10,000 (r) >10,000 (m)
16	MSX-3/MSX-2 (data for MSX-2)	2,500 (h) 900 (r)	5.38 (h) 8.04 (r)	>10,000 (h)	>10,000 (h)
17	Preladenant (SCH-420814)	>1,000 (h) >1,000 (h) 462 (m)	0.9 (h) 0.986 (r) 0.241 (m)	>1,000 (h) >1,000 (m) >1,000 (r)	>1,000 (h) >1,000 (m) >1,000 (r)
A_{2B}-selective antagonists					
18	MRS1754	403 (h) 16.8 (r) 1.45 (m)	503 (h) 612 (r) >10,000 (m)	1.97 (h) 12.8 (r) 3.12 (m)	570 (h) >1,000 (m) >1,000 (r)
19	PSB-603	>10,000 (h) >10,000 (r) 42.4 (m)	>10,000 (h) >10,000 (r) >10,000 (m)	0.553 (h) 0.355 (r) 0.265 (m)	>10,000 (h) >10,000 (r) >10,000 (m)
20	PSB-0788	2,240 (h) 386 (r) 118 (m)	333 (h) 1,730 (r) 235 (m)	0.393 (h) 2.12 (r) 1.90 (m)	>1,000 (h) >10,000 (r) >10,000 (m)
21	PSB-1115	>10,000 (h) 2,200 (r) 591 (m)	3790 (h) 24,000 (r) >10,000 (m)	53.4 (h) 3,140 (r) 1,940 (m)	>10,000 (h) >10,000 (r) >10,000 (m)
22	ATL-802	369 (h) 9,583 (m)	654 (h) 8,393 (m)	2.36 (h) 8.58 (m)	>1,000 (h) >10,000 (m)
A_3-selective antagonists					
23	MRS1523	>10,000 (h) 15,600 (r) >10,000 (m)	3660 (h) 2050 (r) >10,000 (m)	>10,000 (h) >10,000 (r) >10,000 (m)	18.9 (h) 113 (r) 731 (m)
24	MRE3008-F20	1200 (h)	141 (h)	2100 (h)	0.82 (h)
25	PSB-10	1,700 (h) 805 (r)	2,700 (h) 6,040 (r)	30,000 (h)	0.441 (h) 17,000 (r)
26	PSB-11	1,640 (h)	1,280	2,100 (m)	2.34 (h)

[a] *h* human, *m* mouse, *r* rat; data are taken from literature [23, 24] or unpublished data for our laboratory

approved as a drug—by the Japanese Pharmaceutical and Medical Devices Agency (PMDA)—for the treatment of Parkinson's disease (PD) [30]. Its potency and selectivity for the A_{2A}AR is similar in human, rat, and mouse; it is highly selective versus the A_{2B}- and

A_3AR subtypes, but less selective (50–70-fold) versus the A_1AR (*see* Table 2). Other drawbacks are its moderate water solubility and the double bond of its styryl residue which can undergo light-induced E/Z-isomerization in dilute solution, and is prone to light-induced dimerization as a solid [31]. Therefore, it has to be protected from light. The same is true for MSX-3 (**16**), a phosphate prodrug of MSX-2, which is -in contrast to istradefylline- well soluble in water and can therefore be applied to animals in the drinking water [32, 33]. The A_{2A} selectivity of MSX-2 is higher than that of istradefylline (*see* Table 2). One of the most potent and most selective $A_{2A}AR$ antagonists is preladenant (SCH-420814, **17**), which had been evaluated in clinical trials for PD. It was well tolerated, but showed no significant beneficial effects [34].

Recently, a large number of A_{2B}-selective AR antagonists has been described, many of which represent xanthine derivatives [19] (*see* Fig. 4 and Table 2). MRS1754 (**18**) is potent and selective in humans, but not in rat and mice [23, 35]. One of the most potent and selective competitive A_{2B} antagonist in all three species is PSB-603 (**19**) [23, 36]. The compound displays high metabolic stability in human, rat, and mouse (unpublished data). However, the main drawback of **19** is its low water solubility. The related PSB-0788 (**20**) [23, 36] is somewhat more soluble, especially under weakly acidic conditions due to its basicity, but it is less metabolically stable (***unpublished results***). Moreover, it is only moderately selective for A_{2B}- versus A_1ARs in mice (only about 60-fold), whereas it is highly selective in human and rat. PSB-1115 (**21**) was developed as an $A_{2B}AR$ antagonist with high water solubility due to its sulfonate group [37]. It is potent and selective in humans, but not in rat and mice, since it also blocks rodent A_1ARs (*see* Table 2) [23]. ATL 802 (**22**), which can be regarded as an optimized version of **18**, is potent and sufficiently selective in human and mouse [38].

The A_3AR shows the largest species differences for antagonists [5, 39]. Many published antagonists, that are very potent at human A_3ARs, are completely inactive at rodent (rat and mouse) A_3ARs. One of the best A_3AR antagonists for rodent studies is MRS1523 (**23**), which is only moderately potent, but shows high selectivity in human (>100-fold), and at least some selectivity in rat (18-fold versus A_{2A}, >100-fold versus the other AR subtypes), and mice (at least 14-fold versus the other subtypes) [5, 23, 24]. Other potent A_3AR antagonists including MRE3008-F20 (**24**) [40, 41], PSB-10 (**25**), and PSB-11 (**26**) [23, 42] are very potent and selective in humans but virtually inactive at rodent A_3ARs (*see* Table 2).

5 Allosteric Modulators for Adenosine Receptors

The development of allosteric modulator for GPCRs has been an emerging field of research [43–45]. Positive allosteric modulators (PAMs) increasing the potency and/or activity of agonists, as well as negative allosteric modulators (PAMs) acting as noncompetitive antagonists, have been developed for various AR subtypes; however, most compounds developed so far display only moderate potency and/or selectivity [19, 43].

6 P2Y Receptors

P2Y receptors are GPCRs activated by nucleotides. The P2Y receptor family is subdivided into the $P2Y_1$-like receptor subtypes ($P2Y_1$, $P2Y_2$, $P2Y_4$, $P2Y_6$, and $P2Y_{11}$) and the $P2Y_{12}$-like receptor subtypes ($P2Y_{12}$, $P2Y_{13}$, and $P2Y_{14}$) (*see* Fig. 5). The $P2Y_1$-like receptors are coupled to $G_{q/11}$ proteins mediating an activation of phospholipase C_β and subsequent production of IP_3 followed by intracellular calcium release. The $P2Y_{12}$-like receptors are coupled to $G_{i/o}$ proteins inducing an inhibition of adenylate cyclase which results in a reduction of intracellular cAMP levels (*see* Fig. 5). X-ray structures of the $P2Y_{12}$ receptor in complex with agonists and antagonist, all binding to the orthosteric site, have been described [46, 47]. Recently, X-ray structures of the $P2Y_1$ receptor in complex with allosteric antagonists have been obtained [48], a nucleotide that binds close to and partly operlapping with the orthosteric binding site, and a urea derivative which binds in the lateral periphery of the receptor close to the surrounding phospholipids (*see* Fig. 5). Based on homology models and mutagenesis studies, it appears likely, that the orthosteric site for all P2Y receptor subtypes may be in the same area as observed for the $P2Y_{12}$ receptor. Nevertheless, there are significant differences in the orthosteric binding sites of the P2Y receptor subtypes: $P2Y_1$, $P2Y_{12}$ $P2Y_{13}$ receptors are activated by ADP, $P2Y_2$ and $P2Y_{11}$ by ATP (and also by NAD^+ and $NAADP^+$), $P2Y_2$ and $P2Y_4$ by UTP, $P2Y_6$ and $P2Y_{14}$ by UDP, and $P2Y_{14}$ additionally by UDP-sugars such as UDPglucose.

In Table 3, Fig. 6 and 7 useful agonists and antagonists with high potency and selectivity for the various P2Y receptor subtypes are collected. Since radioligand binding assays for all P2Y receptor subtypes are not available, literature data, especially those for agonists, are difficult to compare [6].

The selection was based on recent detailed analyses of P2Y receptor ligands [5, 6, 11, 12, 49] taking into account high affinity across species and selectivity. The majority of the agonists are derivatives of the physiological nucleotides, except for the $P2Y_{11}$

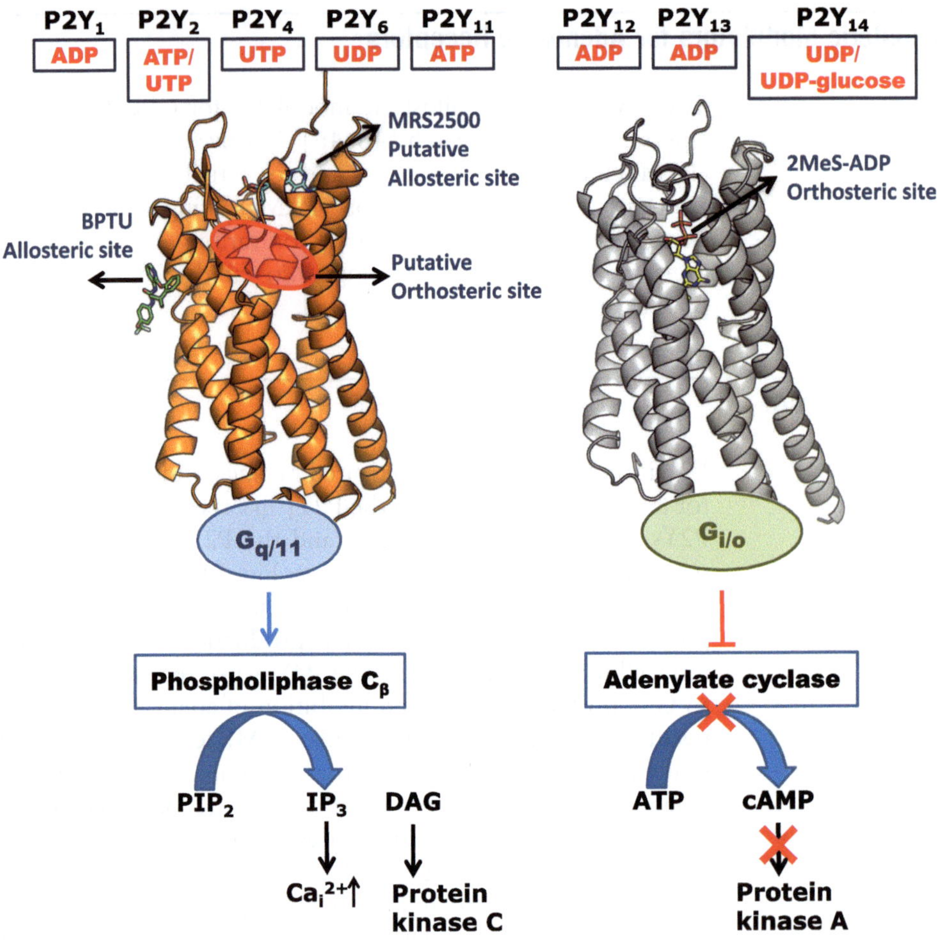

Fig. 5 Physiological agonists, structures, binding sites, and signaling pathways of P2Y receptor subtypes, which are divided into the $P2Y_1$-like and the $P2Y_{12}$-like subgroups. On the left-hand side, the crystal structure of the $P2Y_1$ receptor (orange) in complex with the antagonists MRS2500 (4XNW.pdb) and BPTU (4XNV.pdb) and the putative orthosteric binding site (for ADP) are shown. On the right-hand side, the crystal structure of the $P2Y_{12}$ receptor (gray) in complex with 2MeS-ADP (4PXZ.pdb) is shown. The receptors are represented in cartoon models and the cocrystallized ligands MRS2500 (carbon atoms colored cyan), BPTU (carbon atoms colored green), and 2MeS-ADP (carbon atoms colored yellow) are shown in stick models. Oxygen atoms are colored in red, nitrogen atoms in blue, and phosphorus atoms in orange

receptor agonist NF546, which is derived from the polysulfonate suramin. Its potency is only moderate [50]. Structurally diverse P2Y receptor antagonists have been developed, some of which are competitive antagonists binding to the orthosteric, while others are noncompetitive interacting with an allosteric site of the P2Y receptor (*see* Table 3). The largest variety of potent and selective antagonists is available for the $P2Y_{12}$ receptor subtype, and several of these compounds are therapeutically used as antithrombotic drugs [11], however a $P2Y_{12}$ receptor antagonist with high penetration into the brain is still lacking.

Table 3
Useful agonists and antagonists for P2Y receptor subtypes

P2Y receptor subtype	Agonist	Antagonist
$P2Y_1$	MRS2365 (**27**)	MRS-2500 (**38**, allosteric, but close to the orthosteric site) BPTU (**39**, allosteric)
$P2Y_2$	MRS2698 (**28**) PSB-1114 (**29**)	AR-C118925 (**40**, orthosteric)
$P2Y_4$	MRS4062 (**30**)	PSB-16133 (**41**, allosteric)
$P2Y_6$	5-Methoxy-UDP (**31**) PSB-0474 (**32**) MRS2693 (**33**)	MRS2578 (**42**, allosteric, low solubility)
$P2Y_{11}$	(NF546) (**34**, orthosteric)	NF340 (**43**, orthosteric)
$P2Y_{12}$	2-MeS-ADP (**35**)	Cangrelor (AR-C66931, **44**, orthosteric) Ticagrelor (**45**, orthosteric) PSB-0739 (**46**, orthosteric) AZD 1283 (**47**, orthosteric) Clopidogrel (**48**, prodrug; requires metabolic activation in the liver) (allosteric, covalent)
$P2Y_{13}$		(MRS2211) (**49**, weakly potent, moderate selectivity)
$P2Y_{14}$	MRS2905 (**36**) MRS2690 (**37**, 2-Thio-UDPglucose)	PPTN (**50**, orthosteric, low solubility)

7 P2X Receptors

P2X receptors are ATP-gated ion channels consisting of three subunits, which are permeable for sodium, potassium and calcium ions. Recently, several X-ray structures of P2X receptor, some in complex with orthosteric or allosteric ligands, have been published [7, 51, 52]. The seven different subunits, P2X1–7, can form homomeric or heteromeric channels [4]. Little is known about the P2X5 and P2X6 subtypes, who may be nonfunctional or may require heteromerization for forming a functional channel. So far, only tool compounds for the other five P2X receptor subtypes are available. All P2X receptor subtypes are activated by ATP (Fig. 1), although with different sensitivities (*see* Table 4). The P2X7 receptor subtype is the least sensitive one requiring high, up to millimolar ATP concentrations. Upon prolonged activation, the P2X7 receptor can form large pores allowing molecules with a molecular weight of up to 1 kDa to cross the cell membrane [53]. Due to the high ATP concentrations required for activating the P2X7 receptor,

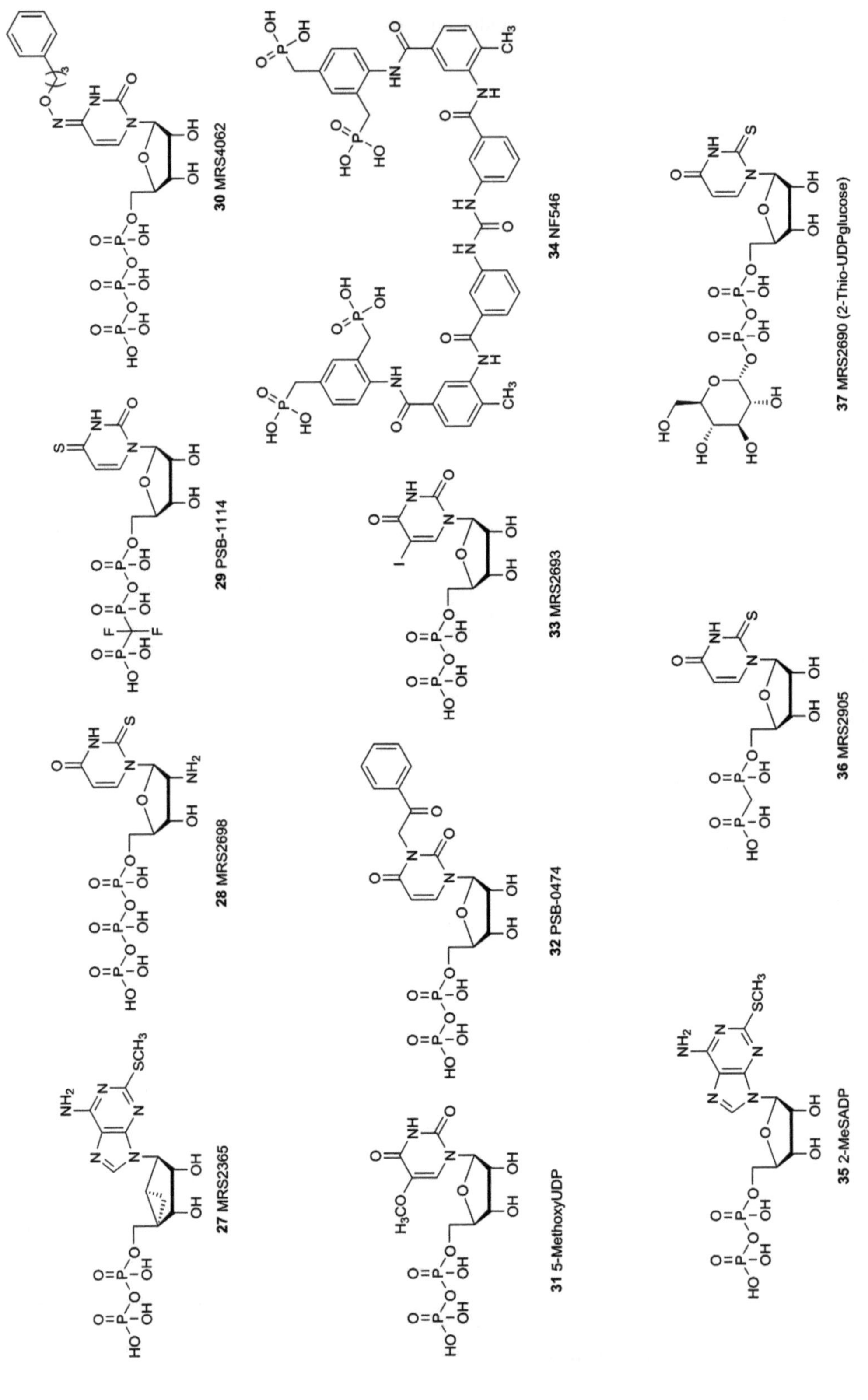

Fig. 6 P2Y receptor agonists

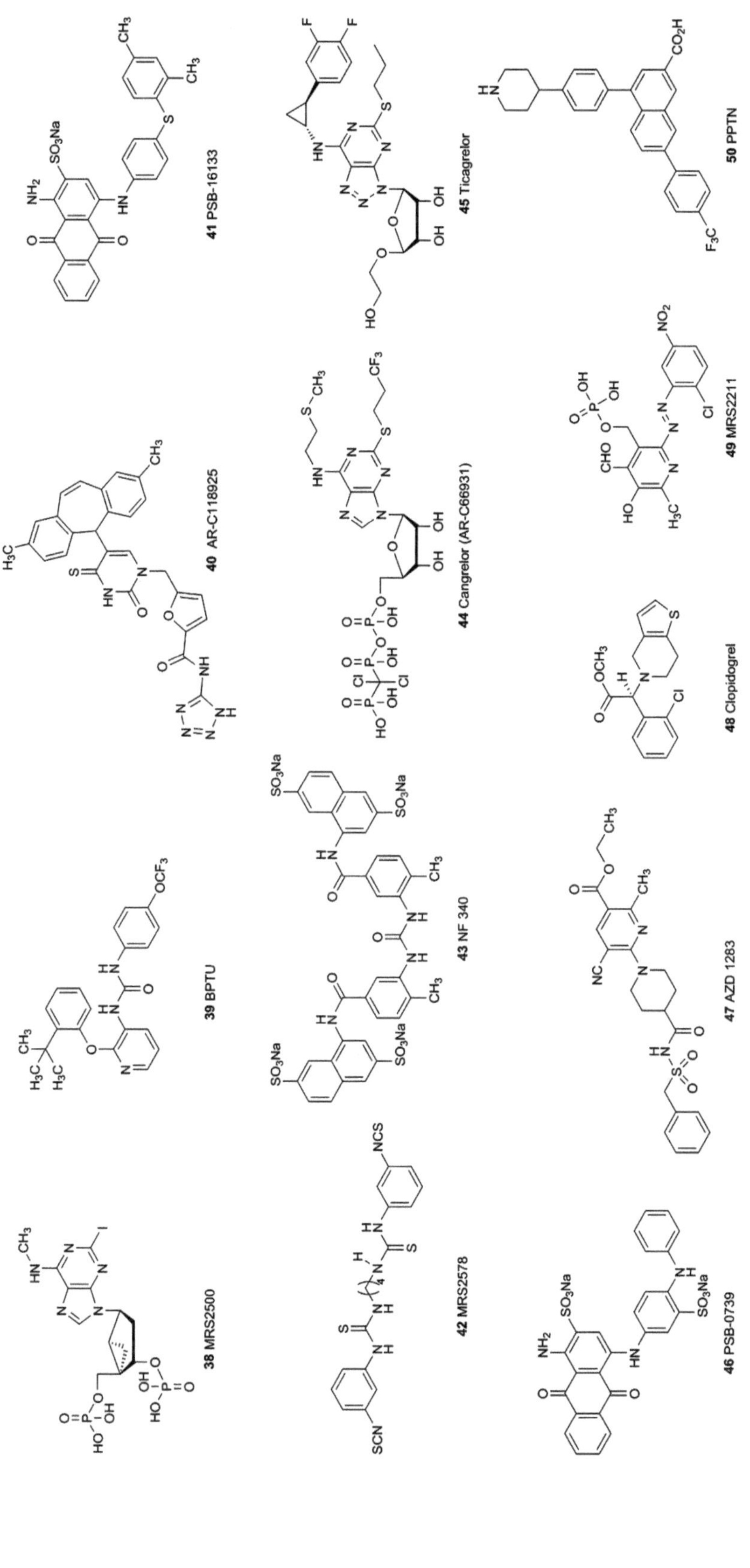

Fig. 7 P2Y receptor antagonists

Table 4
ATP potency and useful antagonists for P2X receptor subtypes

P2X receptor subtype	Agonist	Antagonist
P2X1	ATP (EC_{50} 0.1–0.7 μM, fast desensitization)	NF 279 (**54**) NF 023 (**55**)
P2X2	ATP (EC_{50} 2–8 μM) PSB-10129 (**51**, PAM)	NF 770 (**56**) PSB-10211 (**57**) PSB-1011 (**58**)
P2X3	ATP (EC_{50} ca. 1 μM, fast desensitization)	Gefapixant (**59**, AF-219) AF-353 (**60**, RO-4) AF-906 (**61**, RO-51)
P2X4	ATP (EC_{50} 1–10 μM) Ivermectin (**52**, PAM)	5-BDBD (**62**) PSB-12054 (**63**) PSB-12062 (**64**) BX 430 (**65**, only human, not mouse or rat)
P2X7	ATP (EC_{50} 2–4 mM) BzATP (**53**, about tenfold more potent, but nonselective)	AZ10606120 (**66**) A 740003 (**67**) A 804598 (**68**) JNJ 47965567 (**69**)

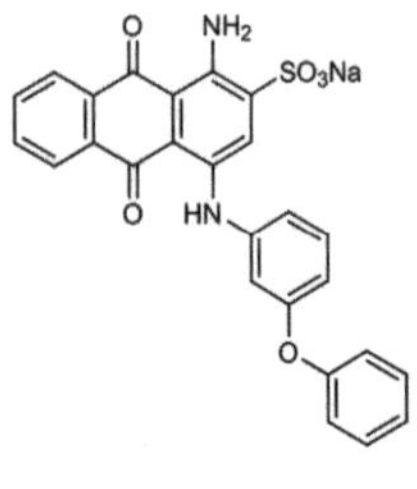

51 PSB-10129

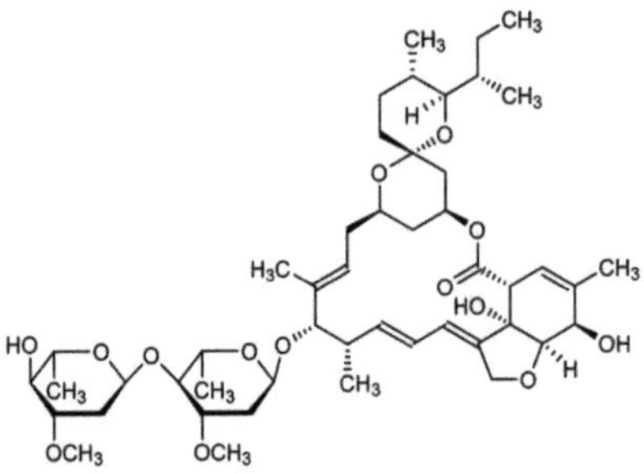

52 Ivermectin

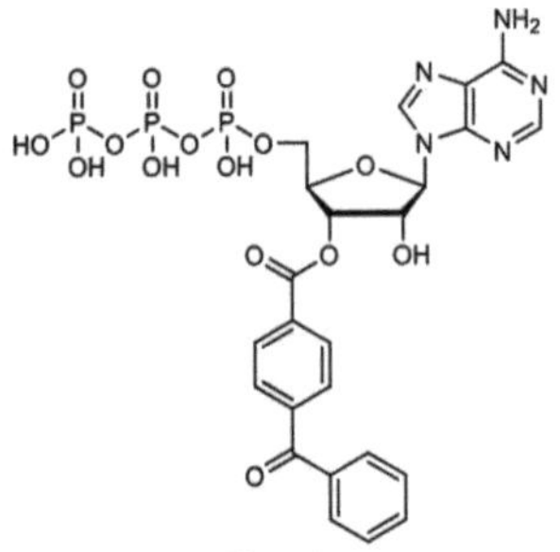

53 BzATP

Fig. 8 P2X receptor agonists and positive allosteric allosteric modulators (PAMs)

which are often already toxic for cells, the more potent, but also nonselective agonist BzATP is often used instead of ATP for activating P2X7 receptors. Selective agonists for P2X receptor subtypes are currently not available, and their development may in fact be difficult due to the high similarity of the positively charged orthosteric ATP-binding site in all subtypes. The development of positive allosteric modulators (PAMs) might be a solution, and in fact, PAMs have been described for the P2X2 (PSB-10129) and the P2X4 receptor (ivermectin). A selection of modulators for the P2X1, P2X2, P2X3, P2X4 and P2X7 receptors is presented in Table 4, Fig. 8 and Fig. 9.

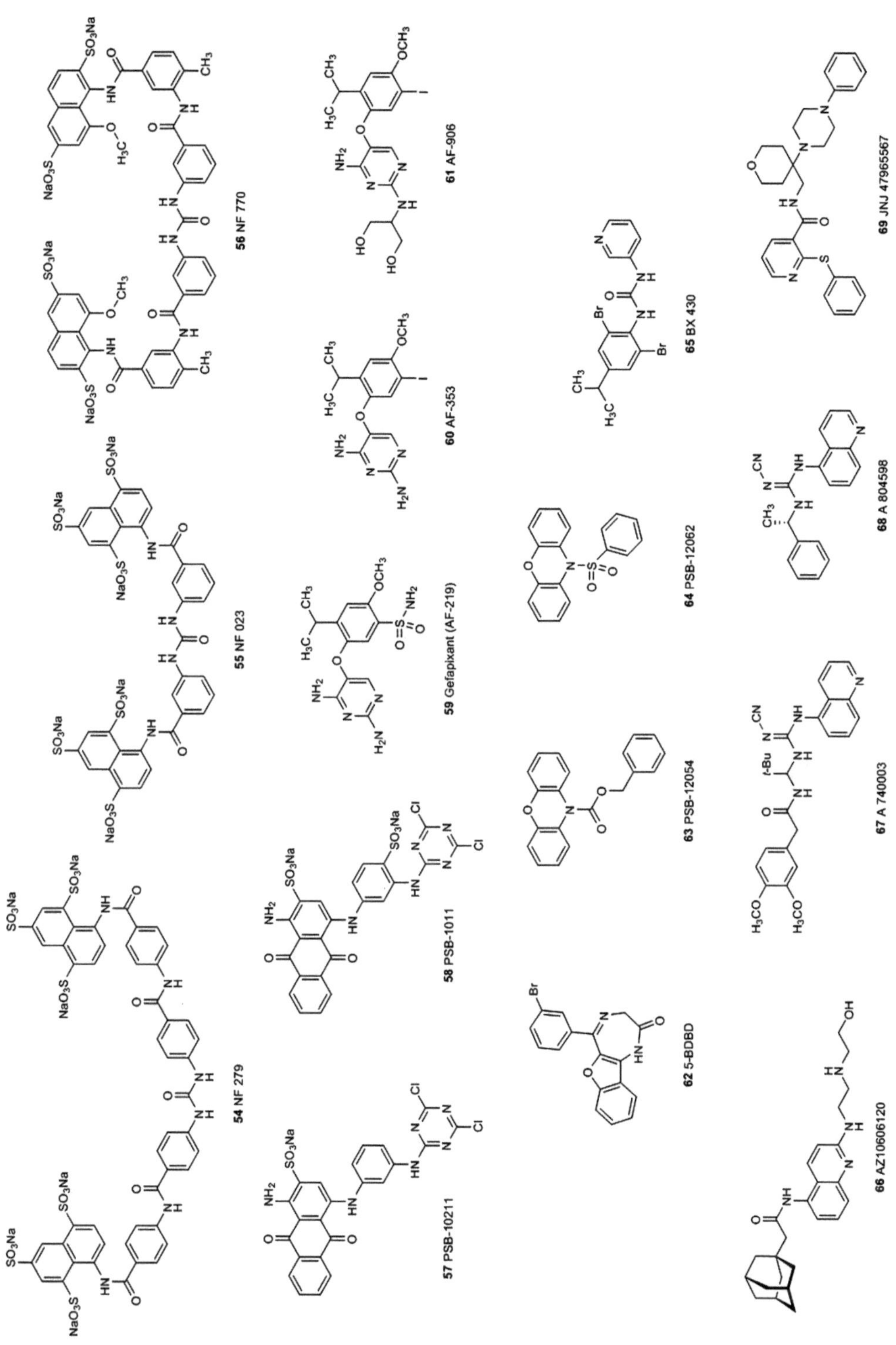

Fig. 9 P2X receptor antagonists

The most intensively studied subtypes are the homomeric P2X3, the heteromeric P2X2/P2X3 and the homomeric P2X7 receptor. Antagonists for these P2X receptors have already been clinically evaluated [13, 54], and further clinical studies are currently being conducted. The majority of the developed antagonists are allosteric modulators (NAMs), which do not bind to the ATP binding site.

8 Conclusions

The interest in purinergic receptors as drug targets has massively increased in recent years. This has led to the development of many useful tool compounds and drugs for adenosine, P2Y and P2X receptors. Despite extensive efforts, however, suitable high-quality tool compounds for a number of scientific questions are still lacking. Due to the recently increased interest in purinergic targets major progress can be expected in the near future.

References

1. Burnstock G (2007) Purine and pyrimidine receptors. Cell Mol Life Sci 64:1471–1483
2. Fredholm BB, IJzerman AP, Jacobson KA et al (2011) International Union of Basic and Clinical Pharmacology. LXXXI. Nomenclature and classification of adenosine receptors – an update. Pharmacol Rev 63:1–34
3. Abbracchio MP, Burnstock G, Boeynaems JM et al (2006) International Union of Pharmacology. LVIII: Update on the P2Y G protein-coupled nucleotide receptors: from molecular mechanisms and pathophysiology to therapy. Pharmacol Rev 58:281–341
4. Khakh BS, Burnstock G, Kennedy C et al (2001) International Union of Pharmacology. XXIV. Current status of the nomenclature and properties of P2X receptors and their subunits. Pharmacol Rev 53:107–118
5. Jacobson KA, Müller CE (2016) Medicinal chemistry of adenosine, P2Y and P2X receptors. Neuropharmacology 104:31–49
6. Rafehi M, Müller CE (2018) Tools and drugs for uracil nucleotide-activated P2Y receptors. Pharmacol Ther 190:24–80
7. Habermacher C, Dunning K, Chataigneau T et al (2016) Molecular structure and function of P2X receptors. Neuropharmacology 104:18–30
8. Thimm D, Knospe M, Abdelrahman A et al (2013) Characterization of new G protein-coupled adenine receptors in mouse and hamster. Purinergic Signal 9:415–426
9. Thimm D, Schiedel AC, Peti-Peterdi J et al (2015) The nucleobase adenine as a signalling molecule in the kidney. Acta Physiol 213:808–818
10. Burnstock G (2017) Purinergic signalling: therapeutic developments. Front Pharmacol 8:661
11. Baqi Y, Müller CE (2018) Antithrombotic $P2Y_{12}$ receptor antagonists: recent developments in drug discovery. Drug Discov Today. https://doi.org/10.1016/j.drudis.2018.09.021
12. von Kügelgen I (2017) Structure, pharmacology and roles in physiology of the $P2Y_{12}$ receptor. Adv Exp Med Biol 1051:123–138
13. Abdulqawi R, Dockry R, Holt K et al (2015) P2X3 receptor antagonist (AF-219) in refractory chronic cough: a randomised, double-blind, placebo-controlled phase 2 study. Lancet 385:1198–1205
14. Meibom D, Albrecht-Küpper B, Diedrichs N et al (2017) Neladenoson bialanate hydrochloride: a prodrug of a partial adenosine A_1 receptor agonist for the chronic treatment of heart diseases. ChemMedChem 12:728–737
15. Congreve M, Brown GA, Borodovsky A et al (2018) Targeting adenosine A_{2A} receptor antagonism for treatment of cancer. Expert Opin Drug Discov 13:997–1003
16. Lukashev D, Sitkovsky M, Ohta A (2007) From "Hellstrom Paradox" to anti-

adenosinergic cancer immunotherapy. Purinergic Signal 3:129–134

17. Vijayan D, Young A, Teng MWL et al (2017) Targeting immunosuppressive adenosine in cancer. Nat Rev Cancer 17:709–724
18. Hinz S, Navarro G, Borroto-Escuela D et al (2018) Adenosine A_{2A} receptor ligand recognition and signaling is blocked by A_{2B} receptors. Oncotarget 9:13593–13611
19. Müller CE, Baqi Y, Hinz S et al (2018) Chapter 6: Medicinal chemistry of A_{2B} adenosine receptors. In: Borea PA et al (eds) The adenosine receptors. Springer Nature Switzerland AG, Switzerland, pp 137–168
20. Jazayeri A, Andrews SP, Marshall FH (2017) Structurally enabled discovery of adenosine A_{2A} receptor antagonists. Chem Rev 117:21–37
21. Glukhova A, Thal DM, Nguyen AT et al (2017) Structure of the adenosine A_1 receptor reveals the basis for subtype selectivity. Cell 168:867–877.e13
22. Franchetti P, Cappellacci L, Vita P et al (2009) N^6-cycloalkyl- and N^6-bicycloalkyl-C5′(C2′)-modified adenosine derivatives as high-affinity and selective agonists at the human A_1 adenosine receptor with antinociceptive effects in mice. J Med Chem 52:2393–2406
23. Alnouri MW, Jepards S, Casari A et al (2015) Selectivity is species-dependent: characterization of standard agonists and antagonists at human, rat, and mouse adenosine receptors. Purinergic Signal 11:389–407
24. Müller CE, Jacobson KA (2011) Recent developments in adenosine receptor ligands and their potential as novel drugs. Biochim Biophys Acta 1808:1290–1308
25. El-Tayeb A, Michael S, Abdelrahman A et al (2011) Development of polar adenosine A_{2A} receptor agonists for inflammatory bowel disease: synergism with A_{2B} antagonists. ACS Med Chem Lett 2:890–895
26. Hinz S, Lacher SK, Seibt BF et al (2014) BAY60-6583 acts as a partial agonist at adenosine A_{2B} receptors. J Pharmacol Exp Ther 349:427–436
27. Müller CE, Jacobson KA (2011) Xanthines as adenosine receptor antagonists. Handb Exp Pharmacol 200:151–199
28. Weyler S, Fülle F, Diekmann M et al (2006) Improving potency, selectivity, and water solubility of adenosine A_1 receptor antagonists: xanthines modified at position 3 and related pyrimido[1,2,3-cd]purinediones. ChemMedChem 1:891–902
29. Kalk P, Eggert B, Relle K et al (2007) The adenosine A_1 receptor antagonist SLV320 reduces myocardial fibrosis in rats with 5/6 nephrectomy without affecting blood pressure. Br J Pharmacol 151:1025–1032
30. Takahashi M, Fujita M, Asai N et al (2018) Safety and effectiveness of istradefylline in patients with Parkinson's disease: interim analysis of a post-marketing surveillance study in Japan. Expert Opin Pharmacother 19:1635–1642
31. Hockemeyer J, Burbiel JC, Müller CE (2004) Multigram-scale syntheses, stability, and photoreactions of A_{2A} adenosine receptor antagonists with 8-styrylxanthine structure: potential drugs for Parkinson's disease. J Org Chem 69:3308–3318
32. Sauer R, Maurinsh J, Reith U et al (2000) Water-soluble phosphate prodrugs of 1-propargyl-8-styrylxanthine derivatives, A_{2A}-selective adenosine receptor antagonists. J Med Chem 43:440–448
33. Faivre E, Coelho JE, Zornbach K et al (2018) Beneficial effect of a selective adenosine A_{2A} receptor antagonist in the APPswe/PS1dE9 mouse model of Alzheimer's disease. Front Mol Neurosci 11:235
34. Stocchi F, Rascol O, Hauser RA et al (2017) Randomized trial of preladenant, given as monotherapy, in patients with early Parkinson disease. Neurology 88:2198–2206
35. Kim YC, Ji X, Melman N et al (2000) Anilide derivatives of an 8-phenylxanthine carboxylic congener are highly potent and selective antagonists at human A_{2B} adenosine receptors. J Med Chem 43:1165–1172
36. Borrmann T, Hinz S, Bertarelli DC et al (2009) 1-Alkyl-8-(piperazine-1-sulfonyl)phenylxanthines: development and characterization of adenosine A_{2B} receptor antagonists and a new radioligand with subnanomolar affinity and subtype specificity. J Med Chem 52:3994–4006
37. Hayallah AM, Sandoval-Ramirez J, Reith U et al (2002) 1,8-Disubstituted xanthine derivatives: synthesis of potent A_{2B}-selective adenosine receptor antagonists. J Med Chem 45:1500–1510
38. Cagnina RE, Ramos SI, Marshall MA et al (2009) Adenosine A_{2B} receptors are highly expressed on murine type II alveolar epithelial cells. Am J Physiol Lung Cell Mol Physiol 297: L467–L474
39. Müller CE (2003) Medicinal chemistry of adenosine A_3 receptor ligands. Curr Top Med Chem 3:445–462
40. Baraldi PG, Preti D, Borea PA et al (2012) Medicinal chemistry of A_3 adenosine receptor modulators: pharmacological activities and

therapeutic implications. J Med Chem 55:5676–5703
41. Borea PA, Varani K, Vincenzi F et al (2015) The A_3 adenosine receptor: history and perspectives. Pharmacol Rev 67:74–102
42. Ozola V, Thorand M, Diekmann M et al (2003) 2-Phenylimidazo[2,1-i]purin-5-ones: structure-activity relationships and characterization of potent and selective inverse agonists at human A_3 adenosine receptors. Bioorg Med Chem 11:347–356
43. Müller CE, Schiedel AC, Baqi Y (2012) Allosteric modulators of rhodopsin-like G protein-coupled receptors: opportunities in drug development. Pharmacol Ther 135:292–315
44. Wootten D, Christopoulos A, Sexton PM (2013) Emerging paradigms in GPCR allostery: implications for drug discovery. Nat Rev Drug Discov 12:630–644
45. Gao ZG, Jacobson KA (2013) Allosteric modulation and functional selectivity of G protein-coupled receptors. Drug Discov Today Technol 10:e237–e243
46. Zhang K, Zhang J, Gao ZG et al (2014) Structure of the human $P2Y_{12}$ receptor in complex with an antithrombotic drug. Nat 509:115–118
47. Zhang J, Zhang K, Gao ZG et al (2014) Agonist-bound structure of the human $P2Y_{12}$ receptor. Nat 509:119–122
48. Zhang D, Gao ZG, Zhang K et al (2015) Two disparate ligand-binding sites in the human $P2Y_1$ receptor. Nature 520:317–321
49. von Kügelgen I, Hoffmann K (2016) Pharmacology and structure of P2Y receptors. Neuropharmacology 104:50–61
50. Meis S, Hamacher A, Hongwiset D et al (2010) NF546 [4,4′-(carbonylbis(imino-3,1-phenylene-carbonylimino-3,1-(4-methyl-phenylene)-carbonylimino))-bis(1,3-xylene-alpha, alpha′-diphosphonic acid)tetrasodium salt] is a non-nucleotide $P2Y_{11}$ agonist and stimulates release of interleukin-8 from human monocyte-derived dendritic cells. J Pharmacol Exp Ther 332:238–247
51. Kawate T (2017) P2X receptor activation. Adv Exp Med Biol 1051:55–69
52. Kasuya G, Yamaura T, Ma XB et al (2017) Structural insights into the competitive inhibition of the ATP-gated P2X receptor channel. Nat Commun 8:876
53. Di Virgilio F, Schmalzing G, Markwardt F (2018) The elusive P2X7 macropore. Trends Cell Biol 28:392–404
54. Rech JC, Bhattacharya A, Letavic MA et al (2016) The evolution of P2X7 antagonists with a focus on CNS indications. Bioorg Med Chem Lett 26:3838–3845

Chapter 4

Homology Modeling of P2X Receptors

Anastasios Stavrou, Sudad Dayl, and Ralf Schmid

Abstract

Since the X-ray structure of the zebra fish P2X4 receptor in the closed state was published in 2009 homology modeling has been used to generate structural models for P2X receptors. In this chapter, we outline how to use the MODELLER software to generate such structural models for P2X receptors whose structures have not been solved yet.

Key words Homology modeling, Protein structure prediction, MODELLER software, P2X receptor, Ion channel

1 Introduction

The determination of high-resolution protein 3D-structures by X-ray crystallography, NMR spectroscopy, and more recently cryo-EM, enables us to understand protein function at an atomistic level. Within a group of evolutionary related proteins their 3D structures tend to be more conserved than their sequences. Once a structure of a protein from any particular protein family is known, the structures of other members of this family can be predicted by using the known structure as a structural template. This protein structure prediction approach is called homology modeling. The availability of a template structure is probed for by sequence similarity searches where the sequence of the protein of interest (target) is compared to a database of sequences of potential templates, that is, proteins with known structures. If a structure shares statistically significant sequence similarity with the target, one can conclude common ancestry (homology). Homology modeling generates one or more 3D models for the target protein based on the structural framework of the identified template structure. This approach works best in scenarios where target and template are closely related and hence share high pairwise sequence identity. Nevertheless, homology modeling can be successfully applied at pairwise sequence identity levels of 30% or even lower, as long as target

Pablo Pelegrín (ed.), *Purinergic Signaling: Methods and Protocols*, Methods in Molecular Biology, vol. 2041, https://doi.org/10.1007/978-1-4939-9717-6_4,

and template are indeed homologous. Several methods, software tools and web servers for homology modeling are available, ranging from the fully automated Swiss-Model server [1, 2] to the highly tuneable MODELLER software [3] (for overview *see* [4]).

P2X receptors are ligand (typically ATP) activated ion-channels which in their functional form occur as homotrimers or heterotrimers. P2X receptors are found ubiquitously in the body with diverse functions such as controlling vascular smooth muscle cells contraction in the heart or involved in the immune system and chronic neuropathic pain (reviewed in [5]). For a long time, high resolution structural information on P2X receptors (P2XR) was elusive, and only limited structural data derived from biochemistry and molecular biology experiments were available. This changed once the X-ray structure of the zebra fish P2X4 receptor (zfP2X4R) in the closed state was published [6]. This structure confirmed the trimeric arrangement of P2XRs, provided detailed insights into the extracellular regions of the receptor, the location of the ATP binding site, the arrangement of the transmembrane helices, and enhanced our understanding of how P2XRs work. In addition to these new insights, this structure also provided the first suitable template for modeling structures of homologous P2XRs in the closed state. In the last few years, additional P2XR structures with higher resolution and in different conformational states have become available and have broadened the scope for P2XR homology modeling. For instance the zfP2X4R [7] and human P2X3R (hP2X3R) [8] structures in the open state with ATP bound allowed the generation of open state homology models. For comprehensive discussion of published P2XR structures and derived models we refer the reader to the reviews by the Young group [9, 10].

Homology modeling of P2XRs comes with specific challenges. For example, most functional interpretation requires a trimeric model rather than a monomer structure (fully automated approaches would typically generate the latter). The selection of the template state is crucial for modeling the required state. A template in the closed state will result a closed state model, and any ATP bound state of P2XR is best modeled based on a template where ATP is present. Other potential pitfalls concern the modeling of regions where no structural information from the template is available. In P2XRs this frequently refers to the intracellular N- and C-termini, and is especially pronounced for the extended intracellular regions in P2X7R. While short insertions and deletions can be modeled successfully (for review on loop modeling *see* [11]), extended stretches modeled without template information frequently will be wrong (*see* **Note 1**). To accommodate these and other challenges we use the MODELLER software [3]. While MODELLER is not the easiest scientific software to use, its flexibility enables the generation of models most relevant to the specific research questions one wants to address.

Homology models in P2XR research have been widely used over the last few years. For example, in our own work we have used homology models for mapping/studying mutations [12, 13], ligand docking [14–16], virtual screening and molecular dynamics simulations [17]. In the next sections, we describe in detail how to generate desired structural models for P2X receptors whose structures have not been solved yet, using the modeling of human P2X2 receptor in the open state as an example.

2 Materials

2.1 MODELLER Installation

The MODELLER software is available free of charge for academic and nonprofit institutions for Linux, Mac, and Windows operating systems [3]. It can be downloaded from the MODELLER project page https://salilab.org/modeller/download_installation.html. The MODELLER installation differs depending on the operating system, for details please refer to the installation guides on the MODELLER download page.

2.2 Test Files

The alignment file (P2X2_P2X3o.pir), template structure file (5svk_t.pdb) and MODELLER instruction file (P2X2o.py) used in this chapter are provided on our GitHub page (https://github.com/RalfSchmid/P2XR-modelling in the directory P2X2R for download, viewing, testing, and modification).

3 Methods

The following instructions guide the reader through the modeling process where we generate a homology model for hP2X2R in the open state. The example files and wording in this section are specific to this particular task. However, the instructions and files provided can easily be adapted to modeling other P2XRs or other states (additional files are provided at https://github.com/RalfSchmid/P2XR-modelling), or in fact for any other protein family. For easier reading, the content of text files, file names and commands to be typed in the terminal are shown in `Courier` font.

3.1 Template Selection

In P2XR homology modeling the first challenge is to identify the most suitable template from the available structures (*see* **Note 2**). The main criteria for template selection are (in descending order) relevance of the conformational state of the template, sequence similarity between target and template, length of the alignment, resolution and quality of the template structure. For all hP2XRs some of this information is provided in Table 1. For modeling hP2X2R in the open state, the zfP2X4R (4DW1) and hP2X3R (5SVK) templates are equivalent in terms of sequence similarity,

Table 1
Human P2XR sequences and some of their potential templates for homology modeling

	PDB identifier (description)		
Sequence *Sequence identifier*	**4DW1 (zfP2X4R, open)**	**5SVK (hP2X3R, open)**	**5U1L (pdP2X7R, closed)**
hP2X1R	150/337 (45%)	171/364 (47%)	155/335 (46%)
sp|P51575|P2RX1_HUMAN	226/337 (67%)	234/364 (64%)	207/335 (62%)
hP2X2R	165/339 (49%)	183/366 (50%)	128/330 (39%)
sp|Q9UBL9|P2RX2_HUMAN	218/339 (64%)	243/366 (66%)	186/330 (56%)
hP2X3R	159/336 (47%)	356/360 (99%)	144/337 (43%)
sp|P56373|P2RX3_HUMAN	219/336 (65%)	358/360 (99%)	194/337 (58%)
hP2X4R	208/338 (62%)	172/369 (47%)	160/340 (47%)
sp|Q99571|P2RX4_HUMAN	260/338 (77%)	249/369 (67%)	216/340 (64%)
rP2X5R (*see* **Note 4**)	175/343 (51%)	172/366 (47%)	140/347 (40%)
sp|P51578|P2RX5_RAT	233/343 (68%)	239/366 (65%)	193/347 (56%)
hP2X6R	145/334 (43%)	154/364 (42%)	128/330 (39%)
sp|O15547|P2RX6_HUMAN	211/334 (63%)	212/364 (58%)	196/330 (59%)
hP2X7R	160/344 (47%)	152/358 (42%)	291/339 (86%)
sp|Q99572|P2RX7_HUMAN	222/344 (65%)	206/358 (58%)	318/339 (94%)

Sequence similarity between P2XR sequences and the templates 4DW1, 5SVK, and 5U1L is derived from BLAST searches with P2XR sequences against the BLAST pdb sequence database [18] (*see* **Note 3**). Results are reported as follows. Row 1: number of identical residues in the alignment/length of the alignment (percentage), and Row 2: "positive" residues/length of the alignment (percentage) where "positive" is defined as sum of identical and similar residues in the alignment

however the length of the respective alignment is different. This is due to the presence of the intracellular cap region in the structure of hP2X3R (5SVK). As we want to include this region in our hP2X2R open state model we opt for 5SVK rather than 4DW1 (zfP2X4R) as template.

3.2 Target-Template Alignment

Once the template has been chosen a sequence alignment between target and template needs to be generated. Such an alignment could be derived from a pairwise BLAST alignment, though we prefer to derive it from multiple sequence alignments. This is because the inclusion of additional data improves the accuracy of such alignments (for details *see* example file `P2XR_sw_pdb.fasta` on https://github.com/RalfSchmid/P2XR-modelling). In addition, we try to rationalize and adjust gaps in the pairwise alignment based on the template structure, that is, we make a judgment on the plausibility of locations for residue insertion or deletion. MODELLER requires the template/target sequence alignment in `pir` format. The content of the example file `P2X2_P2X3o.pir` and explanation of the format is given below.

```
>P1;5svk_t
structureX:5svk_t:7:A:401:C:.:.:.:.
FFTYETPKVIVVKSWTIGIINRVVQLLIISYFVGWVFLHEKAYQVRDTAIESSVVTKVKGSGLYANRVMDVS
DYVTPPQGTSVFVIITKMIVTENQMQGFCPESEE--KYRCVSDSQCG--PERLPGGGILTGRCVNYS-SVLR
TCEIQGWCPTEV-DTVETPIMMEAENFTIFIKNSIRFPLFNFEKGNLLPNLTARDMKTCRFHPDKDPFCPIL
RVGDVVKFAGQDFAKLARTGGVLGIKIGWVCDLDKAWDQCIPKYSFTRLDSVSEKSSVSPGYNFRFAKYYKM
ENGSEYRTLLKAFGIRFDVLVYGNAGKFNIIPTIISSVAAFTSVGVGTVLCDIILLNFLKGADQYKAKKFEE
VNE./
FFTYETPKVIVVKSWTIGIINRVVQLLIISYFVGWVFLHEKAYQVRDTAIESSVVTKVKGSGLYANRVMDVS
DYVTPPQGTSVFVIITKMIVTENQMQGFCPESEE--KYRCVSDSQCG--PERLPGGGILTGRCVNYS-SVLR
TCEIQGWCPTEV-DTVETPIMMEAENFTIFIKNSIRFPLFNFEKGNLLPNLTARDMKTCRFHPDKDPFCPIL
RVGDVVKFAGQDFAKLARTGGVLGIKIGWVCDLDKAWDQCIPKYSFTRLDSVSEKSSVSPGYNFRFAKYYKM
ENGSEYRTLLKAFGIRFDVLVYGNAGKFNIIPTIISSVAAFTSVGVGTVLCDIILLNFLKGADQYKAKKFEE
VNE./
FFTYETPKVIVVKSWTIGIINRVVQLLIISYFVGWVFLHEKAYQVRDTAIESSVVTKVKGSGLYANRVMDVS
DYVTPPQGTSVFVIITKMIVTENQMQGFCPESEE--KYRCVSDSQCG--PERLPGGGILTGRCVNYS-SVLR
TCEIQGWCPTEV-DTVETPIMMEAENFTIFIKNSIRFPLFNFEKGNLLPNLTARDMKTCRFHPDKDPFCPIL
RVGDVVKFAGQDFAKLARTGGVLGIKIGWVCDLDKAWDQCIPKYSFTRLDSVSEKSSVSPGYNFRFAKYYKM
ENGSEYRTLLKAFGIRFDVLVYGNAGKFNIIPTIISSVAAFTSVGVGTVLCDIILLNFLKGADQYKAKKFEE
VNE./*
>P1;P2X2o
sequence:P2X2o:.:.:.:.:.:.:.:.
LWDYETPKVIVVRNRRLGVLYRAVQLLILLYFVWYVFIVQKSYQESETGPESSIITKVKGITTSEHKVWDVE
EYVKPPEGGSVFSIITRVEATHSQTQGTCPESIRVHNATCLSDADCVAGELDMLGNGLRTGRCVPYYQGPSK
TCEVFGWCPVEDGASVSQFLGTMAPNFTILIKNSIHYPKFHFSKGNIAD-RTDGYLKRCTFHEASDLYCPIF
KLGFIVEKAGESFTELAHKGGVIGVIINWDCDLDLPASECNPKYSFRRLDPKHVPA--SSGYNFRFAKYYKI
N-GTTTRTLIKAYGIRIDVIVHGQAGKFSLIPTIINLATALTSVGVGSFLTDWILLTFMNKNKVYSHKKFDK
VTT./
LWDYETPKVIVVRNRRLGVLYRAVQLLILLYFVWYVFIVQKSYQESETGPESSIITKVKGITTSEHKVWDVE
EYVKPPEGGSVFSIITRVEATHSQTQGTCPESIRVHNATCLSDADCVAGELDMLGNGLRTGRCVPYYQGPSK
TCEVFGWCPVEDGASVSQFLGTMAPNFTILIKNSIHYPKFHFSKGNIAD-RTDGYLKRCTFHEASDLYCPIF
KLGFIVEKAGESFTELAHKGGVIGVIINWDCDLDLPASECNPKYSFRRLDPKHVPA--SSGYNFRFAKYYKI
N-GTTTRTLIKAYGIRIDVIVHGQAGKFSLIPTIINLATALTSVGVGSFLTDWILLTFMNKNKVYSHKKFDK
VTT./
LWDYETPKVIVVRNRRLGVLYRAVQLLILLYFVWYVFIVQKSYQESETGPESSIITKVKGITTSEHKVWDVE
EYVKPPEGGSVFSIITRVEATHSQTQGTCPESIRVHNATCLSDADCVAGELDMLGNGLRTGRCVPYYQGPSK
TCEVFGWCPVEDGASVSQFLGTMAPNFTILIKNSIHYPKFHFSKGNIAD-RTDGYLKRCTFHEASDLYCPIF
KLGFIVEKAGESFTELAHKGGVIGVIINWDCDLDLPASECNPKYSFRRLDPKHVPA--SSGYNFRFAKYYKI
N-GTTTRTLIKAYGIRIDVIVHGQAGKFSLIPTIINLATALTSVGVGSFLTDWILLTFMNKNKVYSHKKFDK
VTT./*
```

The file contains two sequence entries, one for the template and one for the target. Each entry starts with two description lines followed by the actual aligned sequence. In the first description line the `5svk_t` corresponds to the name of the template file, in this example `5svk_t.pdb` (the ending ".pdb" is omitted in the description line). For any other modeling project this entry should be changed accordingly. The second line is a little more complex; it contains ten fields separated by colons.

```
structureX:5svk_t:7:A:401:C:.:.:.:.
```

The first field indicates that we are dealing with an X-ray structure, and fields 7–10 contain a "`.`" as place holder. None of these fields needs to be modified by the user. Of importance are fields 2–6. The second field contains the name of the structure file `5svk_t`. This descriptor is best kept the same as in the first description line. Fields 3 and 4 define the first residue of the template structure file that is used in the modeling process. In this example, MODELLER will interpret this as residue `7` of chain `A`. Fields 5 and 6 define the last residue of the template structure. This is read as residue `401` of chain `C`. In the corresponding two lines for the sequence entry, `P2X2o` represents the identifier of the sequence which will be used by MODELLER for naming output files.

For both sequence entries, the description lines are followed by the actual aligned sequences. Note that these are present in triplicate to ensure that we are modeling the hP2X2R as a trimer. The alignment uses the one letter amino acid code with gaps indicated by "`-`". The "*" signifies the end of the respective sequence entry. Another special character "`/`" indicates chain breaks and is used to separate the three chains in the P2X3R template and the P2X2R target. The "`.`" in the alignment stands for any additional, non-amino acid residues to be modeled. In our example, we use it to represent ATP.

Importantly, the start/end numbering for the template as defined in the description line needs to correspond to the actual alignment and to the template file. Hence, in our example the first residue of the template sequence, "F", requires a corresponding phenylalanine residue with residue number "5" and chain identifier "A" to be present in the template structure file (*see* below). The last residue of the template sequence "`.`" corresponds to chain "C" residue "401" in the template structure file (*see* below).

3.3 Template Structure File

Template files are typically downloaded from the PDB (*see* **Note 3**) and occasionally they require editing before being used in homology modeling. As stated above, it is important that the first and last coordinate entry (*see* extract from `5svk_t.pdb` below, highlighted in bold) corresponds to the numbering in the template-target alignment file (chain A, residue 7, F for the first residue and chain C, residue 401 with the "." standing for ATP for the last residue).

```
ATOM 1 N PHE A 7 285.197 176.393 175.589 1.00122.83 N
ATOM 2 CA PHE A 7 285.747 175.102 175.987 1.00134.72 C
ATOM 3 C PHE A 7 284.901 173.972 175.409 1.00136.29 C
ATOM 4 O PHE A 7 283.758 173.766 175.815 1.00134.28 O
<...>
HETATM 8649 H3' ATP C 401 226.438 187.888 125.061 1.00 80.45 L000 H
HETATM 8650 H4' ATP C 401 227.152 185.422 126.161 1.00 77.85 L000 H
HETATM 8651 H5'1 ATP C 401 228.667 186.745 127.581 1.00 75.60 L000 H
HETATM 8652 H5'2 ATP C 401 228.571 187.463 125.964 1.00 75.60 L000 H
END
```

3.4 Instruction File

MODELLER instruction files are written in the Python programming language. Such instruction files can be developed, adapted, and expanded based on examples provided on the MODELLER web page (https://salilab.org/modeller/) without any or with little knowledge of Python programming (*see* **Note 5**). The specific file used for modeling the hP2X2R in the open state (`P2X2o.py`) is available on https://github.com/RalfSchmid/P2XR-modelling in the directory P2X2R. While reading this section we recommend to open this file in parallel with a text editor. Key sections of the file `P2X2o.py` and parameters that might need to be modified in other projects are explained below.

The `automodel` class is MODELLER's standard tool box for modeling. It contains all key methods and data structures required. For P2XR modeling we import this class and expand it with additional features. The `special_patches` block changes the numbering for the models generated. As default the first modeled residue of each chain would be numbered as residue 1. In P2XR homology modeling we normally do not model the N-termini as the corresponding residues are not present in the template. Hence, we want to ensure that the first residue that is included in the model has the correct residue number. In our example, if we base the numbering on the sequence `sp.|P56373|P2RX3_HUMAN`, this would be L25. This is achieved by the setting `25` as value for the `renumber_residues` parameter. This setting can be adjusted for any other P2XR modeling project. For instance, to get the correct numbering for the almost identical, but shorter P2X2R isoform `AAD42947.1` [19] we would change the values of this parameter to `13`. A second addition is the `special restraints` block. This block of code enforces symmetry and ensures that the backbone of our models adhere to trifold symmetry. An important variable to consider is `env.io.hetatm`. By default, MODELLER only reads coordinate entries of the protein chains (*see* Subheading 3.3, lines starting with ATOM). Setting the value of env.io.hetatm to True enables MODELLER to read hetero atom entries from the template structure file (*see* Subheading 3.3, lines starting with HETATM), and hence is a requirement for including ATP molecules in the modeling process. In the next section a user may need to adapt the values of the settings alnfile and knowns. The values"P2X2_P2X3o.pir" and "5svk_t" for these variables need to correspond to the respective alignment and template structure files. Note that for the template structure file the ending .pdb is not required to occur in the entry. In the same block of code, sequence is specified as "P2X2o". This entry should match the corresponding sequence identifier in the alignment file, and will also be used in the naming of the output files. The two entries a. starting_model and a.ending_model define the index of the first and last model generated. This setting can be used to define the number of models generated. For instance, to generate 100 instead of the five models in the example file one would set the value of a. ending_model to 100.

3.5 Running MODELLER

The default MODELLER installation does not have a GUI, and is run from a terminal window. We recommend starting with a new project directory for each round of modeling as MODELLER will generate many output files. To run the example files and generate five models of hP2X2R in the open state, the project directory should only contain the alignment file (P2X2_P2X3o.pir), the instruction file (P2X2o.py) and the template structure file (5svk_t.pdb). Once these three files are present (and MODELLER is installed) the modeling process can be launched by typing the command below in a terminal window (*see* **Note 6**).

```
mod9v20 P2X2o.py
```

The precise name of the MODELLER executable may differ as depending on the version that has been installed. The command above is valid for MODELLER version 9.20.

3.6 The MODELLER Output

The sheer number of output files generated by MODELLER can be confusing. To make life easier for the first-time user we focus on the most important files in this section. Assuming the modeling process has been successful, the files ending with .pdb are the most interesting as they contain the coordinates of the generated models. The number of models generated depends on the settings defined in P2X2o.py. The "ranked_models.txt" file should be consulted to find the "best" model. In case things go wrong (typically this means MODELLER exits directly after being started and there are not any models generated) the logfile (in this example P2X2o.

`log`) is the user's best friend. There is no need to read the full logfile, typically the last ~20 lines give an indication of what went wrong and how to fix it (*see* **Note 7**). In our discussion of MODELLER output we have ignored the `P2X2o.ini`, `P2X2o.rsr`, `P2X2o.V99990001`, and `P2X2o.D00000001` etc. output files and refer the interested user to the MODELLER manual (https://salilab.org/modeller/manual/).

3.7 Ranking Structures and Interpretation

There are two measures that MODELLER mainly uses to rank its output homology models, the DOPE score and the modpdf score. The DOPE, or Discrete Optimized Protein Energy, score, is an atomic distance-dependent statistical potential designed to distinguish native states, or native-like states in case of homology modeling, from nonnative states [20]. DOPE scores are specific to a target sequence. This means that scores for models of different proteins (e.g., P2X2R and P2X1R) are not directly comparable. An alternative measure is the molecular pdf, or molpdf score. This score is based on the sum of all spatial constraints defined in the model, so it again cannot be used to compare different proteins. Like for the DOPE score, the lowest score defines the "best" model. The DOPE score is considered the better measure of the models' rank, but in practice, one may consider both scores. Critical thinking and visual inspection of models is advisable when making use of such models.

4 Notes

1. Regions of a target protein for which no template is available can be modeled by other means such as loop modeling for shorter regions, or ab initio modeling techniques. This is, however, beyond the scope of this chapter.
2. MODELLER also allows the use of multiple templates. This goes beyond the scope of this chapter, for details *see* the MODELLER manual (https://salilab.org/modeller/manual/node21.html). An example where multiple templates could be employed in a meaningful way is the modeling of the hP2X4R in the open form based on the open state structures of zfP2X4R (4DW1) and hP2X3R (5SVK) as templates. Here the hP2X4R is orthologous to zfP2X4R, hence we would expect higher structure similarity compared to hP2X4R/hP2X3R pair. However, the zfP2X4R (4DW1) template is lacking structural information for the intracellular cap. Adding hP2X3R (5SVK) as second template would allow to benefit from both, the presumed higher structure similarity between the two P2X4R orthologs, and the structural information solely available from hP2X3R (5SVK).

3. The term "PDB" normally refers to the Protein Data Bank (https://www.rcsb.org/), a database where all publicly available *protein structures* are stored, and from where we typically retrieve our template structures for homology modeling. In the context of a BLAST search the term "pdb" refers to a BLAST-formatted *protein sequence* database derived from the Protein Data Bank, which contains all sequences of the structures that have been deposited in the Protein Data Bank.
4. For Table 1 we chose the rat P2X5R sequence for comparison. The canonical human P2X5R sequence in SwissProt (sp|Q93086|P2RX5_HUMAN) is a nonfunctional splice variant and would not allow for meaningful comparisons of sequence similarities. It would make very little sense to generate a homology model for sp|Q93086|P2RX5_HUMAN, unless perhaps to demonstrate that for protein structural reasons this splice variant is extremely unlikely to form a functional P2XR.
5. A few tips on how to read Python code:
   (a) Python code is structured in lines and blocks. Blocks are multiple lines of code that are interpreted as a single unit. Blocks of code starts with a line ending with " : " followed by one or more lines of indented code.
   (b) "#" indicates a comment, any text after a "#" character is not read as code.
   (c) Text or string arguments, for example file names are in single or double quotes.
6. When launching MODELLER users may encounter the somewhat confusing warning regarding platform dependent and platform independent libraries. In brief, it is safe to ignore it (*see* release notes https://salilab.org/modeller/release.html#issues).
7. In a Linux or Mac OSX terminal one can use the `tail` command to selectively look at the end of a file (for example: `tail -20 P2X2o.log`). Typical errors one might encounter are files not being not found (look out for typos in file names and make sure you launch MODELLER from the directory where your input files are present), or alignment errors. The latter occur when the template sequence in the alignment file does not fully correspond to the residues present in the template structure file.

References

1. Biasini M, Bienert S, Waterhouse A, Arnold K, Studer G, Schmidt T, Kiefer F, Gallo Cassarino T, Bertoni M, Bordoli L, Schwede T (2014) SWISS-MODEL: modelling protein tertiary and quaternary structure using evolutionary information. Nucleic Acids Res 42: W252–W258
2. Bordoli L, Schwede T (2012) Automated protein structure modeling with SWISS-MODEL workspace and the protein model portal. Methods Mol Biol 857:107–136
3. Webb B, Sali A (2014) Comparative protein structure modeling using MODELLER. Curr Protoc Bioinformatics 47:5.6.1–5.6.32

4. Schmidt T, Bergner A, Schwede T (2014) Modelling three-dimensional protein structures for applications in drug design. Drug Discov Today 19:890–897
5. Burnstock G, Kennedy C (2011) P2X receptors in health and disease. Adv Pharmacol 61:333–372
6. Kawate T, Michel JC, Birdsong WT, Gouaux E (2009) Crystal structure of the ATP-gated P2X (4) ion channel in the closed state. Nature 460:592–598
7. Hattori M, Gouaux E (2012) Molecular mechanism of ATP binding and ion channel activation in P2X receptors. Nature 485:207–212
8. Mansoor SE, Lu W, Oosterheert W, Shekhar M, Tajkhorshid E, Gouaux E (2016) X-ray structures define human P2X(3) receptor gating cycle and antagonist action. Nature 538:66–71
9. Grimes L, Young MT (2015) Purinergic P2X receptors: structural and functional features depicted by X-ray and molecular modelling studies. Curr Med Chem 22:783–798
10. Pasqualetto G, Brancale A, Young MT (2018) The molecular determinants of small-molecule ligand binding at P2X receptors. Front Pharmacol 9:58
11. Fiser A, Do RK, Sali A (2000) Modeling of loops in protein structures. Protein Sci 9:1753–1773
12. Allsopp RC, El Ajouz S, Schmid R, Evans RJ (2011) Cysteine scanning mutagenesis (residues Glu52-Gly96) of the human P2X1 receptor for ATP: mapping agonist binding and channel gating. J Biol Chem 286:29207–29217
13. Roberts JA, Allsopp RC, El Ajouz S, Vial C, Schmid R, Young MT, Evans RJ (2012) Agonist binding evokes extensive conformational changes in the extracellular domain of the ATP-gated human P2X1 receptor ion channel. Proc Natl Acad Sci U S A 109:4663–4667
14. Allsopp RC, Dayl S, Bin Dayel A, Schmid R, Evans RJ (2018) Mapping the allosteric action of antagonists A740003 and A438079 reveals a role for the left flipper in ligand sensitivity at P2X7 receptors. Mol Pharmacol 93:553–562
15. Allsopp RC, Dayl S, Schmid R, Evans RJ (2017) Unique residues in the ATP gated human P2X7 receptor define a novel allosteric binding pocket for the selective antagonist AZ10606120. Sci Rep 7:725
16. Huo H, Fryatt AG, Farmer LK, Schmid R, Evans RJ (2018) Mapping the binding site of the P2X receptor antagonist PPADS reveals the importance of orthosteric site charge and the cysteine-rich head region. J Biol Chem 293:12820–12831
17. Fryatt AG, Dayl S, Cullis PM, Schmid R, Evans RJ (2016) Mechanistic insights from resolving ligand-dependent kinetics of conformational changes at ATP-gated P2X1R ion channels. Sci Rep 6:32918
18. Johnson M, Zaretskaya I, Raytselis Y, Merezhuk Y, McGinnis S, Madden TL (2008) NCBI BLAST: a better web interface. Nucleic Acids Res 36:W5–W9
19. Roberts JA, Digby HR, Kara M, Ajouz SE, Sutcliffe MJ, Evans RJ (2008) Cysteine substitution mutagenesis and the effects of methanethiosulfonate reagents at P2X(2) and P2X (4) receptors support a core common mode of ATP action at P2X receptors. J Biol Chem 283:20126–20136
20. Shen M, Sali A (2006) Statistical potential for assessment and prediction of protein structures. Protein Sci 15:2507–2524

Chapter 5

Using RNA Interference for Purinoceptor Knockdown In Vivo

Rebeca Padrão Amorim, Iscia Teresinha Lopes Cendes, and Maria Jose da Silva Fernandes

Abstract

RNA interference (RNAi) is a powerful post-transcriptional gene silencing (PTGS) induced by small double-stranded RNA (dsRNA). The method allows silencing of genes of interest by translation inhibition or by mRNA degradation. In this chapter, we provide a brief overview of the mechanisms involved in each step of gene silencing. A nonviral infusion of short siRNA into ventricular system of rats was used to study purinoceptor in the rat brain.

Key words RNAi, Hippocampus, Purinergic receptor, Brain, Rat

1 Introduction

1.1 The RNA Interference (RNAi)

The RNAi is a post-transcriptional gene silencing (PTGS) cascade described in the 1990s and that can be used as a method to modulate, usually decreasing, gene expression. It can be induced by the use of small (21–23 bases long) double-stranded RNA molecule, small-interfering RNA (siRNA), that specifically recognizes a target mRNA sequence, triggering its silencing, either by inhibiting the translation of the gene, or by mRNA cleavage by the action of an RNAse [1–5].

In 2006, Andrew Z. Fire and Craig C. Mello received the Nobel Prize in Medicine and Physiology for their participation in the elucidation of the mechanisms involved in gene silencing by RNAi. The authors observed that double-stranded RNA (dsRNA) was able to silence the gene expression of target molecules in the nematode *Caenorhabditis elegans* [3].

In eukaryote cells, small RNAs (~19–25 nucleotides, nt) or micro-RNAs (miRNA) are generated during endogenous gene-silencing. In this pathway, miRNA is processed to generate the primary miRNA (pri-miRNA) which contains a hairpin, stem-loop structures that are cleaved by a nuclear microprocessor complex Drosha-DGCR8 to form an intermediate hairpin RNA

Pablo Pelegrín (ed.), *Purinergic Signaling: Methods and Protocols*, Methods in Molecular Biology, vol. 2041, https://doi.org/10.1007/978-1-4939-9717-6_5, © Springer Science+Business Media, LLC, part of Springer Nature 2020

(~60–70 nt), the precursors of miRNA (pre-miRNA). The pre-miRNAs are exported to the cytoplasm by the enzyme exportin 5, and they are processed by an RNase III Dicer to form a small miRNA duplex. After this, a single strand is incorporated into the RNA-induced silencing complex (RISC) allowing its binding and silencing of target transcripts [6].

The miRNAs are small endogenous dsRNAs (approximately 22 nucleotides), while the siRNAs are synthetic dsRNAs (19–30 nucleotides) used to silence target genes. The shRNAs are synthetic dsRNAs that generate siRNAs and have a similar structure to miRNAs [4, 7].

Dicer is an essential enzyme for both endogenous and exogenous RNAi [4]. Into the cytoplasm of the cell, the dsRNA molecules are recognized and cleaved by Dicer to generate dsRNA with 21–23 nucleotides, with two unpaired nucleotides at the 3′ ends [8]. A Dicer RNA binding domain PAZ is responsible for the size of the generated dsRNA [4]. The miRNA or siRNA is loaded into RISC and cleaved by an endonuclease Argonaute 2 (AGO2) which releases one of the strands, leading the stranded-guide RNA molecule in activated RISC, that will direct the target sequence recognition for pairing and silencing [9–14]. The thermodynamic stability of the siRNA terminus governs the selectivity of which strand will be loaded into RISC [15, 16]. Strands with less thermodynamic stability bind more easily to AGO2 to be cleaved.

Usually, when the complementarity is perfect, the silencing occurs by mRNA degradation, but when it is partial silencing occurs by translation inhibition followed by mRNA degradation [4]. After silencing the target mRNA, is released and the RISC is recycled to another silencing cycle [17]. Figure 1 shows a diagram of posttranscriptional gene silencing by siRNA and miRNA.

Currently, RNAi is a widely used technique to study specific gene functions in different species [2, 18–20]. RNAi has also been used in pathogen control, in plant and animal generation, and in genetic disease and tumor control [21]. The RNAi method is the most appropriate strategy to specifically silencing gene expression in mammalian cells due to its efficiency and specificity [22]. Silencing of the mRNA by RNAi results in lower expression of the target protein, allowing studies to be developed with greater practicality, low cost, and efficiency [23].

When designing and selecting a siRNA, one should consider the potency, specificity and stability of the nucleases, since the siRNAs are designed to be perfectly complementary to the target mRNA sequences [4, 24, 25]. The siRNA has a small size (21–23 nucleotides) and does not cause immunostimulation. However, the use of slightly longer siRNAs results in greater efficiency, since they need to be processed by Dicer before being incorporated into the RISC [22].

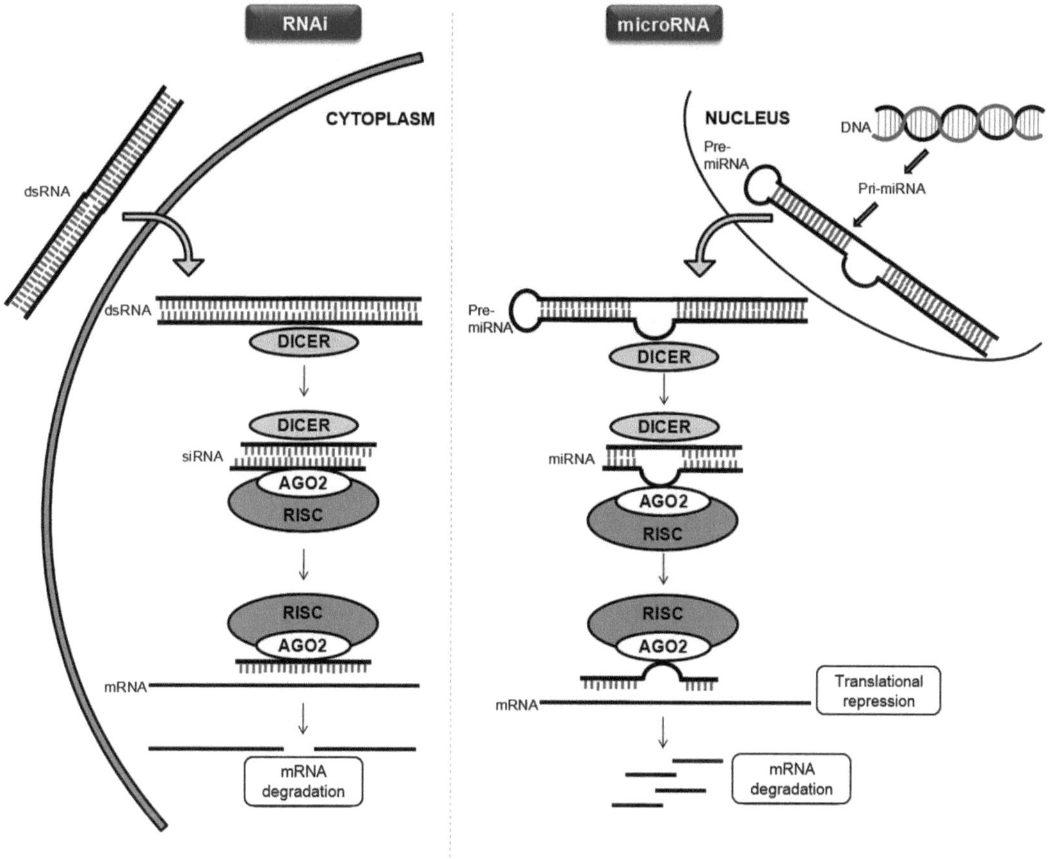

Fig. 1 Diagram of posttranscriptional gene silencing by siRNA and miRNA

The delivery of siRNA in most cells is only possible by transfection agents or high-pressure [26, 27]. Insert the siRNA into the cell consists in the main limitation of the method mainly in the central nervous system (CNS) a privileged organ protected by a blood-brain barrier [4, 6, 28, 29]. Focal delivery into the parenchyma or intracerebroventricular routes can be used to study genes located in the CNS. However, this insertion must be done preserving siRNA complexes stability [6]. Vectors with low immunostimulation capacity are the most appropriate for the use in RNAi research or therapy. Liposomes, nanoparticles, exosomes, and peptides are widely used [6, 30, 31].

Kumar et al. proposed a method siRNA transfection into mice CNS, which consists of conjugating the siRNA with a peptide sequence derived from the rabies virus glycoprotein plus nine carboxy-terminal arginine residues (RVG-9dR, YTIWMPENPRP GTPCDIFTNSRGKRASNGGGGRRRRRRRRR) [32].

Binding of siRNA to RVG is possible through chimera formation obtained by the nona-D-arginine residues addition at the carboxy terminus of RVG. This chimera facilitates binding with siRNA

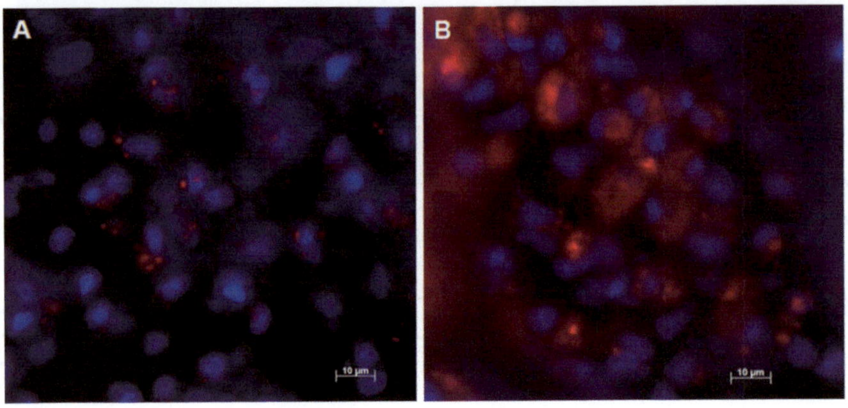

Fig. 2 Red fluorescence in (**a**), indicates the presence of unconjugated BLOCK-IT™ and in (**b**), a conjugated form of BLOCK-IT™ with RVG-9dR is shown into the cell. The data show that RVG-9dR is a very efficient vehicle to deliver siRNA into the cell in the brain. Calibration bar 10 μm

due to the difference of charges between the molecules (arginine residues have positive charge and siRNA negative charge) and allows the siRNA to be inserted into the cells via specific binding to acetylcholine receptors [32].

SiRNA:RVG-9dR complex when administered intravenously crosses the blood–brain barrier, where siRNA can be recognized and degraded by effectors of genetic silencing [32]. Now, it is known that this transfection method is not selective for neuronal cells and is also capable of transfecting macrophages and microglia [33].

In a previous study by Amorim et al., RVG-9dR-mediated transfection was used to study purinergic P2X7 receptors in rat brain [34]. As these receptors are expressed peripherally [35], the route of administration of the conjugated RVG:siRNA to silencing the P2X7 receptor in the brain was intracerebroventricular (icv).

To show the siRNA transfection, a fluorescently labeled RNA oligonucleotide, BLOCK-IT™, conjugated with RVG-9dR was performed (Fig. 2). Unconjugated BLOCK-IT™ was used as a control of the reaction. Using this methodology, we were able to show the presence of BLOCK-IT™ in the cell showing the important role of the RVG in the insertion of the siRNA (Fig. 2b). The low fluorescence of the unconjugated form of BLOCK-IT™ is shown in Fig. 2a. These results show the efficiency of the peptide RVG-9dR to deliver siRNA into the hippocampal cells using the icv. route of delivery.

1.2 Purinoceptor as a Target for RNAi

Purinergic P2 receptors belong to a family of receptors activated by nucleotides. ATP has been recognized as an important neurotransmitter in the peripheral and CNS and is involved in a wide range of physiologic functions in higher mammals [36]. When released in the extracellular medium, ATP activates P2X receptors and is rapidly hydrolyzed by ectonucleotidases and its products activate other signaling pathways [37]. To date, there are seven cloned and characterized subtypes of P2X receptors (P2X1–P2X7). They are homotrimeric or heterotrimeric ionotropic receptors, with a fast-synaptic response when activated by ATP. The P2Y receptor subfamily are G-protein-coupled receptors, with eight subtypes described until now, and present slow synaptic transmission when activated by ADP, UTP, or UDP [37].

Distinct from the other P2X receptors, the P2X7 is activated under a high level of ATP (mM), since it has a lower affinity for ATP. In this condition, P2X7 can form a pore to permeabilize the cell membrane to molecules up to 900 Da in size [38]. This mechanism can involve the participation of other channels, such as pannexin-1 [39].

In the CNS, the P2X7 receptor is expressed in glial cells and neurons [40, 41], although controversial data are reported [42]. In neurons, P2X7 receptors are located mainly in presynaptic terminals and modulate the neurotransmitter release, including GABA and glutamate [43, 44]. Since P2X7 has been described in 1996, there are a growing number of studies aimed at elucidating the role of this receptor in neurodegenerative diseases in which inflammation plays a prominent role [45]. This chapter presents a methodological design employing siRNA used for study the role of the purinoceptor P2X7 in the rat brain, during the epileptogenesis. A nonviral infusion of siRNA into the ventricular system was performed [34]. In addition to being a method used in preclinical investigation protocols, RNAi is a therapeutic modality, since it allows delivery to block specific targets in diseased tissues.

2 Materials

1. Diethylpyrocarbonate water: 0.1% diethylpyrocarbonate (DEPC) distilled water (*see* **Note 1**).
2. siRNA probe.
3. RVG-9dR: peptide sequence YTIWMPENPRPGTPCDIFTN SRGKRASNGGGGRRRRRRRRR.
4. Vehicle solution: 5% glucose, 0.9% sodium chloride (*see* **Note 2**).
5. RNase-free microtubes.

3 Methods

1. Search the target genomic sequence (e.g., siRNA designed to silencing P2X7 receptors in the hippocampus) in the target organism (e.g., *Rattus norvegicus*) (*see* **Note 3**).

3.1 Design, Synthesis, and Preparation of siRNA

2. Design 21 nucleotides sequence in the target mRNA. Look for the identity of the target oligo by the Basic Local Alignment Search Tool (BLAST) program (*see* **Note 4**), from the antisense oligo of siRNA (which is the functional molecule) against EST bank of the target organism.
3. Selection of the siRNA probe (*see* **Note 5**) with the perfect match of the antisense target oligo (100% complementarity) (*see* **Note 6**).
4. Synthesize the sequences in a specialized company (*see* **Note 7**).
5. Suspend the lyophilized sequences in diethylpyrocarbonate water.
6. In a new microtube, make the equimolar combination of the complementary RNAs, to obtain the double-stranded siRNA. Heat at 95 °C for 5 min for denaturation and alignment of the molecules. Keep the tube closed in vacuo, for slow temperature drop and annealing between the RNA strands, obtaining the siRNA, at a 50 μM concentration.

3.2 siRNA:RVG-9dR Complex Preparation for In Vivo Application (See Note 8)

1. Start the preparation of the siRNA: RVG-9dR complex, 20 min prior to animal application.
2. SiRNA should be previously conjugated with RVG-9dR at 1:10 molar ratio, in RNAse free microtubes, protected from light and kept at room temperature, for 15 min. After this, dilute the complex in vehicle solution (*see* **Note 9**).
3. Administer 1 μg of siRNA per animal (intracerebroventricular, icv) (*see* **Notes 9** and **10**).

4 Notes

1. Add 1 ml of 0.1% diethylpyrocarbonate (DEPC) to 1000 ml distilled water. Mix well and let set at room temperature for 1 h. Autoclave it. Let cool to room temperature prior to use.
2. Add 5% glucose in 0.9% normal saline.
3. Genomic sequence: we suggest using a nucleotide database, which can be found on the NCBI server at: https://www.ncbi.nlm.nih.gov/nuccore/ [46].
4. Basic Local Alignment Search Tool: we suggest using BLAST, which can be found on the NCBI server at: www.ncbi.nlm.nih.gov/BLAST [47].
5. The siRNAs design is not limited to choosing a 21 nucleotides sequence based on the target mRNA. siRNA targets identification is based on the protocol proposed by Tuschl [2] with some modifications. Some aspects should be considered when choosing siRNA:

- The target must be located in the coding region, excluding the first 100 nucleotides after the start codon, and the last 100 nucleotides before the stop codon, since this region may be complexed to a regulatory translation protein making access difficult for the siRNA. Considering this, 5′ and 3′ untranslated regions (UTR) should not be considered.
- The GC content should be between 30% and 55%.
- The siRNA must present TT at the 3′ ends since they are more stable.
- The free energy of the double-strands of the siRNA, sense and antisense strand, must be analyzed. This is one more parameter to ensure the efficiency of the target's silencing. The analysis can be done by a free online software tool Strand Analysis (SA) program that uses thermodynamic features of the siRNA termini, defined as Gibbs free energy, to choose the guide strand [48].
- Must not have secondary structures (hairpins, duplexes or loops). The analysis can be done using the Gene Runner program, oligo subprogram, working with the RNA sequence format;
- siRNA should have no identity with any other sequence in the genome, even if they are as low as 16/21 nt.

6. Cleavage of target RNA by slicer occurs when there is a perfect match of the antisense oligo with the target (100% complementarity). Therefore, the existence of only one imbalance drastically reduces the slicer activity (less than 10%), impairing gene silencing.
7. Specialized companies have siRNAs previously designed and/or design siRNAs of interest.
8. The siRNA: RVG-9dR complexation protocol was established based on prior protocols developed by Kumar et al. and Pascoal et al. [32, 49].
9. The vehicle amount (5% glucose) should be calculated according to the amount of prepared siRNA:RVG-9dR complex, considering that 1 μg of siRNA will be delivered into the brain using a volume not exceeding tissue capacity.
10. The half-life of siRNA is determined by means of a time curve after siRNA administration, with successive measurement of the presence of the gene target (PCR) or protein target (Western blot) on tissue samples of interest. If necessary, the siRNA can be applied repeatedly.

Acknowledgments

The authors thank the Brazilian agencies Fundação de Amparo à Pesquisa do Estado de São Paulo (FAPESP), Conselho Nacional de Desenvolvimento Científico e Tecnológico (CNPq), and Coordenação de Aperfeiçoamento de Pessoal de Nível Superior (CAPES) for financial support.

The authors declare no competing interests.

References

1. Corey DR (2007) RNA learns from antisense. Nat Chem Biol 3:8–11. https://doi.org/10.1038/nchembio0107-8
2. Elbashir SM, Harborth J, Lendeckel W et al (2001) Duplexes of 21-nucleotide RNAs mediate RNA interference in cultured mammalian cells. Nature 411:494–498. https://doi.org/10.1038/35078107
3. Fire A, Xu S, Montgomery MK et al (1998) Potent and specific genetic interference by double-stranded RNA in Caenorhabditis elegans. Nature 391:806–811. https://doi.org/10.1038/35888
4. França NR, Mesquita D Jr, Lima AB et al (2010) Interferência por RNA: uma nova alternativa para terapia nas doenças reumáticas. Rev Bras Reumatol 50:695–702. https://doi.org/10.1590/S0482-50042010000600008
5. Zamecnik PC, Stephenson ML (1978) Inhibition of Rous sarcoma virus replication and cell transformation by a specific oligodeoxynucleotide. Proc Natl Acad Sci U S A 75:280–284
6. Boudreau RL, Rodríguez-Lebrón E, Davidson BL (2011) RNAi medicine for the brain: progresses and challenges. Hum Mol Genet 20:21–27. https://doi.org/10.1093/hmg/ddr137
7. Bagga S, Bracht J, Hunter S et al (2005) Regulation by let-7 and lin-4 miRNAs results in target mRNA degradation. Cell 122:553–563. https://doi.org/10.1016/j.cell.2005.07.031
8. Bernstein E, Caudy AA, Hammond SM et al (2001) Role for a bidentate ribonuclease in the initiation step of RNA interference. Nature 409:363–366. https://doi.org/10.1038/35053110
9. Ameres SL, Martinez J, Schroeder R (2007) Molecular basis for target RNA recognition and cleavage by human RISC. Cell 130:101–112. https://doi.org/10.1016/j.cell.2007.04.037
10. Kim DH, Rossi JJ (2008) RNAi mechanisms and applications. Biotechniques 44:613–616. https://doi.org/10.2144/000112792
11. Meister GM, Landthaler A, Patkaniowska Y et al (2004) Human Argonaute2 mediates RNA cleavage targeted by miRNAs and siRNAs. Mol Cell 15:185–197. https://doi.org/10.1016/j.molcel.2004.07.007
12. Rand TA, Ginalski K, Gridhin NV et al (2004) Biochemical identification of argonaute 2 as the sole protein required for RNA-induced silencing complex activity. Proc Natl Acad Sci U S A 101:14385–14389. https://doi.org/10.1073/pnas.0405913101
13. Rand TA, Petersen S, Du F et al (2005) Argonaute2 cleaves the anti-guide strand of siRNA during RISC activation. Cell 123:621–629. https://doi.org/10.1016/j.cell.2005.10.020
14. Tang G (2005) siRNA and miRNA: an insight into RISCs. Trends Biochem Sci 30:106–114. https://doi.org/10.1016/j.tibs.2004.12.007
15. Khvorova A, Reynolds A, Jayasena SD (2003) Functional siRNAs and miRNAs exhibit strand bias. Cell 115:209–216
16. Schwarz DS, Hutvagner G, Du T et al (2003) Asymmetry in the assembly of the RNAi enzyme complex. Cell 115:199–208
17. Jinek M, Doudna JA (2009) A three-dimensional view of the molecular machinery of RNA interference. Nature 457:405–412. https://doi.org/10.1038/nature07755
18. Fraser AG, Kamath RS, Zipperlen P et al (2000) Functional genomic analysis of C. elegans chromosome I by systematic RNA interference. Nature 408:325–330. https://doi.org/10.1038/35042517
19. Gönczy P, Echeverri C, Oegema K et al (2000) Functional genomic analysis of cell division in C. elegans using RNAi of genes on chromosome III. Nature 408:331–336. https://doi.org/10.1038/35042526
20. Mccaffrey AP, Kay MA (2002) A story of mice and men. Gene Ther 9:1563. https://doi.org/10.1038/sj.gt.3301890

21. Dogini DB, Pascoal VD, Avansini SH et al (2014) The new world of RNAs. Genet Mol Biol 37:285–293
22. Grimm D (2009) Small silencing RNAs: state-of-the-art. Adv Drug Deliv Rev 61:672–703. https://doi.org/10.1016/j.addr.2009.05.002
23. Pascoal VDB (2010) O papel da interleucina-1 beta na fase aguda do modelo de epilepsia do lobo temporal induzido pela pilocarpina. Dissertação. Universidade Estadual de Campinas, São Paulo
24. Hartmann R, Justesen J, Sarkar SN et al (2003) Crystal structure of the 2′-specific and double-stranded RNA-activated interferon-induced antiviral protein 2′-5′-oligoadenylate synthetase. Mol Cell 12:1173–1185
25. Pei Y, Tuschl T (2006) On the art of identifying effective and specific siRNAs. Nat Methods 3(9):670–676. https://doi.org/10.1038/nmeth911
26. Fedorov Y, Anderson EM, Birmingham A et al (2006) Off-target effects by siRNA can induce toxic phenotype. RNA 12:1188–1196. https://doi.org/10.1261/rna.28106
27. Song E, Zhu P, Lee SK et al (2005) Antibody mediated in vivo delivery of small interfering RNAs via cell-surface receptors. Nat Biotechnol 23:709–717. https://doi.org/10.1038/nbt1101
28. Hassani Z, Lemkine GF, Erbacher P et al (2005) Lipid-mediated siRNA delivery down-regulates exogenous gene expression in the mouse brain at picomolar levels. J Gene Med 7:198–207. https://doi.org/10.1002/jgm.659
29. Pardridge WM (2007) Blood–brain barrier delivery. Drug Discov Today 12:54–61. https://doi.org/10.1016/j.drudis.2006.10.013
30. Boorn JGVD, Schlee M, Coch C et al (2011) SiRNA delivery with exosome nanoparticles. Nat Biotechnol 29:325–326. https://doi.org/10.1038/nbt.1830
31. Lu M, Xing H, Xun Z et al (2018) Exosome-based small RNA delivery: progress and prospects. Asian J Pharm Sci 13:1–11. https://doi.org/10.1016/j.ajps.2017.07.008
32. Kumar P, Wu H, McBride JL et al (2007) Transvascular delivery of small interfering RNA to the central nervous system. Nature 448:39–43. https://doi.org/10.1038/nature05901
33. Kim SS, Ye C, Kumar P et al (2010) Targeted delivery of siRNA to macrophages for anti-inflammatory treatment. Mol Ther 18:993–1001. https://doi.org/10.1038/mt.2010.27
34. Amorim RP, Araújo MGL, Valero J et al (2017) Silencing of P2X7R by RNA interference in the hippocampus can attenuate morphological and behavioral impact of pilocarpine-induced epilepsy. Purinergic Signal 13:467–478. https://doi.org/10.1007/s11302-017-9573-4
35. Burnstock G (2013) Purinergic mechanisms and pain—an update. Eur J Pharmacol 716:24–40. https://doi.org/10.1016/j.ejphar.2013.01.078
36. Burnstock G (2008) Purinergic signalling and disorders of the central nervous system. Nat Rev Drug Discov 7:575–590. https://doi.org/10.1038/nrd2605
37. Abbracchio MP, Burnstock G, Verkhratsky A et al (2009) Purinergic signalling in the nervous system: an overview. Trends Neurosci 32:19–29. https://doi.org/10.1016/j.tins.2008.10.001
38. Sperlágh B, Illes P (2014) P2X7 receptor: an emerging target in central nervous system diseases. Trends Pharmacol Sci 35:537–547. https://doi.org/10.1016/j.tips.2014.08.002
39. Pelegrin P, Surprenant A (2006) Pannexin-1 mediates large pore formation and interleukin-1β release by the ATP-gated P2X7 receptor. EMBO J 25:5071–5082. https://doi.org/10.1038/sj.emboj.7601378
40. Engel T, Gomez-Villafuertes R, Tanaka K et al (2012) Seizure suppression and neuroprotection by targeting the purinergic P2X7 receptor during status epilepticus in mice. FASEB J 26:1616–1628. https://doi.org/10.1096/fj.11-196089
41. Xiang Z, Burnstock G (2005) Expression of P2X receptors on rat microglial cells during early development. Glia 52:119–126. https://doi.org/10.1002/glia.20227
42. Sim JA, Young MT, Sung HY et al (2004) Reanalysis of P2X7 receptor expression in rodent brain. J Neurosci 24:6307–6314. https://doi.org/10.1523/JNEUROSCI.1469-04.2004
43. Armstrong JN, Brust TB, Lewis RG et al (2002) Activation of presynaptic P2X7-like receptors depresses mossy fiber-CA3 synaptic transmission through p38 mitogen-activated protein kinase. J Neurosci 22:5938–5945. doi:20026618
44. Sperlágh B, Köfalvi A, Deuchars J et al (2002) Involvement of P2X7 receptors in the regulation of neurotransmitter release in the rat hippocampus. J Neurochem 81(6):1196–1211
45. Di Virgilio F, Ben DD, Sarti AC et al (2017) The P2X7 receptor in infection and inflammation. Immunity 47:15–31. https://doi.org/10.1016/j.immuni.2017.06.020

46. National Center for Biotechnology Information, US National Library of Medicine, Rockville Pike. https://www.ncbi.nlm.nih.gov/nuccore/. Accessed Sept 2018
47. National Center for Biotechnology Information, US National Library of Medicine, Rockville Pike. www.ncbi.nlm.nih.gov/BLAST. Accessed Sept 2018
48. Pereira TC, Pascoal VDB, Secolin R et al (2007) Strand analysis, a free online program for the computational identification of the best RNA interference (RNAi) targets based on Gibbs free energy. Genet Mol Biol 30 (4):1206–1208. https://doi.org/10.1590/S1415-47572007000600030
49. Pascoal VDB, Marchesini RB, Matos AHB et al (2010) The il1β have a protective action in the acute phase of the pilocarpine-induced epilepsy model. J Epilepsy Clin Neurophysiol 16 (3):97–99. https://doi.org/10.1590/S1676-26492010000300003

Chapter 6

Developmental Expression of Ectonucleotidase and Purinergic Receptors Detection by Whole-Mount In Situ Hybridization in *Xenopus* Embryos

Camille Blanchard and Karine Massé

Abstract

Xenopus embryos are one of the most used animal models in developmental biology and are well suited for apprehending functions of signaling pathways during embryogenesis. To do so, it is necessary to be able to detect expression pattern of the key genes of these signaling pathways. Here we describe the whole-mount in situ hybridization technique to investigate the expression pattern of ectonucleotidases and purinergic receptors during embryonic development.

Key words P2X receptors, Ectonucleotidases, Purinergic signaling pathway, Whole-mount in situ hybridization, Expression pattern, *Xenopus* embryo

1 Introduction

The whole-mount in situ hybridization technique allows the detection of specific mRNA sequences in preserved whole embryo. Its principle relies on the hybridization of a labeled complementary probe (antisense strand) to the sequence of mRNA of interest (sense strand), which is performed within the cell. This technique is therefore used to establish the spatiotemporal expression profile of newly identified genes in wild-type embryos but can also be used to analyse any potential effects on gene expression (downregulation or upregulation) in manipulated embryos.

The in situ hybridization was first described in 1969 in frog oocytes [1]. At that time, radioactive isotopes were used to label the probe, and the first in situ hybridization using nonradioactive probes was published 20 years later in Drosophila model [2]. Non-radioactive probes are usually labeled with the digoxigenin (DIG) hapten, a steroid derived from the plant *Digitalis purpurea*, which is not found in animal tissues. DIG-labeling occurs during in vitro transcription reaction: the DIG-11-UTP replaces UTP in that

Pablo Pelegrín (ed.), *Purinergic Signaling: Methods and Protocols*, Methods in Molecular Biology, vol. 2041, https://doi.org/10.1007/978-1-4939-9717-6_6, © Springer Science+Business Media, LLC, part of Springer Nature 2020

enzymatic reaction in a ratio of 35–65%, therefore the riboprobe is composed of ATP, GTP, CTP, UTP, and DIG-11-UTP. The nucleotide analogue can be then detected with an anti-DIG antibody. This technique was later adapted to vertebrates, especially to *Xenopus* embryos in 1990 [3]. Since then, several improved protocols have been published, allowing detection of low-expressed transcripts, improved staining specificity, and reduced background staining [4–7].

Xenopus laevis embryos have been one of the most productive models for cell and developmental biology. Indeed, *X. laevis* has many advantages, namely large, robust and abundant embryos, a rapid external development cycle (3 days when embryos are cultured at 23 °C), facilitating micromanipulation and misexpression studies. The other characteristic of this model is its temperature dependant development cycle, with a fourfold increase length when embryos are cultured at 12 °C. This unique feature allows to access easily to all embryonic stages and to study the early phases of development. In particular, frog neurulation phase is well distinct from the other embryonic phases, and *Xenopus* has been of great importance to decipher mechanisms regulating neurulation [8, 9]. Moreover, the recent sequencing of its allotetraploid genome has widened its use and placed *Xenopus* as a genetic animal model [10–12]. Furthermore, it is nowadays possible to design easily and generate quickly specific probes for any mRNA; even probes which are able to distinguish homeologs. More details on this animal model can be found on Xenbase site (www.xenbase.org).

We already have demonstrated that *X. laevis* embryos are particularly suited for the study of the roles of the different components of the purinergic signaling pathway during embryogenesis. Using whole mount in situ hybridization, we have established the first comparative spatiotemporal expression map of several key actors of this pathway, purinergic receptors and ectonucleotidases [13, 14, 15]. Expression profiles for these different actors have been established using this technique in other animal models (*see* [16] for review), but our studies are unique as the expression pattern was obtained for all members of the family and during early phases of development such as neurulation. Furthermore, we provided the first in vivo evidence that this pathway regulates the early steps of eye formation [17]. Indeed, using in situ hybridization we showed that extracellular ADP induces the expression of the EFTF (eye field transcription factors), such as *pax6*, at the time of eye field specification.

Whole mount in situ hybridization is a time consuming technique, with multiple and long incubation steps. Here we describe a protocol that might last longer than others, but which was worked out to be efficient with any probe, on any purinergic signaling member. An experienced person can easily process 20 tubes simultaneously corresponding to different probes or embryo stages tested. Embryos are first needed and can be obtained by in vitro fertilization. The endogenous mRNAs are then fixed with an aldehyde cross linking reagent such as MEMFA and hybridized with a labeled radioprobe, most often a DIG-labeled probe. The main advantage of using riboprobes is that RNA-RNA hybrids are very thermostable and resistant to digestion to RNases. This allows the removal of free unincorporated RNA probe during the RNase treatment and therefore reduces background nonspecific staining risks. Digoxigenin is subsequently detected by an alkaline phosphate coupled anti DIG antibody. The presence of the enzyme in a cell/tissue/organ is then detected by adding a substrate in the buffer. A precipitate will form during the chromogenic reaction and will reveal the activity of the enzyme, and therefore the presence of the transcripts. This in situ hybridization method indirectly allows the localization of the mRNA of interest. Precise expression profile can then be established by cryostat sectioning the stained embryos. Moreover, it is also possible to detect in the same embryo transcripts from two different genes. This is possible by the use of two probes labeled with different nucleotide analogues, such as DIG and Fluorescein [18], and the variety of alkaline phosphate substrates available [19].

2 Materials

All solutions for embryo culture and fixation, for RNA probe preparation and in situ hybridization experiments are prepared with ultrapure deionized water (18 MΩ-cm at 25 °C). Concentrated embryo stock solutions (10× *see* below) and in situ hybridization buffer are filtered.

2.1 Embryo Fertilization and Culture

1. 450–750 U human chorionic gonadotropin hormone (hCG).
2. NaCl/gentamicin solution: 20 mM NaCl, 5 μg/ml gentamicin.
3. 90 mm petri dish.
4. Lethal anesthetic solution: 0.35 g/l 4-aminobenzoic acid ethyl ester, 0.35 g/l Ethyl 4-aminobenzoate also called Benzocaine. This is a lethal dose for the frog.
5. 1× Egg collection media or High Salt MBS (HSB): 108 mM NaCl, 1 mM KCl, 0.7 mM $CaCl_2$, 1 mM $MgSO_4$, 2.5 mM $NaHCO_3$, 5 mM Hepes (pH 7.8). The solution is prepared from filtrated 10× HBS solution.
6. 10× MBS: 88 mM NaCl, 10 mM KCl, 10 mM $MgSO_4$, 25 mM $NaHCO_3$, 50 mM HEPES pH 7.8.

7. 10× HSB solution: for 1 l add 100 ml of filtered 10× MBS, 7 ml of 0.1 M $CaCl_2$, and 4 ml of 5 M NaCl, pH is adjusted to 7.7.
8. 1× MMR solution: 100 mM NaCl, 2 mM KCl, 2 mM $CaCl_2$, 1 mM $MgSO_4$, 5 mM Hepes (pH 7.8). The solution is prepared from filtrated 10× MMR stock solution. The pH is adjusted to 7.4.
9. Culture media: 0.1× MMR with 50 μg/ml of gentamycin.
10. Cysteine solution: 3% L-cysteine solution in water. The pH is adjusted to 7.6 with the addition of NaOH (*see* **Note 1**).
11. Fixation buffer (MEMFA): 100 mM MOPS pH 7.4, 2 mM EGTA, 1 mM MgSO4, 4% formaldehyde. MEMFA buffer needs to be freshly prepared (*see* **Note 2**).
12. Ethanol solutions: (1) 25% ethanol, 75% water; (2) 50% ethanol, 50% water; (3) 75% ethanol, 25% water; (4) 100% ethanol.
13. Stereo-microscope with LED lamp (we use a Nikon SM2645 with Schott lamp KL200).

2.2 In Situ Hybridization

1. PBw solution: 1× PBS, 0.1% Tween 20.
2. Ethanol–PBw solutions: (1) 75% ethanol, 25% PBw; (2) 50% ethanol, 50% PBw; (3) 25% ethanol, 75% PBw.
3. PBw–Proteinase K solution: 10 μg/ml proteinase K in PBw.
4. PBw-formaldehyde solution: 4% formaldehyde in PBw.
5. Formaldehyde solution: 4% formaldehyde in PBS.
6. Stop solution: 2 μl of EDTA 0.5 M, 5 μl of LiCl 8 M and 300 μl of precooled ethanol 100%.
7. Triethanolamine solution: 0.1 M Triethanolamine. This solution is prepared by diluting 1 ml of stock solution from any manufacturer (i.e., Sigma-Aldrich) into 72.6 ml of water. The pH is then adjusted to 7.5 with HCl. This solution is very corrosive for the pH electrode, which requires intense wash with water after use.
8. Acetic anhydride.
9. Hybridization buffer: 50% formamide, 5× SSC, 1 mg/ml torula RNA, 1× Denhardt, 0.1% Tween 20, 10 mM EDTA, 100 μg/ml heparin (*see* **Note 3**). The hybridization buffer is stored at −20 °C after filtration (0.45 μm). For Torula solution preparation, *see* **Note 4**.
10. MAB buffer: 0.1 M Maleic Acid (pH 7.5), 150 mM NaCl (*see* **Note 5**).
11. MAB-blocking solution: 2% Blocking reagent (we use the one from Roche), 5% goat serum in MAB solution. The Blocking Reagent is prepared as 10% solution in MAB solution and is

kept in −20 °C. Dissolution of the blocking reagent requires heating but the solution remains cloudy.

12. Alkaline phosphate staining buffer: 0.1 M Tris–HCl (pH 9.5), 0.1 M NaCl, 50 mM $MgCl_2$, 0.1% Tween 20, 5 mM Levamisole. The staining buffer is freshly made.
13. Bouin's solution: 5% acetic acid, 9. 25% formaldehyde. The Bouin's fixative solution is freshly made (*see* **Notes 2** and **5**).
14. 20× SSC stock buffer: 3 M NaCl and 300 mM trisodium citrate (adjusted to pH 7.0 with HCl).
15. 2× SSC/CHAPS buffer: 2× SSC + 0.1% CHAPS.
16. RNAse A solution: 2X SCC+0.1% CHAPS+RNAse A (10 μg/ml).
17. 0.2× SSC/CHAPS buffer: 0.2× SSC + 0.1% CHAPS.
18. Bleaching solution: 1% H_2O_2, 5% formamide, 0.5× SSC made in PBS 1× (*see* **Note 6**).
19. Ethanol–PBS solution: 50% ethanol, 50% PBS.
20. 70 ethanol–PBS solution: 70% ethanol, 30% PBS.
21. 100% methanol.
22. Glycine/Tween solution: 1% glycine, 0.1 M HCl pH 2, 1% Tween 20.
23. All molecular reagents, RNA polymerases, DIG RNA labeling kit, BM Purple, mouse anti DIG coupled to alkaline phosphatase, are purchased from Roche, but they could be also purchased from other vendors.
24. Orbital shaker and a rocker to produce a three dimensional action.
25. Stereo-microscope and LED lamp.

3 Methods

3.1 Egg Collection

Egg collection can be performed either by manual collection or by natural laying into a high-salt buffer. Manual collection is performed by holding and gently but firmly massaging the female belly with one thumb while holding the females above a 90 mm petri dish. This first method is though to mimic the actions of the male frog to encourage the female to lay eggs. The second method is though to be less stressful for the frogs than manual collection. However, the number of eggs collected can be lower and skin irritation can occur if the pH of the solution is not correct. Here, we describe this last method. Be careful to always wear gloves when manipulating the frogs.

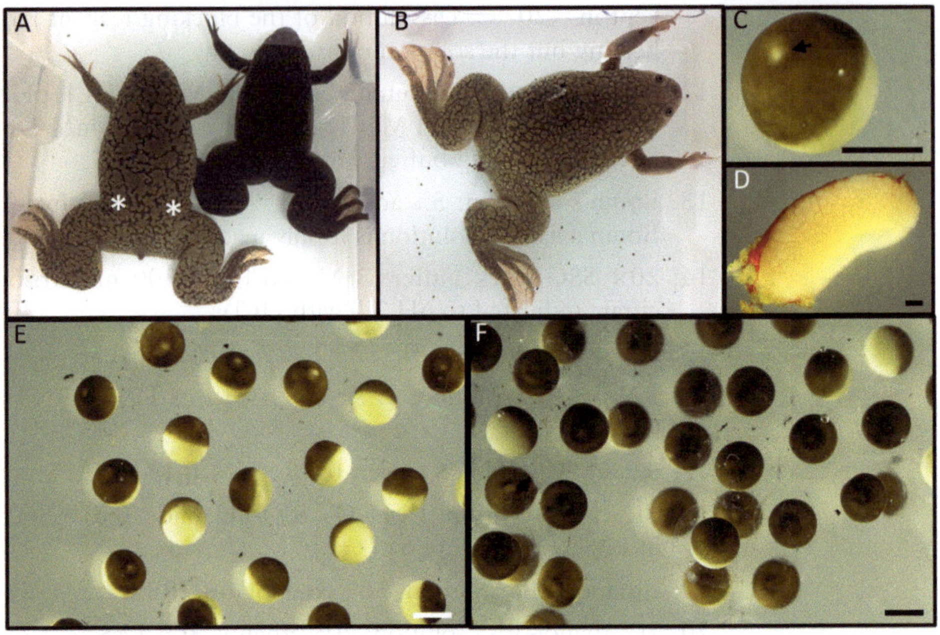

Fig. 1 (**A**) Photography of *Xenopus laevis* female and male adults. The female is on the left and the male on the right. The asterisks indicate the injection points for hormonal stimulation. (**B**) Photography of a female laying in the tank. The eggs are laid through the cloaca, located between the hind limbs. (**C**) Photography of a laid egg, a stage VI oocyte. The white spot (arrow) is the maturing spot indicating that germinal vesicle breakdown and meiosis I have occurred. Note that the jelly surrounding the egg is hardly visible before water addition. (**D**) Photography of a freshly isolated testis. Note the presence of capillaries and yellow fat tissues connected to the testis. (**E, F**) Photography of *Xenopus* eggs just after (**E**) and 30 min after fertilization (**F**). Note the difference of orientation of the eggs; the eggs are positioned animal pole up (**F**), indicating that the gravitational rotation has occurred. The scale represents 1 mm

1. *Xenopus laevis* female (Fig. 1a) ovulation is stimulated by injection of 450–750 U of hCG (depending of the female size) into the dorsal lymph sac, using a fine needle attached to a 1 ml syringe (*see* **Note 7**).
2. Keep the frog isolated in water overnight (approx. 16 h).
3. Place the egg-laying females in HSB 1× buffer and keep them isolated throughout the day (*see* **Note 8**).
4. Collect eggs every 45–60 min and place them into a petri dish with HSB 1×, keep the eggs at 18 °C until fertilization (Fig. 1c, *see* **Note 9**).
5. When egg collection is over, transfer the females into a water tank containing NaCl/gentamicin solution, and keep them isolated overnight as the frogs may keep laying eggs.
6. Before returning the females to the colony, check for any signs of illness as ovulation induction and egg collection are the most common cause of disease. Females can be stimulated two to three times a year with a minimal period of 4 months between hormone stimulation.

3.2 Testis Isolation

1. In the morning of egg collection, place a *Xenopus laevis* male (Fig. 1a) in a 2 l beaker containing 500 ml of room temperature lethal anesthetic solution during at least 30 min.
2. Before dissection, check the animal death by the absence of heart beating.
3. Place the frog on the back on a foil covered by clean paper towel. Make a small cut in the loose belly skin using scissors and then cut a large flap of skin in order to expose the lower belly. Using forceps, lift the abdominal muscles, and cut in order to expose viscera. Note that it is important to cut on either side of the midline to prevent any blood vessels damage.
4. Using forceps, pull out the abdominal yellow fat. Testes are attached to the fat body and are whitish, about 1 cm long and characterized by the presence of capillaries (Fig. 1d).
5. Using forceps and scissors, remove both testes and place them into a petri dish containing 1× MMR. Remove any fat tissues that may be attached to the testes and place them into a new 1× MMR petri dish in order to rinse them from any blood.
6. Place each testis into 1× MMR small petri dish and store at 4 °C. Testis can be kept under these conditions for 48–72 h without affecting sperm viability.
7. Wrap the frog carcass and freeze at −20 °C.
8. It is possible to check the quality of the sperm by crushing a small piece and realizing sperm observation by phase-contrast microscopy.

3.3 Fertilization

1. Transfer the eggs into a 60 mm petri dish and remove all solution.
2. Cut a small piece of testis and rub it over the eggs. Make sure that each egg is touched in order to maximize fertilization percentage.
3. Incubate for 7–10 min to let the sperm to bind to the oocytes.
4. Add water or 0.1× MMR solution into the petri dish (Fig. 1e).
5. Fertilization can be monitored by observation of the eggs using binocular. The cortical contraction (pigment movement toward the animal pole), the gravitational rotation (due to perivitelline space swelling), the appearance of a black mark (scar of the sperm penetration), and the disappearance of the maturation spot are signs of fertilization (Fig. 1f, *see* **Note 10**).
6. Fertilization efficiency should reach 80–100%. If efficiency is poor, repeat the fertilization of another batch of eggs, as egg quality can vary during the day. If it remains poor, prepare a new batch of testes, as males can be sterile.

7. Thirty minutes after fertilization, dejelly the embryos by removing solution and adding the cysteine solution.
8. Gently swirl the eggs for a few minutes until the jelly membranes are visible in the solution and fertilized eggs pack together. Dejellying usually only takes 2–5 min, but this time is frog-dependent. As prolonged exposure to cysteine will damage the embryos, it is safer to monitor dejellying process by binocular observation.
9. When the dejellying is complete, remove the cysteine solution and rinse the fertilized eggs 5–7 times with 0.1× MMR. It is essential to act quickly and remove any traces of cysteine.
10. Place the fertilized eggs into a clean dish containing culture media.
11. Remove any unfertilized eggs and dead embryos using a pipette and place the dish in the incubator at 12 or 18 °C depending of the desired embryonic stage.
12. Embryos need to be sorted and placed into a clean petri dish filled with new culture media every day of culture, until they reach the desired embryonic stage. Embryo staging is done using the Niewkoop and Faber Tables [20].

3.4 Embryo Preparation

Embryos are prepared for in situ hybridization procedure by fixation and dehydration. For the early organogenesis stages before hatching, it is best to remove the vitelline membrane, otherwise the embryos will be fixed while being curled inside this membrane. There is no need to remove this membrane before these stages, as dehydration into ethanol and the proteinase K treatment during the in situ hybridization procedure (*see* Subheading 3.6, **step 3**) will remove the vitelline membrane of early stages embryos.

Every manipulation of the embryos is critical, as they are very delicate and can break easily: do not hesitate to leave a small volume of the previous liquid with the embryos to avoid snapping them.

1. Transfer the embryos into a 2 or 5 ml tube, depending of the number of embryos to be fixed.
2. Let the embryos sank to the bottom of the tube, then remove most of the culture media.
3. Add 0.5 or 2 ml MEMFA fixation solution (depending of the tube size).
4. Remove MEMFA/culture media mixed solution.
5. Add MEMFA fixation solution to fill up the tube in order to prevent any embryo damage.
6. Fix the embryos for 1–2 h at room temperature or overnight at 4 °C, with agitation on the nutator.

7. Rinse the fixed embryos with water 4–6 times with agitation on the nutator over a period of 2 h. Typically, do a quick wash, then two times 15 min wash followed by two washes of 30 min and a final one of 1 h.
8. Gradually dehydrate the embryos into ethanol by washing 5 min with agitation on the nutator in each of the following solutions: 25% ethanol–75% water; 50%ethanol–50%water; 75%ethanol–25%water; 100% ethanol.
9. Change the ethanol solution and store the embryos in ethanol 100% at −20 °C until use. The dehydrated embryos can be stored for months under these conditions.

3.5 RNA Probe Synthesis

Commonly, digoxigenin-labeled RNAs are used as probes for in situ hybridization. These antisense probes are produced by in vitro transcription using RNA polymerase (T3, T7 or Sp6) using a *Xenopus* DNA sequence. This DNA template is most often cloned into a bacteriophage promoter carrying vector, which needs linearized and purified prior probe synthesis. It is preferable to linearize the vector just after the *Xenopus* DNA insert to avoid any bacterial sequences incorporated into the probe (*see* **Note 11**). An alternative to generate the template is to amplify by PCR the DNA fragment, with primers containing the T3, T7 or Sp6 promoter at their 5′ ends [21]. It is also necessary to prepare the sense probe, which will act as control for nonspecific hybridization and staining, for example due to probe trapping. Note that riboprobes can be very easily degraded (highly sensitive to RNases) and special caution is required when manipulating them: wearing gloves, using sterile filtered buffer and keeping the probes on ice after their purification. However, they remain the most used probe type used in laboratory. The following centrifugation steps are carried out in a tabletop centrifuge with a 24-microtubes rotor, precooled at 4 °C.

1. Prewarm the buffer tube at 37 °C.
2. Defrost the DIG RNA labeling mix and keep on ice.
3. Take the RNA Polymerase enzyme from the freezer and keep in a benchtop cooler. Add the following reagents to a 1.5 ml centrifuge tube in the order specified: (1) water (to adjust the final volume up to 20 μl); (2) DNA template (500–700 ng of linearized plasmid or 100 ng of PCR product); (3) Buffer (5× SP6/T3/T7 buffer to a final 1× concentration); (4) 10× DIG RNA labeling mix to a 1× final concentration; (5) 40 U of RNase inhibitor; (6) 40 U of RNA polymerase (T3/T7/Sp6). Mix every solution prior adding a new reagent. Because of the presence of spermidine in the reagent, it is essential to assemble this enzymatic reaction at room temperature.
4. Mix well without vortexing the tube, and spin the tube.

5. Incubate for 2–3 h at 37 °C (*see* **Note 12**).
6. Add 10 U of RNase-free DNaseI.
7. Incubate for 10 min at 37 °C.
8. Stop the enzymatic reaction and precipitate the probes by adding 105 μl of stop solution.
9. Mix well by inverting several times the tube.
10. Incubate at −20 °C for 15–45 min.
11. Centrifuge the tube 15 min at 13,400 × *g* at 4 °C.
12. Carefully remove the supernatant and add 500 μl of cooled ethanol 70%. The pellet is translucent and can be hard to see. Make sure to note the orientation of the tube when placing it in the centrifuge in order to identify the future pellet side. Pipette the supernatant on the opposite side.
13. Centrifuge the tube 10 min at 12,000 rpm at 4 °C.
14. Carefully remove the supernatant and let the pellet dry for 15–30 min at room temperature, with the tube opened.
15. Add 20–30 μl of water, depending of the size of the pellet (*see* **Note 13**).
16. Incubate the tube 1 h on ice or overnight at 4 °C for probe resuspension.
17. Homogenize the solution by gently pipetting up and down, and spin the tube. The probe can be stored at −20 °C for a few weeks and at −80 °C for a longer storage (*see* **Note 14**).

3.6 Probe Hybridization for In Situ Hybridization

The different steps are performed under agitation on the nutator at room temperature unless stated otherwise. Place the hybridization buffer at 60 °C at the start of the day. This first day involves several steps and incubations, so it can easily last for 10 h (*see* **Note 15**).

1. Gradually rehydrate the embryos into PBw solution by washing 5 min with agitation on the nutator in each of the following solutions: 75% ethanol–25% PBw; 50%ethanol–50%PBw; 25% ethanol–75%PBw; 100% PBw.
2. Transfer the embryos into glass vial tubes of 2 ml (*see* **Note 16**).
3. Remove the PBw solution.
4. Add 2 ml of PBw-Proteinase K solution and incubate for 5 min (*see* **Note 17**).
5. Wash the embryos twice for 5 min in 2 ml of triethanolamine solution (*see* **Note 18**).
6. Add 5 μl of acetic anhydride directly to each tube, and incubate for 5 min. Repeat this step once (*see* **Note 19**).
7. Wash twice the embryos 10 min with PBw.
8. Refix the embryos in PBw-formaldehyde solution for 20 min.

9. Wash the embryos five times during 5 min in PBw.
10. Remove most of the PBw solution.
11. Add 1 ml of prewarmed hybridization buffer in each tube.
12. Once the embryos have sunk to the bottom of the tube, replace this solution by 1 ml of clean prewarmed hybridization buffer.
13. Incubate for 6 h at 60 °C in a water bath. The length of this prehybridization step can be shortened to 4.5 h if needed. However we strongly advise to keep the 6 h incubation time.
14. Just before its use, defrost the RNA probe stock on ice. Pipette 1–5 μl and mix with 10 μl of ultra-pure water (*see* **Note 20**).
15. Denature the diluted RNA probe by heating at 65 °C for 5 min.
16. Put on ice immediately to prevent any renaturation. Dilute this solution into 1 ml of hybridization buffer, prewarmed at 60 °C.
17. Replace hybridization buffer of the embryos with 500 μl of hybridization buffer containing the RNA probe.
18. Incubate overnight at 60 °C in a water bath.

3.7 Antibody Incubation for In Situ Hybridization

The different buffers need to be prewarmed at 37 or 60 °C before their use. The length of each step can be increased if necessary, but in any case, they cannot be shortened. All washes are done by filling up the tube (2 ml of buffer) except **step 10**. This second day involves long incubations, so it can easily last for 8 h.

1. Recover the hybridization buffer containing the probe (*see* **Note 21**).
2. Wash twice the embryos with hybridization buffer at 60 °C for 30 min.
3. Wash three times the embryos in 2× SSC/CHAPS buffer at 60 °C during 20 min.
4. Replace 2× SSC/CHAPS buffer with the RNase A solution and incubate at 37 °C for 30 min (*see* **Note 22**).
5. Wash twice the embryos with 2× SSC for 10 min at 60 °C.
6. Wash the embryos with 0.2× SSC/CHAPS buffer at 60 °C for 1 h. Repeat this step.
7. Wash the embryos with MAB buffer at room temperature for 5 min.
8. Wash the embryos with MAB buffer at room temperature for 15 min.
9. Replace MAB buffer with MAB-blocking solution and incubate for 1 h at room temperature.
10. Replace MAB-blocking solution with fresh MAB-blocking solution containing the mouse anti-DIG antibody diluted

1:2000 and place the tubes at 4 °C vertically under agitation and incubate overnight. Use the minimal volume needed (350–500 μl) to cover the embryos (*see* **Note 23**).

3.8 Chromogenic Reaction for In Situ Hybridization

All steps are performed under agitation on the nutator at room temperature.

1. Recover the antibody solution and store at 4 °C. This solution can be reused 3–4 times.
2. Wash the embryos with 2 ml of MAB buffer for 1 h. Transfer the embryos into 12 or 24-wells plate depending on the number of embryos per tube.
3. Repeat the wash step with 1 or 2 ml of MAB buffer at least five times (*see* **Note 24**).
4. Rinse the embryos in PBS for 5 min while preparing the alkaline phosphate staining buffer.
5. Wash the embryos twice in the staining buffer for 15 min.
6. Replace the last wash solution with BM purple, the alkaline phosphate substrate, and incubate at room temperature until staining becomes apparent. As this step is light sensitive, the plate needs to be wrapped in foil (*see* **Note 25**).
7. Monitor the staining process by observation under the binocular (*see* **Note 26**).
8. When staining is complete, rinse the embryos in PBS for 10 min. Repeat this step twice.
9. Wash the embryos for 10 min in ethanol–PBS solution (*see* **Note 27**).
10. Wash the embryos for 10 min in methanol 100% (two washes maximum).
11. Wash the embryos for 10 min in ethanol–PBS solution.
12. Wash the embryos for 10 min in PBS.
13. Fix the embryos with Bouin's solution for at least 1 h, without any rocking. The embryos will clump together and agitation may damage them. This fixation step can be performed overnight.
14. Remove background and Bouin's solution with several washes (at least 6) of 1 h in 70 ethanol–PBS solution (*see* **Note 28**).
15. Wash three times the embryos in PBS: first 5 min, then 10 min, and finally 20 min.
16. Replace the PBS solution with the bleaching solution.
17. Place the plate, covered with foil, on a light box. Bleaching process can take a few hours, depending of the pigmentation of the embryos. Monitor this process by regular observation under binocular.

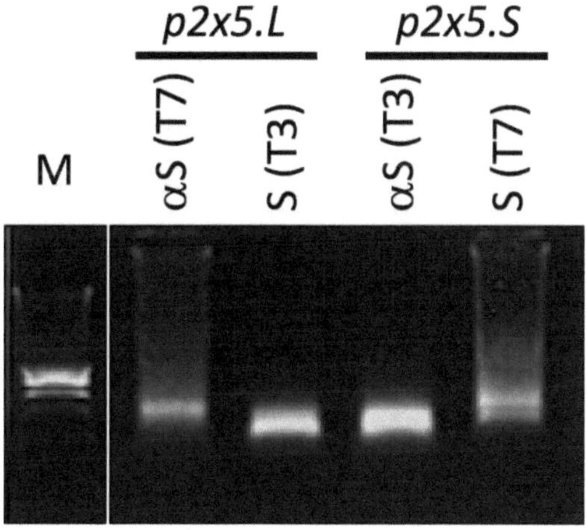

Fig. 2 Scan of the *p2x5.S* and *p2x5.L* antisense (αS) and sense (S) riboprobe electrophoresis. Note the smear in the wells corresponding to the riboprobes synthetized with T7 RNA polymerase (T7), indicating that this precise in vitro transcription reaction did not give a perfect result. The intensity of the riboprobe bands synthetized with the T3 RNA polymerase (T3) corresponds to the expected yield (1 μl/20 μl loaded on the gel). As an indication, 1 μl of each T3 RNA polymerase probe and 3 Placeholder Textμl of each T7 RNA polymerase probe per ml of hybridization buffer were used during the corresponding in situ hybridization. *M* Marker

18. Rinse the embryos twice in PBS.
19. Refix in formaldehyde solution and store at 4 °C until observation. Rinse again the embryos in PBS before observation, taking care that you do not inhale formaldehyde during observation, and always restore them in formaldehyde solution at 4 °C.
20. To observe deep stains embryos can be cleared using 2:1 BB/BA clearing solution (*see* **Note 29**).
21. Stained embryo can be sectioned at 10 μm with a cryostat (*see* **Note 30**). In Fig. 2, an example of a stained embryo for *entpd2* and *p2x1* is presented.

3.9 Double In Situ Hybridization

This procedure allows the detection of two different transcripts in the same embryo without any major changes in the previous protocol. Two probes will be labeled with different UTP analogs (classically digoxigenin and fluorescein). The embryos will be sequentially incubated with two different alkaline phosphate-coupled antibodies against digoxigenin and fluorescein and the staining step performed with two different alkaline phosphate substrates, such as BM purple (light purple stain), NBTP/BCIP (dark blue stain), BCIP (turquoise stain), Magenta-phos (magenta stain), and Fast Red (red stain). The first antibody to add is the one binding to the labeling nucleotide analog associated to the strongest expressed receptor. An alkaline phosphate substrate, which will give a light stain, such as BCIP or Magenta, should be used for detection of the highly expressed receptor (*see* **Note 31**).

1. Probes synthesis: one probe is DIG labeled whereas the second probe is fluorescein labeled. The protocol to generate these probes is identical (*see* Subheading 3.5) except that DIG-UTP is replaced by fluorescein-UTP for the second probe synthesis.
2. Follow steps in Subheading 3.6, except that the two probes are added simultaneously in the hybridization buffer.
3. Follow steps in Subheading 3.7. The embryos will be incubated overnight with only one antibody, anti-DIG antibody (1:2000) or anti-fluorescein (1:5000).
4. Follow steps in Subheading 3.8 up to the **step 8** without any changes.
5. Rinse the embryos three times in PBw for 5 min.
6. Inactivate the alkaline phosphatase by incubating the embryos for 40 min at room temperature in glycine/Tween solution.
7. Wash four times the embryos in MAB for 10 min.
8. Replace the last MAB wash by MAB-blocking solution for 5 min.
9. Incubate the embryos with the second antibody diluted in MAB-blocking solution.
10. Place the tubes at 4 °C vertically under agitation and incubate overnight. Use the minimal volume needed (350–500 μl) to cover the embryos (*see* **Note 23**).
11. Then continue with the **step 1** of the Subheading 3.8.

4 Notes

1. Cysteine is an irritant to the eyes, skin, and respiratory tract. Wear appropriate protection while handling it. pH will be first very acidic, between 1 and 1.5, use NaOH 10 N to adjust pH.
2. Formaldehyde is highly toxic and volatile. Avoid breathing the vapors and wear appropriate gloves and safety glasses when handling it. Always use in a chemical fume hood.
3. Formamide is teratogenic. Wear appropriate protection while handling it. Always use in a chemical fume hood.
4. Torula RNA solution is prepared at 10 mg/ml. Dissolution of the powder needs agitation with moderate heating. Heat first the solution in the microwave (2× 15 s, low-power with careful

mixing between the two pulses) then place on a magnetic stirrer with heater for a few minutes.

5. Maleic acid is toxic and harmful if inhaled, ingested or absorbed through skin. Wear appropriate protection. Acetic acid is harmful if inhaled, ingested or absorbed through skin. Wear appropriate protection and use in a chemical fume hood.
6. H_2O_2 is toxic and damaging to the skin. Wear appropriate protection while handling it. Always use in a chemical fume hood.
7. Injection is done near the lateral line sense organs on both side of the females. It is important to penetrate the skin and the lymph sac with an angle not superior to 30° to prevent any bleeding caused by muscle damage. To insure egg laying, especially if the females have not been hormonally stimulated for more than 4 months or if they are newly purchased, it is preferable to prime the females with a small hormone dose (50 U) the week before egg collection.
8. Egg-laying onset is variable. Hormonal stimulation response can be checked by cloaca observation. After induction of ovulation, cloaca must be red, swollen and open (*see* Fig. 1b). If females do not present these signs, keep them into water until hormone stimulation response. In some cases, females will not respond to hormonal stimulation. It is preferable to wait for females laying and check the egg quality before sacrificing the male. Laid eggs should present a pigmented animal pole with the maturating spot.
9. It is preferable to fertilize immediately the collected eggs. However, we have already fertilized eggs that have been kept for a few hours (up to 4 h) at 18 °C. Fertilize the earliest layings first as egg quality and competence for fertilization decreases during the day. Time of collection should be indicated on the petri dish.
10. Pigmentation of the egg could make the observation of some of these fertilization signs, such as sperm entry scar, difficult. The major sign for fertilization success is the gravitational rotation, occurring in the 20 min after fertilization.
11. Constructs in order to synthetize specific probes for each *Xenopus laevis p2x* receptor or ectonucleotidases can be requested to the corresponding author.
12. We advise to take 1 μl of the reaction mix before the DNase step in order to check the enzymatic yield by DNA electrophoresis (1% agarose 1× TBE gel).
13. It is preferable to concentrate than to dilute the probe. We do not resuspend the probe in water volume superior to 30 μl for its use (*see* **Note 20**) and storage reasons.

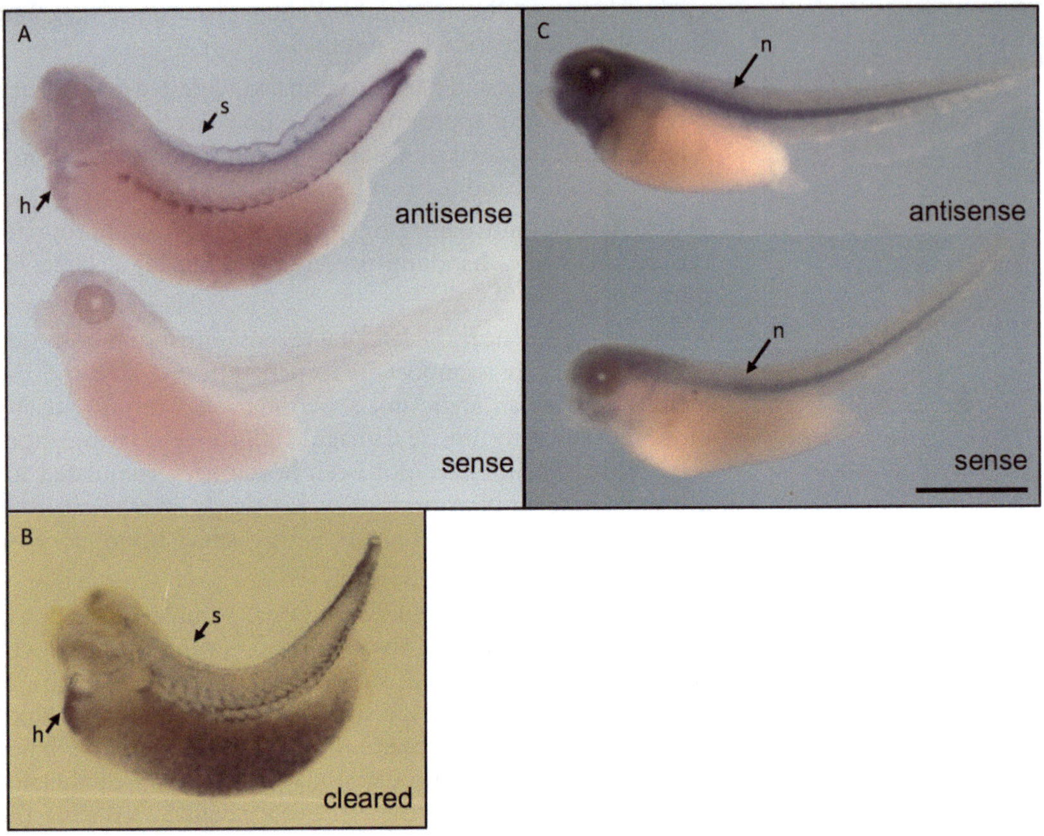

Fig. 3 (**A**) Photography of a representative embryo stained by in situ hybridization with the ectonucleotidase *entpd2* antisense and sense probes. No staining is visible with the sense probe, confirming that staining in the muscle tissues (*s* somites) is specific. (**B**) Photography of a representative cleared embryo stained by in situ hybridization with the ectonucleotidase *entpd2* antisense. Note that the staining in the heart (h) is more visible after clearing step. (**C**) Photography of a representative embryo stained by in situ hybridization with the *p2x1.L* antisense and sense probes. Note the notochord nonspecific staining obtained with the sense probe, due to probe trapping. The scale represents 1 mm

14. Size of the probe is critical for correct penetration inside the embryos. The probes we designed are around 500 bp long. For probes longer than 1000 bp, partial alkaline hydrolysis is performed by adding 2 volumes of carbonate buffer (60 mM Na_2CO_3, 40 mM $NaHCO_3$, pH 10.2) and incubation for 5 min at 60 °C. Reaction is stopped by the addition of 1 volume of neutralization buffer (1 M Tris–HCl pH 8.0, 1.5 M NaCl). Fragmented probes are purified by ethanol precipitation and pellet resuspended in RNase-free water.

    Run 1 μl of the probe in a 1% agarose TBE 1× gel to check the probe quality and get an estimation of the yield. The probe should appear as one band (Fig. 3). No DNA template should be visible, if so, it is recommended to start again the protocol at the DNase I step even if the probe yield will be reduced. A smear without any visible band on the gel indicates that the probe is partially or totally degraded and should not be used.

15. When the in situ hybridization is performed for the first time, we advise to test this procedure using a neuronal or neural marker, strongly expressed in *Xenopus* embryos, such as *tubb2b* (N-tubulin [22]), *sox2* [23], or *sox3* [24].
16. The number of embryos per glass tube depends of the stage of the embryos, but up to 50 embryos can be processed in the same tube. Embryos from different stages can also be mixed into the same tube. However, because it may be hard to differentiate certain stages, especially gastrula and neurula stages after bleaching step, we do not mix in the same tube stage close embryos. As staining is embryo dependent, it is crucial to analyze at least ten embryos per probe and per stage.
17. The proteinase K treatment aims to permeabilize the embryos and improve probe penetration inside the embryos, and will also reduce background staining. However, proteinase K treatment has to be timed, as excessive treatment will induce embryos to break up or embryo epidermis to peel.
18. Triethanolamine is highly toxic, inflammable, corrosive to skin, eyes, and respiratory tract and is harmful if inhaled, ingested, or absorbed through skin. Wear appropriate protection and use in a chemical fume hood.
19. This treatment acetylates and neutralizes free amines. This helps to prevent any electrostatic interaction between the RNA probe and basic proteins, which may induce background staining.
20. The estimation of the probe volume to use could be quite complicated and can be estimated to 0.5 μg/ml of hybridization buffer. However, the quality of the probe determined by electrophoresis is as important as probe quantity to estimate the volume to use (Fig. 2).
21. Probes can be reused at least three times and stored in the hybridization buffer at −20 or −80 °C. If used probe is reused, there is no need to denature it as the probe is stored into the hybridization buffer containing formamide.
22. This step increases the stringency of the hybridization. Even if it reduces the signal, this step cannot be omitted because of the existence of homologs in the purinergic receptors and ectonucleotidase families.
23. To reduce background staining due to nonspecific binding, a prediluted antibody solution (1:50 in MAB solution) is prepared in advance and incubated at 4 °C overnight with fixed wild type embryos of different stages. This prediluted

antibody solution is then used at 1:40 in the antibody hybridization step, and should be stored at 4 °C.

24. The last MAB wash can be done overnight if needed, as the antibody complex is extremely stable. This option is even recommended if staining step takes less than 8 h, or if the staining duration is unknown (probe never used in the laboratory).
25. Other alkaline phosphate substrates can be used. NBT/BCIP is another widely used substrate in laboratory, especially for detecting transcripts strongly expressed, which gives a darker blue staining. This substrate needs to be diluted 1:100 in alkaline phosphate buffer. However, we prefer to use BM Purple as it reduces background staining when long staining process is needed as for *p2x6* expression detection.
26. The duration of the staining step is variable and is probe and temperature dependent. For strongly expressed transcript, such as *N-tubulin*, the staining can be done in 2 h. Regarding the p2x receptors, the length of the staining step is from 1 day (*p2x5*) to 1 week (*p2x1.S*).
27. These washes, especially the methanol wash, remove background staining. Repeat the methanol wash once if needed (methanol solution will become pink if unfixed stains are removed).
28. This step is critical to remove chromogenic components before bleaching. If the wash solution remains pink after six washes, keep washing until the solution is transparent. Residual Bouin fixative can also affect the bleaching process. The last wash can be performed overnight, or even over weekend.
29. Dehydrate the embryos into methanol (two washes of 10 min each) and transfer the embryos into depression slides. Clear the embryos by replacing the methanol by 2:1 BB (benzyl benzoate) /BA (benzyl alcohol) (Murray's clear) clearing solution. After observation, embryos can be stored in BB/BA in the dark. For longer storage, transfer the embryos in 70% methanol–30% PBS. Note that this procedure is irreversible.
30. Stained embryos are gradually infiltrated in gelatin–sucrose solution (7.5 g sucrose, 17.5 ml fish gelatin, up to 50 ml with PBS). This solution stock needs several hours under gentle agitation at room temperature until complete dissolution and is then stored at 4 °C. It needs to be brought to room temperature at least 1 h before use. Embryos are washed 30 min at room temperature in the following solutions: 10% gelatin/sucrose–PBS, 25% gelatin/sucrose–PBS, 50% gelatin/sucrose–PBS. Embryos are then incubated for 1 h in 75% gelatin/sucrose–PBS and 100% gelatin/sucrose solution before an incubation overnight in 100% gelatin/sucrose. Drop each

identified embryo into a Tissue-Tek OCT (Optimal cutting temperature compound) solution filled mold placed on dry ice, and orient the embryo as polymerization proceeds. It is recommended to proceed with a binocular. Store the sample at −80 °C until use.

31. We routinely use combination of Magenta and BM Purple (or NBT/BCIP), or NBT/BCIP and Fast Red. Note that Fast Red is soluble in methanol and not stable in clearing agents. Avoid the combination of BM Purple and BCIP or BMP Purple and NBTP/BCIP as the double staining (blue/purple shade) can be confusing.

References

1. Gall JG, Pardue ML (1969) Formation and detection of RNA-DNA hybrid molecules in cytological preparations. Proc Natl Acad Sci U S A 63:378–383
2. Tautz D, Pfeifle C (1989) A non-radioactive in situ hybridization method for the localization of specific RNAs in Drosophila embryos reveals translational control of the segmentation gene hunchback. Chromosoma 98:81–85
3. Hemmati-Brivanlou A, Frank D, Bolce ME, Brown BD, Sive HL, Harland RM (1990) Localization of specific mRNAs in Xenopus embryos by whole-mount in situ hybridization. Development 110:325–330
4. Harland RM (1991) In situ hybridization: an improved whole-mount method for Xenopus embryos. Methods Cell Biol 36:685–695
5. Sive HL, Grainger RM, Harland RM (2000) Whole-mount in situ hybridization. Cold Spring Harbor Laboratory Press, Cold Spring Harbor, NY
6. Monsoro-Burq AH (2007) A rapid protocol for whole-mount in situ hybridization on Xenopus embryos. CSH Protoc 2007. https://doi.org/10.1101/pdb.prot4809
7. Saint-Jeannet JP (2017) Whole-mount in situ hybridization of Xenopus embryos. Cold Spring Harb Protoc. https://doi.org/10.1101/pdb.prot097287
8. Spemann H, Mangold H (1924) Induction of embryonic primordia by implantation of organizers from different species. In: Willier BH, Oppen-heimer JM (eds) Foundations of experimental embryology. Hafner, New York, pp 144–184
9. Gould SE, Grainger RM (1997) Neural induction and antero-posterior patterning in the amphibian embryo: past, present and future. Cell Mol Life Sci 53:319–338
10. Session AM, Uno Y, Kwon T et al (2016) Genome evolution in the allotetraploid frog Xenopus laevis. Nature 538:336–343
11. Sater AK, Moody SA (2016) Using Xenopus to understand human disease and developmental disorders. Genesis 55. https://doi.org/10.1002/dvg.22997
12. Tandon P, Conlon F, Furlow JD, Horb ME (2017) Expanding the genetic toolkit in Xenopus: approaches and opportunities for human disease modeling. Dev Biol 426:325–335
13. Massé K, Eason R, Bhamra S, Dale N, Jones EA (2006) Comparative genomic and expression analysis of the conserved NTPDase gene family in Xenopus. Genomics 87:366–381
14. Massé K, Bhamra S, Allsop G, Dale N, Jones EA (2010) Ectophosphodiesterase/nucleotide phosphohydrolase (Enpp) nucleotidases: cloning, conservation and developmental restriction. Int J Dev Biol 54:181–193
15. Blanchard C, Boué-Grabot E, Massé K (2019) Comparative embryonic spatio-temporal expression profile map of the Xenopus P2X receptor family Frontiers in Cellular Neuroscience. https://doi.org/10.3389/fncel.2019.00340
16. Massé K, Dale N (2012) Purines as potential morphogens during embryonic development. Purinergic Signal 8:503–521
17. Massé K, Bhamra S, Eason R, Dale N, Jones EA (2007) Purine-mediated signalling triggers eye development. Nature 449:1058–1062
18. Jowett T, Lettice L (1994) Whole-mount in situ hybridizations on zebrafish embryos using a mixture of digoxigenin- and fluorescein-labelled probes. Trends Genet 10:73–74
19. Koga M, Kudoh T, Hamada Y, Watanabe M, Kageura H (2007) A new triple staining method for double in situ hybridization in

combination with cell lineage tracing in whole-mount Xenopus embryos. Dev Growth Differ 49:635–645

20. Nieuwkoop F (1994) Normal table of Xenopus laevis (Daudin). Garland Publishing Inc., New York. ISBN: 0-8153-1896-0
21. David R, Wedlich D (2001) PCR-based RNA probes: a quick and sensitive method to improve whole mount embryo in situ hybridizations. Biotechniques 30:769–772
22. Oschwald R, Richter K, Grunz H (1991) Localization of a nervous system-specific class II beta-tubulin gene in Xenopus laevis embryos by whole-mount in situ hybridization. Int J Dev Biol 35:399–405
23. Mizuseki K, Kishi M, Shiota K, Nakanishi S, Sasai Y (1998) SoxD: an essential mediator of induction of anterior neural tissues in Xenopus embryos. Neuron 21:77–85
24. Penzel R, Oschwald R, Chen Y, Tacke L, Grunz H (1997) Characterization and early embryonic expression of a neural specific transcription factor xSOX3 in Xenopus laevis. Int J Dev Biol 41:667–677

Chapter 7

Histochemical Approach for Simultaneous Detection of Ectonucleotidase and Alkaline Phosphatase Activities in Tissues

Karolina Losenkova, Marius Paul, Heikki Irjala, Sirpa Jalkanen, and Gennady G. Yegutkin

Abstract

Studies on pathophysiology and the therapeutic potential of extracellular ATP and other purines represent an important and rapidly evolving field. The integral response of the cell is determined by multiple factors, including the release of endogenous ATP, co-expression of different types of nucleotide- and adenosine-selective receptors, as well as the specific makeup of ectoenzymes governing the duration and magnitude of purinergic signaling. Current findings support the presence of an extensive network of purine-converting ectoenzymes that are co-expressed to a variable extent among the mammalian tissues and share similarities in substrate specificity. Here, we describe a histochemical approach for simultaneous detection of ecto-nucleotidase and tissue-nonspecific alkaline phosphatase (TNAP) activities in the same tissue slice. Further employment of this technique for staining human palatine tonsil cryosections revealed selective distribution of the key ectoenzymes within certain tonsillar structures, including germinal centers and connective tissues (ecto-5′-nucleotidase/CD73), as well as interfollicular area (TNAP and NTPDase1/CD39).

Key words Enzyme histochemistry, NTPDase1/CD39, Ecto-5′-nucleotidase/CD73, Tissue-nonspecific alkaline phosphatase, Human tonsils

1 Introduction

Extracellular ATP, adenosine, and other purines are important signaling molecules implicated in an array of cell-specific responses in virtually all organs and tissues. Most models of purinergic signaling depend on functional interactions between distinct processes, including (1) the release of endogenous ATP; (2) triggering of signaling events via a series of nucleotide- and nucleoside-selective receptors; (3) ectoenzymatic inactivation of nucleotides and finally, (4) metabolism and/or re-uptake of nucleotide-derived adenosine into the cell [1, 2]. The last decade has seen a great increase in publications concerning nucleoside triphosphate diphosphohydrolase-1 (NTPDase1, otherwise known as CD39),

Pablo Pelegrín (ed.), *Purinergic Signaling: Methods and Protocols*, Methods in Molecular Biology, vol. 2041,
https://doi.org/10.1007/978-1-4939-9717-6_7,

ecto-5′-nucleotidase/CD73 (*e*N/CD73), tissue-nonspecific alkaline phosphatase (TNAP) and other nucleotide-inactivating ectoenzymes [3, 4]. These key ectonucleotidases have been extensively investigated in regard to their implications in such pathological states as inflammation [5], vascular remodeling [6–8], brain diseases [4, 7, 9], as well as tumor growth and metastasis [3, 10, 11].

Different analytic approaches have been employed for the measurement of ecto-nucleotidase activities, including colorimetric P_i-liberating assays, capillary electrophoresis, chromatography-based assays, and lead nitrate-based enzyme histochemistry [3]. The latter technique is of particular importance in defining the distribution of ectonucleotidase activities within a tissue, taking advantage of the abilities of these enzymes to generate inorganic phosphorus (P_i) when incubated with appropriate nucleotide substrates in the presence of lead nitrate $Pb(NO_3)_2$ (Fig. 1a). This technique was originally employed by Wachstein and Meisel for the histochemical characterization of hepatic phosphatases [12]. Using this approach, tissue localization of different ectonucleotidases has been characterized in the murine brain [9], dorsal root ganglion and spinal cord [13], thoracic aortas and other blood vessels [6, 14], peripheral lymph nodes [8], rat meninges [15], human endometrium [16] and oviducts [17], as well as other human and rodent tissues [3]. The activity of TNAP can also be detected histochemically by using 5′-bromo-4-chloro-3-indolyl phosphate/nitro blue tetrazolium (BCIP/NBT) as artificial chromogenic substrates (Fig. 1b) [9, 13]. Here, we describe a convenient method for investigating the expression levels, catalytic activities and tissue distribution of NTPDase1/CD39, *e*N/CD73 and TNAP, using human tonsil cryosections as a representative source of the ectoenzymes studied.

2 Materials

2.1 Buffers and Other Working Solutions

1. Trizma–maleate buffer (TMB): weigh Trizma–maleate powder, dissolve it in MQ-water at the concentration of 9.5 g/L (40 mM) and adjust pH of the mixture to 7.3 by using 1 M NaOH. TMB should be stored at +4 °C and warmed up at room temperature (RT) before the experiment.
2. Trizma–maleate sucrose buffer (TMSB): dissolve 40 g sucrose in 0.5 L of TMB. For assaying ectonucleotidase and TNAP activities, adjust the pH values of TMSB to 7.3 and 9.0, respectively (*see* **Note 1**). TMSB should be stored at +4 °C and warmed up at room temperature (RT) before the experiment.
3. Blocking buffer for immunofluorescence staining: Phosphate buffered saline (PBS; comprising 137 mM NaCl, 2.7 mM KCl,

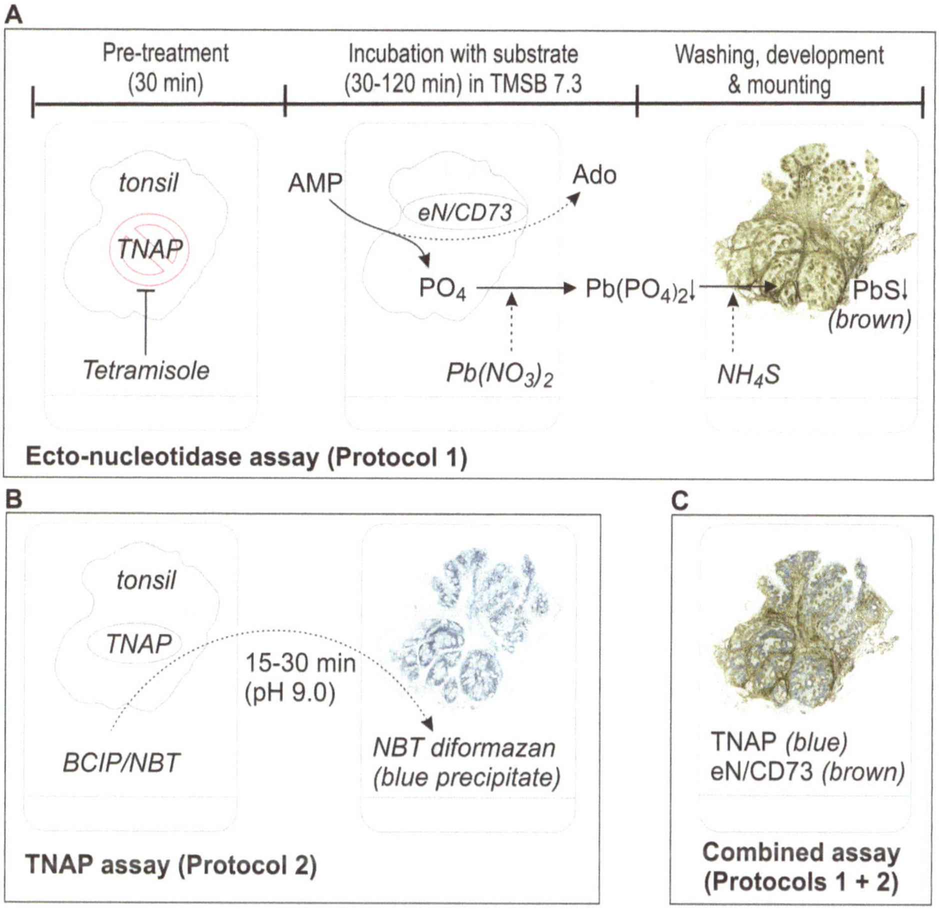

Fig. 1 Enzyme histochemistry assays for detecting ecto-nucleotidase and TNAP activities in tissues. (**a**) Lead nitrate-based ecto-nucleotidase histochemistry is comprised of several steps, including: pre-treatment of the tissue slices with TNAP inhibitor tetramisole, incubation with appropriate nucleotide substrate (in this particular case, AMP) and finally, chemical conversion of the reaction product into insoluble brown precipitate. In the presence of $Pb(NO_3)_2$, the inorganic phosphorus generated in the course of the ecto-nucleotidase reaction is converted into insoluble lead diphosphate $Pb(PO_4)_2$ (white precipitate). Subsequent addition of the developing solution (0.5% $(NH_4)_2S$) triggers the further conversion of $Pb(PO_4)_2$ to a lead salt PbS, which can be detected as a brown staining by light microscopy. (**b**) The principle of the TNAP activity assay is based on the ability of alkaline phosphatases to remove the phosphate group of BCIP, an artificial chromogenic substrate. The resulting blue dye, 5,5′-dibromo-4,4′-dichloro-indigo is further oxidized by NBT, which forms an insoluble dark blue diformazan precipitate. (**c**) Both ecto-nucleotidase and TNAP can be assayed simultaneously by combining the two approaches above, as described in Subheading 3

4.3 mM Na_2HPO_4, and 1.47 mM KH_2PO_4; pH 7.4), supplemented with 2% bovine serum albumin (BSA) and 0.2% Triton X-100.

4. Ammonium sulfide solution: Dilute 1 mL of 20% $(NH_4)_2S$ stock solution in 40 mL of MQ-water. The obtained working solution of 0.5% $(NH_4)_2S$ has to be prepared and stored at RT under a fume hood (*see* **Note 2**).

5. Calcium chloride solution: Prepare stock solution of 0.5 M $CaCl_2$ in MQ-mater and store them at RT.
6. Magnesium sulfate solutions: Prepare stock solution of 0.5 M $MgSO_4$ in MQ-mater and store them at RT.
7. Lead nitrate solution: Weigh $Pb(NO_3)_2$ and dissolve it in MQ-water at a concentration of 120 mg/mL. A freshly prepared solution of lead nitrate has to be used for each staining assay (*see* **Note 3**).

2.2 Enzyme Substrates and Inhibitors

1. Inhibitor of TNAP activity tetramisole (otherwise known as levamisole): Weigh 240 mg of tetramisole and dissolve the powder in 1 mL of TMB (pH 7.3) in order to achieve the final concentration of 1 M. A freshly prepared stock solution of tetramisole has to be used for each assay.
2. Substrates for assaying ecto-nucleotidase activities: Prepare stock solutions of ATP, ADP and AMP dissolved in MQ-water at concentrations of 40–100 mM. The obtained stock solutions can be aliquoted and stored for several months at −20 °C.
3. Substrates for assaying TNAP activity: Weigh 4.3 mg of BCIP and 8.2 mg of NBT and dissolve them in 50 mL TMSB (pH 9.0) at final concentrations of 0.2 mM for both substrates.

2.3 Antibodies

1. Polyclonal guinea pig anti-human NTPDase1/CD39 ($hN1\text{-}1_c$) antibody (provided by Prof. Jean Sevigny, Quebec, Canada; http://ectonucleotidases-ab.com).
2. Secondary Cy™3-conjugated donkey anti-guinea pig IgG.

3 Methods

All staining experiments were conducted by using human tonsil cryosections as the appropriate enzyme source. Palatine tonsils were obtained from adult patients with chronic tonsillitis undergoing routine tonsillectomy. The tonsils were washed with physiological salt solution, embedded in the cryo-mold with Tissue-Tek® O.C.T. compound (Sakura Finetek Europe B.V., The Netherlands), cut using a cryostat and stored at −80 °C (*see* **Note 4**). Prior to the experiment, the slides were taken from the freezer and equilibrated for 15–20 min at RT (*see* **Note 5**). Subsequent staining procedures were carried out in a glass chamber at RT, unless otherwise specified. Multiple images of adjacent tissue areas were captured using Pannoramic 250 and Pannoramic Midi FL slide scanners (3DHistech Ltd., Budapest, Hungary), and further stitched to a larger overview using the accompanying Pannoramic Viewer 1.15.4 software.

3.1 In Situ Ecto-Nucleotidase Activity Assay

1. For blocking TNAP activity, pre-incubate tissue cryosections for 30 min with 5 mM tetramisole dissolved in 50 mL TMB.
2. Dissolve the following compounds from stock solutions at 1:100–1:500 dilutions in 50 mL TMSB (pH 7.3): ATP, ADP, AMP or other nucleotides (at final concentration ranging from 0.3 to 1 mM, as determined based on pilot assays), tetramisole (2.5 mM), $CaCl_2$ (0.5 μM), and lead nitrate (0.24 mg/mL). The concentrations of nucleotide substrates, incubation times and other experimental settings have to be optimized for each particular tissue studied (*see* **Note 6**). Specifically, for assaying tonsillar *e*N/CD73 activity, tissue slices were incubated for 60 min with 1 mM AMP (Fig. 2a), while for determining NTPDase1/CD39 activity the samples were incubated for 30 min with 300 μM ATP and ADP as preferred enzyme substrates (Fig. 3a).
3. Wash the slides with TMB (3 × 5 min) and then add 40 mL of ammonium sulfide (0.5%) to a glass chamber with samples under a fume hood and wait for 30 s.
4. Wash the slides again with TMB (3 × 5 min) and rinse once with MQ-water (1–2 min). Mount the slides with stained tissue using an aqueous mounting agent Aquatex (Merck), and store them at RT for subsequent microscopic examination.

3.2 Evaluation of TNAP Activity in the Tissue

1. Incubate the slides with artificial TNAP substrates (BCIP/NBT, 0.2 mM each) dissolved in 50 mL of TMSB (pH 9.0) supplemented with 5 mM $MgSO_4$. Generation of the insoluble reaction product NBT diformazan can be directly visualized by progressively increased blue staining intensity. In this particular experiment, clearly detectable signal has been observed after 15 min of incubation of the human tonsils with BCIP/NBT mixture.
2. Wash the slides with TMB (3 × 5 min), rinse with MQ-water and proceed with mounting and microscopic examination, as specified in Subheading 3.1, **step 4** (Fig. 2b).

3.3 Combined Assay for Simultaneous Detection of Ecto-Nucleotidase and TNAP Activities

By combining the above-described protocols, co-localization of ecto-nucleotidase and TNAP activities may be revealed within the same tissue.

1. Follow the development of evolving blue stain during incubation of the tissue with BCIP/NBT, as described in Subheading 3.2, **step 1**.
2. Rinse the slide with TMB.
3. Transfer the slide into another chamber with the appropriate "nucleotidase-designed mixture" containing $Pb(NO_3)_2$ and certain nucleotide substrate, followed by a step-by-step

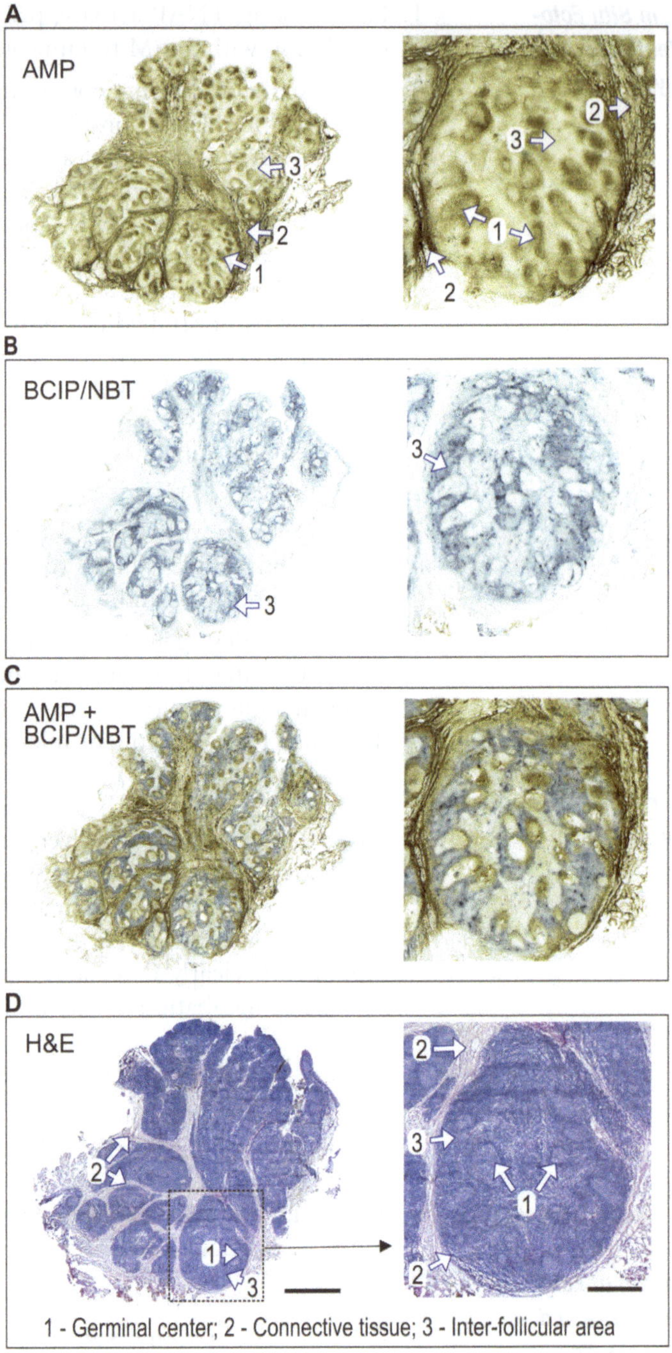

Fig. 2 Histochemical analysis of the distribution of eN/CD73 and TNAP activities in human tonsils. (**a**) Tonsillar eN/CD73 (AMPase) activity was assayed by incubating tissue slices with AMP in the presence of $Pb(PO_4)_2$, followed by microscopic detection of the nucleotide-derived P_i as a brown precipitate. (**b**) The activity of TNAP was measured by using the artificial chromogenic enzyme substrates BCIP/NBT and subsequent monitoring the development of the blue color reaction. (**c**) Co-staining analysis of eN/CD73 (brown) and TNAP (blue)

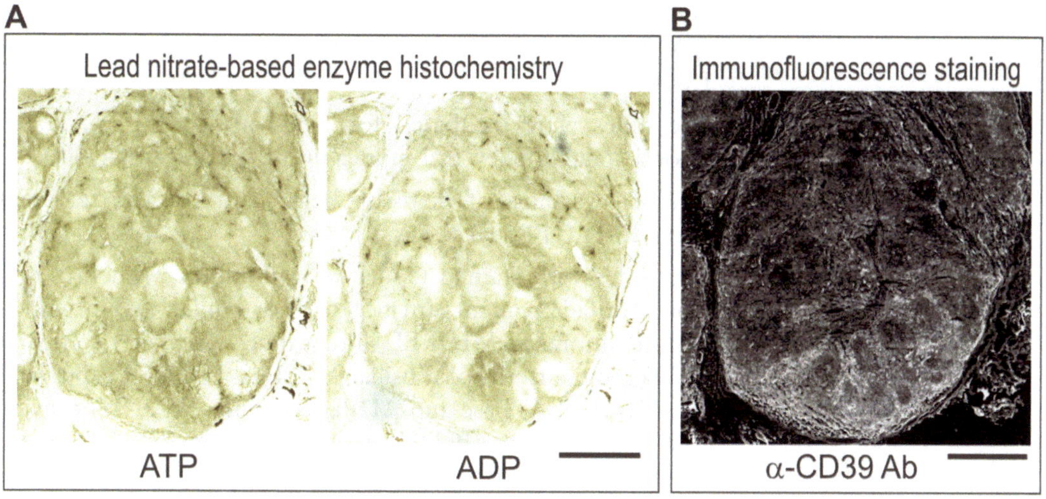

Fig. 3 Distribution of ATPase and ADPase activities and the expression levels of NTPDase1/CD39 in human tonsils. (**a**) Tonsillar ecto-nucleotidases were assayed by incubating tissue slices with ATP and ADP in the presence of $Pb(PO_4)_2$, as indicated. (**b**) For immunofluorescence staining, human tonsils were sequentially incubated with anti-CD39 antibody and isotype-matched fluorochrome-conjugated second-stage Ig. The images were captured using Pannoramic 250 (**a**) and Pannoramic Midi FL (**b**) slide scanners as tile scans of adjacent areas stitched into larger mosaics. Scale bar: 2 mm

determination of P_i generated in the course of the ecto-nucleotidase reaction (*see* Subheading 3.1). Figure 2c depicts a representative co-staining image of tonsillar TNAP (blue) and eN/CD73-mediated AMPase (brown) activities, showing their selective and spatially distinct localization in the germinal centers and connective tissues (eN/CD73), as well as blood vessels and inter-follicular area (TNAP) (*see* **Note 7**).

3.4 Immunofluorescence Staining

Given the ability of ecto-nucleotidases and other purinergic enzymes to share similarities in substrate specificity, additional immunohistochemical analysis may serve as an important auxiliary tool for identifying the exact nature of ectoenzyme responsible for hydrolysis of certain nucleotide substrate(s) [3].

1. For the immunofluorescence analysis of NTPDase1/CD39 expression, human tonsil sections were pre-incubated for 30 min with the appropriate blocking buffer in Shandon Sequenza Staining System.

Fig. 2 (continued) activities was performed using the combination of the above protocols "**a**" and "**b**", as described in the Methods section. (**d**) Tissue samples were also stained with hematoxylin and eosin (H&E). All images were captured as tile scans of adjacent areas by using the Pannoramic 250 slide scanner. Scale bars: 3 mm (left images) and 1 mm (right)

2. Overnight incubation at +4 °C with guinea pig anti-human CD39 antibody (diluted in 200 μL of blocking buffer at 1:300).
3. The slides were incubated for 1 h with Cy™3-conjugated second-stage donkey anti-guinea pig IgG diluted at 1:600.
4. Intensive washings of the tissues before and after the treatment using 200 μL PBS (3 × 5 min).
5. The slides with stained tissues were mounted with ProLong Gold Antifade reagent (Invitogen™) or similar, and examined using Pannoramic MIDI FL slide scanner. The fairly comparable staining patterns of tonsillar NTPDase1/CD39 expression (Fig. 3b) and ATPase and ADPase activities (Fig. 3a) provide clear evidence that CD39 represents a major nucleotide-inactivating ectoenzyme responsible for sequential dephosphorylation of ATP and ADP in human tonsils.

   In summary, the (immuno)histochemical approach described here provides a useful tool for comprehensive analysis of the expression levels and catalytic activities of key nucleotide-inactivating enzymes in different tissues, as well as their spatial compartmentalization and potential cross-talk with other components of the purinergic signaling cascade at various (patho) physiological settings.

4 Notes

1. Since Tris-based buffers have high degree of temperature sensitivity, the pH values for TMB and TMSB have to be set for the ambient temperature at which the buffers will be used in the experiment. Moreover, during the whole staining protocol, avoid using PBS and other phosphorus-containing buffers, which may interfere with specific detection of P_i liberated in the course of ecto-nucleotidase reaction.
2. The diluted 0.5% solution of ammonium sulfide can be stored in a fume hood, protected from light. The same developing solution could be re-used for staining multiple slides in several experiments. However, in the case of long-term storage, make sure that this solution still has a "normal" bright-yellow color. Otherwise, prepare a fresh one.
3. Lead nitrate is poorly dissolved in TMB and TMSB. Therefore, it is important to use MQ-water for preparation of $Pb(NO_3)_2$ stock solution.
4. Short-term fixation of the dissected tissue for 30–60 min with cold paraformaldehyde diluted in PBS (4%) does not affect the catalytic activities of ecto-nucleotidases in subsequent enzyme histochemistry assays. In the case of relatively large tissue

specimens, thawed slides with non-fixed tissue slices can be fixed for 5–10 min with 4% paraformaldehyde just before their transfer into the glass chamber. Noteworthy, this maneuver may also prevent potential tissue detachment from the slide during subsequent staining procedures.

5. Collect the exact amount of tissue specimens required for each staining experiment. Avoid re-freezing of the thawed slides.
6. These experimental settings were chosen based on the preliminary staining experiments performed with human tonsils, and therefore may serve only as a guidance. Additional pilot experiments may be required for further optimizing the assay's conditions for each particular tissue and enzyme studied.
7. Basically, this methodological paper only describes the overall patterns of ecto-enzyme distribution within major tonsillar structures. If deemed necessary, additional quantification of the measured staining intensities and more thorough cytometric analysis of the expression levels of ecto-nucleotidases among different cell types can be performed using appropriate image analysis software.

References

1. Ralevic V, Burnstock G (1998) Receptors for purines and pyrimidines. Pharmacol Rev 50 (3):413–492
2. Yegutkin GG (2008) Nucleotide- and nucleoside-converting ectoenzymes: important modulators of purinergic signalling cascade. Biochim Biophys Acta 1783(5):673–694
3. Yegutkin GG (2014) Enzymes involved in metabolism of extracellular nucleotides and nucleosides: Functional implications and measurement of activities. Crit Rev Biochem Mol Biol 49(6):473–497
4. Zimmermann H, Zebisch M, Strater N (2012) Cellular function and molecular structure of ecto-nucleotidases. Purinergic Signal 8 (3):437–502
5. Eltzschig HK, Sitkovsky MV, Robson SC (2012) Purinergic signaling during inflammation. N Engl J Med 367(24):2322–2333
6. Kauffenstein G, Drouin A, Thorin-Trescases N, Bachelard H, Robaye B, D'Orleans-Juste P, Marceau F, Thorin E, Sevigny J (2010) NTPDase1 (CD39) controls nucleotide-dependent vasoconstriction in mouse. Cardiovasc Res 85(1):204–213
7. Buchet R, Millan JL, Magne D (2013) Multisystemic functions of alkaline phosphatases. Methods Mol Biol 1053:27–51
8. Yegutkin GG, Auvinen K, Rantakari P, Hollmen M, Karikoski M, Grenman R, Elima K, Jalkanen S, Salmi M (2015) Ecto-5′-nucleotidase/CD73 enhances endothelial barrier function and sprouting in blood but not lymphatic vasculature. Eur J Immunol 45 (2):562–573
9. Langer D, Hammer K, Koszalka P, Schrader J, Robson S, Zimmermann H (2008) Distribution of ectonucleotidases in the rodent brain revisited. Cell Tissue Res 334 (2):199–217
10. Maj T, Wang W, Crespo J, Zhang H, Wei S, Zhao L, Vatan L, Shao I, Szeliga W, Lyssiotis C, Liu JR, Kryczek I, Zou W (2017) Oxidative stress controls regulatory T cell apoptosis and suppressor activity and PD-L1-blockade resistance in tumor. Nat Immunol 18(12):1332–1341
11. Allard B, Longhi MS, Robson SC, Stagg J (2017) The ectonucleotidases CD39 and CD73: novel checkpoint inhibitor targets. Immunol Rev 276(1):121–144
12. Wachstein M, Meisel E (1957) Histochemistry of hepatic phosphatases of a physiologic pH; with special reference to the demonstration of bile canaliculi. Am J Clin Pathol 27(1):13–23
13. Street SE, Kramer NJ, Walsh PL, Taylor-Blake-B, Yadav MC, King IF, Vihko P, Wightman RM, Millan JL, Zylka MJ (2013) Tissue-nonspecific alkaline phosphatase acts redundantly with PAP and NT5E to generate

adenosine in the dorsal spinal cord. J Neurosci 33(27):11314–11322

14. Mercier N, Kiviniemi TO, Saraste A, Miiluniemi M, Silvola J, Jalkanen S, Yegutkin GG (2012) Impaired ATP-induced coronary blood flow and diminished aortic NTPDase activity precede lesion formation in apolipoprotein E-deficient mice. Am J Pathol 180 (1):419–428
15. Yegutkin GG, Guerrero-Toro C, Kilinc E, Koroleva K, Ishchenko Y, Abushik P, Giniatullina R, Fayuk D, Giniatullin R (2016) Nucleotide homeostasis and purinergic nociceptive signaling in rat meninges in migraine-like conditions. Purinergic Signal 12 (3):561–574
16. Aliagas E, Vidal A, Torrejon-Escribano B, Taco Mdel R, Ponce J, de Aranda IG, Sevigny J, Condom E, Martin-Satue M (2013) Ecto-nucleotidases distribution in human cyclic and postmenopausic endometrium. Purinergic Signal 9(2):227–237
17. Villamonte ML, Torrejon-Escribano B, Rodriguez-Martinez A, Trapero C, Vidal A, Gomez de Aranda I, Sevigny J, Matias-Guiu X, Martin-Satue M (2018) Characterization of ecto-nucleotidases in human oviducts with an improved approach simultaneously identifying protein expression and in situ enzyme activity. Histochem Cell Biol 149 (3):269–276

Chapter 8

Flow Cytometry of Membrane Purinoreceptors

Nicole Schwarz, Marten Junge, Friedrich Haag, and Friedrich Koch-Nolte

Abstract

Mammalian purinoreceptors respond to extracellular nucleotides and their metabolites, for example, following the release of ATP or NAD^+ from cells and their hydrolysis by ectonucleotidases. Membrane purinoreceptors are expressed as ionotropic ligand-gated ion channels designated P2X receptors, or as metabotropic G-protein coupled receptors designated P1 or P2Y receptors, on the cell surface of different cell types. In this chapter, we provide protocols to monitor the expression and activity of purinoreceptors on the cell membrane of living cells by flow cytometry.

Key words Purinoreceptors, Purinergic signaling, Flow cytometry, Ca^{2+}-influx, Externalization of phosphatidylserine, Ectodomain shedding, Inflammasome, P2X7

1 Introduction

Flow cytometry is a rapid multiparametric analysis of single cells in suspension [1]. A flow cytometer measures fluorescence of cells excited by light of specific wavelengths generated by a laser. Fluorescent probes can expand fluorescent properties of the cell. In contrast to fluorescence immunohistochemistry, flow-cytometric analyses cannot provide any information about the three-dimensional organization of cells within tissue. However, a modern flow cytometer can assess multiple parameters for each cell, thus allowing fine differentiation between cell subsets. Flow cytometry allows the quantification of protein expression levels at the time of measurement and thus provides a different level of information than gene expression analyses on the RNA level (e.g., RT-PCR or RNA sequencing). Moreover, flow cytometry also permits real-time monitoring of cellular responses to extracellular nucleotides that cannot be visualized on a transcriptional level (e.g., shedding of cell surface proteins, externalization of phosphatidylserine, permeabilization of the cell membrane to Ca^{2+} and to fluorescent dyes,

Nicole Schwarz and Marten Junge contributed equally to this work.

Pablo Pelegrín (ed.), *Purinergic Signaling: Methods and Protocols*, Methods in Molecular Biology, vol. 2041, https://doi.org/10.1007/978-1-4939-9717-6_8, © Springer Science+Business Media, LLC, part of Springer Nature 2020

alterations of the mitochondrial membrane potential, swelling and shrinking of the cell, and assembly of the inflammasome). While multiparametric analyses, that is, the simultaneous detection of numerous cell surface proteins using monoclonal antibodies conjugated to different fluorochromes, represents a major advantage of flow cytometry, the potential spectral overlap of the fluorochromes utilized demands careful experimental design and compensation.

A major limitation of flow cytometry is the necessity to prepare a suspension of single cells. This is relatively straightforward for cell lines that grow in suspension (e.g., many lymphoma cell lines) and for primary cells prepared from blood. Preparing a cell suspension is more complex in case of adherent cell lines and primary cells from solid tissues. In such cases, the procedures required to achieve a cell suspension (e.g., treatment with trypsin or collagenase at 37 °C and/or mechanical agitation) can profoundly affect the integrity of membrane proteins and the cellular vitality. These limitations make flow cytometry more useful for the analysis of some tissues and cells than others. A particular concern for studies of cells expressing purinoreceptors and purine metabolizing ectoenzymes is the fact that mechanical stress can induce the release of nucleotides from cells during cell preparation [2]. Nucleotides released from cells and metabolites generated by ectoenzymes in turn can trigger purinoreceptor-mediated cellular responses, even under carefully controlled conditions. Recently, experimental tools and protocols have been developed to alleviate the inadvertent activation of purinoreceptors during cell preparation, including the use of small-molecule and antibody-based inhibitors [3–5].

This chapter provides protocols and advice to apply flow cytometry for monitoring the expression and function of purinoreceptors on cell lines and primary cells, using the ATP-gated P2X7 ion channel as an example [6–8].

2 Materials

1. Purinoreceptor-specific antibodies: for flow cytometry use validated monoclonal antibodies that recognize purinoreceptors in native conformation [9, 10] (*see* **Note 1**).
2. Purinoreceptor-transfected cells: purchase a synthetic gene block encompassing the full length open reading frame of the desired purinoreceptor flanked by suitable restriction enzyme sites and clone the gene block into a eukaryotic expression vector (e.g., pcDNA6) (*see* **Note 2**). Transfect HEK293T cells or lymphoma cells using a suitable transfection system (e.g., jetPEI for HEK293T cells or electroporation for lymphoma cells) (*see* **Note 3**). Optional: in order to facilitate visualization of transfected vs. nontransfected cells, cotransfect

cells with a cDNA expression vector encoding GFP. Similarly, cells can be cotransfected with expression vectors encoding ARTC2.2 or CD62L (*see* **Note 4**). Harvest cells 24–48 h post transfection. For stable selection, replace medium 24–48 h after transfection with cell culture medium containing a suitable selection drug (e.g., blasticidin in case of pCDNA6). Monitor expression and activity of the respective purinoreceptor by flow cytometry as described below.

3. P2X7/ARTC2.2 antagonists and agonists: KN-62 (1-[*N,O*-bis (5-isoquinolinesulfonyl)-*N*-methyl-L-tyrosyl]-4-phenylpiperazine) [11], A-438079 [12]; P2X7-specific nanobodies 13A7, 14D5, and Dano1 [13]; ARTC2.2-specific nanobody s + 16a (Treg-Protector™, Biolegend) [14] (*see* **Note 5**).

2.1 Monitoring the Cell Surface Expression of Purinoreceptors

1. Antibodies: titrate each lot of fluorochrome-conjugated antibodies (final concentrations typically are in the range of 0.1–1 μg/ml): anti-P2X7 (Hano44), anti-CD19 (B4-1D3), anti-CD3 (145-2C11), anti-CD4 (RM4-5), anti-CD8 (RPA-T8), anti-CD25 (PC61), anti-F4/80 (BM8), and anti-CD11b (M1/70). Use anti-CD45 (30F11) for intravascular staining of leukocytes [15] (*see* **Note 6**). Use anti-CD16/CD32 (2.4G2) as FcR-block to prevent unspecific binding of Abs to Fc-receptor-expressing leukocytes (*see* **Note 7**). Use Dynabead-immobilized goat anti-mouse IgG for B cell depletion of splenocytes.
2. FACS buffer: PBS pH 7.4, 1 mM EDTA, 0.5% bovine serum albumin (BSA).
3. P2X7-expression vector: standard eukaryotic expression vector (e.g., pcDNA6-blasticidin (Addgene #V22120) containing the full-length open reading frame of P2X7 (NCBI Gene ID: 11872); jetPEI transfection reagent (Q-Biogen) or similar transfection reagent; and 15 μg/ml solution of blasticidin.
4. Cells: P2X7-expressing lymphoma cell lines Yac-1 (ATCC: TIB160), P2X7-transfected HEK293T cells (ATCC: CRL11268).
5. Cell culture media: RPMI or DMEM tissue culture medium containing 10% fetal calf serum (FCS).
6. Cell culture CO_2 incubator set at 5% CO_2 and 37 °C.
7. Primary cells from murine blood, thymus, spleen, lymph nodes (*see* **Notes 4–6**): scissors, forceps, a preparation board, 70% ethanol, 15 and 50 ml centrifuge tubes, 70 μm cell strainers, 10 cm petri dishes, 10 ml syringe piston, serological pipettes, pipette boy, cell centrifuge, Neubauer chamber, cell microscope, Dynabead-immobilized sheep anti-mouse IgG, and a suitable magnet for depletion of B cells.

8. Propidium iodide: 100× stock solution at 20 μg/ml for discrimination of dead cells.
9. Refrigerated centrifuge for 15 and 50 ml conical tubes.
10. Flow cytometer (e.g., FACS-Canto II (BD)).
11. Appropriate data analysis software (e.g., FlowJo (Tristar)).

2.2 Monitoring ATP-Induced Ca^{2+} Influx in Transfected HEK293T Cells

1. Cells: transfected HEK293T cells, primary macrophages or T cells (*see* **Notes 4–6**).
2. 4 mM ATP stock solution in PBS pH 7.4, store in aliquots at −80 °C.
3. RPMI 1640 medium.
4. $PBS^{+/+}$ containing Ca^{2+} and Mg^{2+} pH 7.4 with 0.1% bovine serum albumin (BSA).
5. FACS buffer: PBS pH 7.4, 1 mM EDTA, 0.5% bovine serum albumin (BSA).
6. Ca^{2+} probe: Fluo4-acetoxymethyl (AM) ester (*see* **Note 8**).
7. Antibodies: monoclonal antibody anti-CD16/CD32 (2.4G2) as FcR-block, fluorochrome-conjugated monoclonal antibody against cell surface protein of interest (e.g., anti-CD11b (M1–70)), and P2X7-specific nanobodies (*see* **Note 5**).
8. Water bath.
9. Infrared lamp (*see* **Note 9**).

2.3 Monitoring ATP-Induced Uptake of DNA-Staining Dyes

1. Cells: transfected HEK293T cells, primary macrophages or T cells (*see* **Notes 4–6**).
2. RPMI 1640 medium.
3. FACS buffer: PBS pH 7.4, 1 mM EDTA, 0.5% bovine serum albumin (BSA).
4. NaCl buffer: 10 mM HEPES, 140 mM NaCl, 5 mM KCl in ddH_2O pH 7.4 for Yo-Pro1 uptake.
5. DNA staining dyes: 4′,6-diamidino-2-phenylindole (DAPI, MW 277.32 g/mol), alternatively Yo-Pro1 iodide (MW 629.3216 g/mol), or ethidium bromide (EtBr, MW 394.32 g/mol).
6. Antibodies: monoclonal antibody anti-CD16/CD32 (2.4G2) as FcR-block, fluorochrome-conjugated monoclonal antibody against cell surface protein of interest (e.g., anti-CD11b (M1-70)) (*see* **Notes 1** and 7), and P2X7-specific nanobodies (*see* **Note 5**).
7. Water bath.
8. Infrared lamp (*see* **Note 9**).

2.4 Monitoring ATP-Induced Externalization of Phosphatidylserine

1. Cells: transfected HEK293T cells, primary macrophages, or T cells (*see* **Notes 4–6**).
2. 4 mM ATP stock solution in PBS pH 7.4, store in aliquots at −80 °C.
3. RPMI 1640 medium.
4. Annexin V binding buffer: 10 mM HEPES, 140 mM NaCl, 2.5 mM $CaCl_2$ in ddH_2O pH 7.4.
5. Phosphatidylserine detector: fluorochrome-conjugated Annexin-V.
6. Antibodies: monoclonal antibody anti-CD16/CD32 (2.4G2) as FcR-block, fluorochrome-conjugated monoclonal antibody against cell surface protein of interest (e.g., anti-CD4 (RM4–5) and anti-CD8 (RPA-T8)) (*see* **Notes 1** and **7**), and P2X7-specific nanobodies (*see* **Note 5**).
7. Water bath (*see* **Note 9**).

2.5 Monitoring ATP-Induced Ectodomain Shedding of Cell Surface Proteins

1. Cells: cotransfected HEK293T cells, primary macrophages, or T cells (*see* **Notes 4–6**).
2. 4 mM ATP stock solution in PBS pH 7.4, store in aliquots at −80 °C.
3. RPMI 1640 medium.
4. FACS buffer: PBS pH 7.4, 1 mM EDTA, 0.5% bovine serum albumin (BSA).
5. Antibodies: monoclonal antibody anti-CD16/CD32 (2.4G2) as FcR-block, fluorochrome-conjugated monoclonal antibody against cell surface protein of interest (e.g., FITC-labeled anti-CD62L (MEL-14), and anti-CD27 (LG.3A 10)) (*see* **Notes 1** and 7), and P2X7-specific nanobodies (*see* **Note 5**).
6. Water bath (*see* **Note 9**).

2.6 Monitoring ATP-Induced Inflammasome Assembly (ASC Specks)

1. Cells: monocytes or macrophages.
2. 4 mM stock solution of ATP in PBS pH 7.4, store in aliquots at −80 °C.
3. PBS pH 7.4.
4. RPMI 1640 medium.
5. FACS buffer: PBS pH 7.4, 1 mM EDTA, 0.5% bovine serum albumin (BSA).
6. Permeabilization buffer: PBS pH 7.4, 0.2% saponin, 1% bovine serum albumin (BSA).
7. 2% paraformaldehyde solution (PFA) in PBS pH 7.4
8. 10 mM stock solution of nigericin.
9. 5 μg/ml stock solution of LPS.

10. Antibodies: monoclonal rabbit anti-ASC antibody (rb IgG, Santa Cruz Biotechnology), fluorochrome-conjugated secondary antibody (e.g., PE-conjugated anti-rabbit secondary antibody) (*see* **Notes 1** and 7), and P2X7-specific nanobodies (*see* **Note 5**).
11. Water bath (*see* **Note 9**).

2.7 Monitoring ATP-Induced Processing of IL-1β

1. Cells: monocytes or macrophages.
2. 4 mM stock solution of ATP in PBS pH 7,4, store in aliquots at −80 °C.
3. PBS pH 7.4.
4. RPMI 1640 medium.
5. FACS buffer: PBS pH 7.4, 1 mM EDTA, 0.5% bovine serum albumin (BSA).
6. Permeabilization buffer: PBS pH 7.4, 0.3% saponin, 1% bovine serum albumin (BSA).
7. 4% Paraformaldehyde solution (PFA) in PBS pH 7.4.
8. 5 μg/ml stock solution of LPS.
9. Antibodies: fluorochrome-conjugated monoclonal anti-proIL-1β antibody (NJTEN3), monoclonal antibody anti-CD16/CD32 (2.4G2) as FcR-block, fluorochrome-conjugated monoclonal antibody against cell surface protein of interest (e.g., anti-CD11b (M1–70), anti-Ly-6C (HK1.4), and anti-Ly-6G (1A8)) (*see* **Notes 1** and 7), and P2X7-specific nanobodies (*see* **Note 5**).
10. Water bath (*see* **Note 9**).

3 Methods

3.1 Monitoring Cell Surface Expression of Purinoreceptors by Flow Cytometry

1. Conjugate purinoreceptor-specific monoclonal antibody to a suitable fluorochrome (e.g., Alexa 488) according to the manufacturer's instructions.
2. Resuspend purinoreceptor-expressing lymphoma cells, transfected HEK293T cells or primary lymphocytes from peripheral blood, spleen, lymph nodes, etc. in FACS buffer at 2×10^7 cells/ml.
3. Transfer 50 μl aliquots of cells (1×10^6 cells) into 5 ml polystyrol tubes. In case of primary lymphocytes: centrifuge cells at $300 \times g$ for 10 min at 4 °C, resuspend cells in 50 μl twofold concentrated anti-CD16/CD32 FcR-block in FACS buffer and incubate for 10 min at 4 °C.
4. Prepare twofold concentrated cocktails of fluorochrome-conjugated antibodies in FACS buffer (e.g., anti-P2X7, anti-

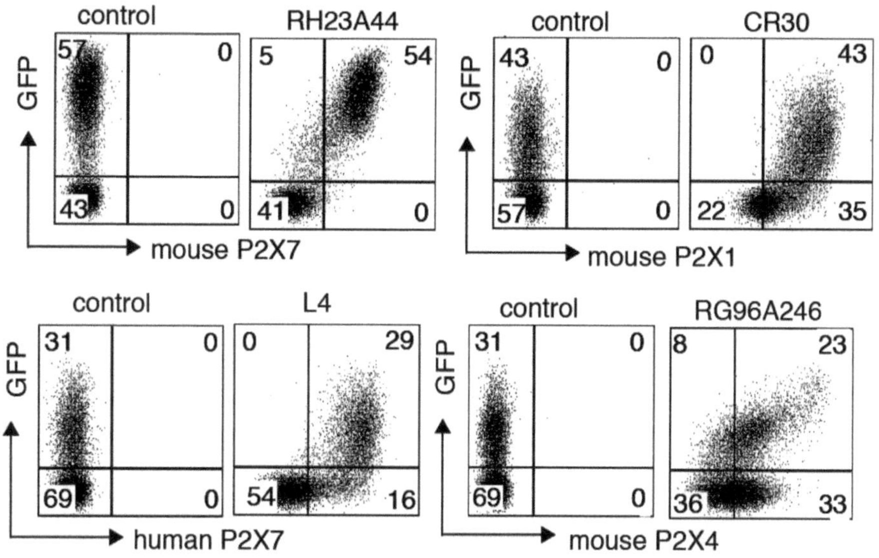

Fig. 1 Monitoring cell surface expression of P2X receptors on transiently transfected HEK293T cells. HEK293T cells were transiently cotransfected with expression constructs for green fluorescent protein (GFP) and either mouse P2X7, mouse P2X1, mouse P2X4, or human P2X7. Forty-eight hours post transfection, cells were harvested by trypsinization (*see* **Notes 4** and **5**) and stained with the indicated antibodies directed against mouse P2X7 (rat mAb RH23A44), mouse P2X1 (rabbit pAb CR30), mouse P2X4 (rat mAb RG96A246), or human P2X7 (mouse mAb L4) followed by appropriate Alexa 647-conjugated secondary antibodies. Control stainings were performed with secondary antibodies only. Flow cytometric analyses were performed with a BD FACS CantoII and the FlowJo software (Reproduced by permission of The American Association for the Advancement of Science ©2016 [Danquah et al., Sci. Transl. Med. 8, 366ra162, Fig. 1C])

CD3, anti-CD4, anti-CD8, and anti-CD25). Add 50 μl of antibody solution per tube and incubate in the dark for 30 min at 4 °C (*see* **Note 7**).

5. Wash samples twice with 2 ml FACS buffer by centrifugation at 300 × *g* for 10 min at 4 °C.
6. Resuspend cells in 250 μl FACS buffer, optional: add 10 μl propidium iodide for dead cell discrimination.
7. Analyze stained cells by flow cytometry and appropriate data analysis software (Figs. 1 and 2).

3.2 Monitoring ATP-Induced Ca^{2+} Influx in Transfected HEK293T Cells

1. Resuspend transfected HEK293T cells in RPMI 1640 medium at 2 × 10^7 cells/ml.
2. Transfer 50 μl aliquots of cells (1 × 10^6 cells) into 5 ml polystyrol tubes.
3. Prepare twofold concentrated Fluo4-AM/antibody mix solutions in RPMI 1640. Add 50 μl of Fluo4-AM/Ab mix solution per tube. Allow cells to load with Ca^{2+}-sensitive dye in the dark for 20 min at 37 °C (*see* **Note 8**).

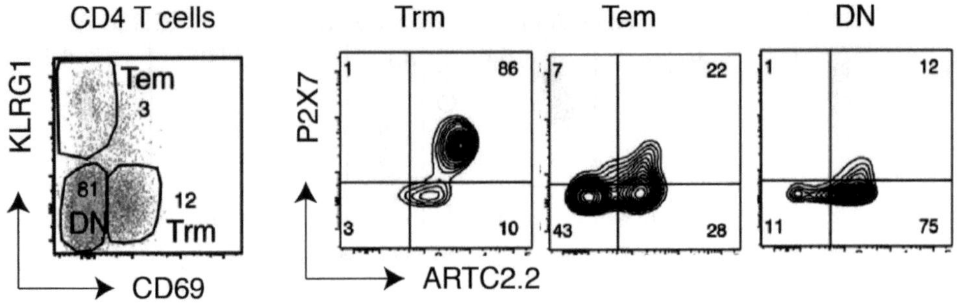

Fig. 2 Monitoring cell surface expression of P2X7 by mouse liver CD4 T cells. In order to prevent ADP-ribosylation of P2X7 in response to NAD^+ released during cell preparation, the ARTC2.2-blocking nanobody s + 16a (50 μg/100 μl NaCl) was injected intraperitoneally 30 min before sacrifice (*see* **Note 5**). In order to stain intravascular leukocytes, the PerCP-conjugated CD45-specific mAb (2 μg/100 μl NaCl) was injected intravenously 3 min before sacrifice (*see* **Note 6**). The liver was gently mashed through a metal sieve using a syringe piston, and liver leukocytes were enriched by Percoll density gradient centrifugation. Eythrocytes were lysed with ACK lysis buffer; cells were washed and resuspended in FACS buffer. Cells were stained with a panel of fluorochrome-conjugated antibodies directed against the following cell surface markers: ARTC2.2 (clone Nika109), CD3 (clone 145-2C11), CD4 (clone RM4–5), CD8 (clone 53-6.7), CD45 (clone 30-F11), CD69 (clone H1.2F3), KLRG1 (clone 2F1), and P2X7 (clone RH23A44) and PE-labeled CD1d-tetramer (PBS-57-loaded) (*see* **Note 7**). Gating was performed on helper T cells ($CD4^+/CD3^+/CD1d\text{-}tet^-$). Staining with KLRG1 and CD69 was used to identify effector memory T cells (Tem), resident memory T cells (Trm), and double negative naïve T cells (DN) (panel 1). Panels 2–4 show the cell surface expression of P2X7 and ARTC2.2 on the subpopulations of helper T cells. Flow cytometric analyses were performed with a BD FACS CantoII and the FlowJo software (Reproduced by permission of Frontiers in Immunology ©2018 [Rissiek et al. Front. Immunol. 9, 1580, Fig. 1])

4. Wash cells twice in 2 ml cold RPMI 1640 by centrifugation at 300 × *g* for 10 min at 4 °C.
5. Resuspend cells in 300 μl in cold $PBS^{+/+}$ with 0.1% BSA.
6. Keep samples on ice until measurement.
7. To monitor Ca^{2+} uptake in real time: adjust sample temperature to 37 °C in a water bath and maintain sample temperature at 37 °C by using an infrared lamp while measuring (*see* **Note 9**).
8. Equilibrate until baseline before adding 300 μl twofold concentrated ATP in warm $PBS^{+/+}$ (*see* **Note 10**) and continue measuring the sample by flow cytometry.
9. Analyze with appropriate data analysis software (Fig. 3).
10. For endpoint measurements, incubate cells in 300 μl $ATP/PBS^{+/+}$ solution with 0.1% BSA in the dark for 20 min at 37 °C. Add 300 μl of $PBS^{+/+}$ with 0.1% BSA and analyze cells by flow cytometry.

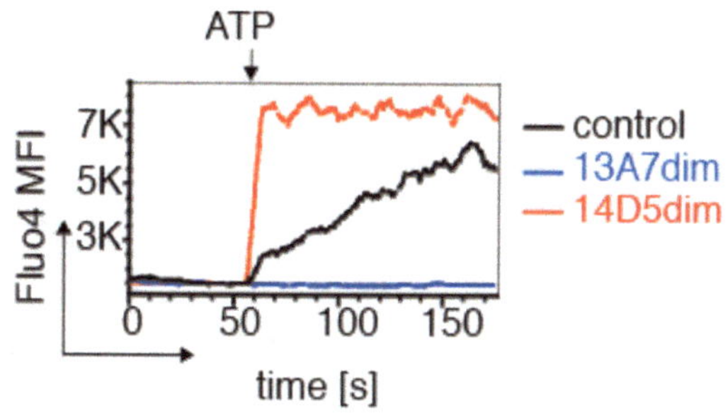

Fig. 3 Monitoring ATP-induced Ca^{2+} influx in mouse P2X7-transfected HEK293T cells. HEK293T cells stably transfected with mouse P2X7 were harvested by trypsinization, washed, resuspended in PBS and loaded with 2 μM of the Ca^{2+} indicator Fluo-4 for 20 min at 37 °C (*see* **Notes 4–6**). Cells were washed, resuspended in a FACS tube in FACS buffer without (control) or with 100 nM of the P2X7 antagonizing nanobody 13A7dim or the P2X7 potentiating nanobody 14D5dim. Cells were incubated for 10 min at 37 °C before real-time analysis by flow cytometry on a BD FACS-Canto II. An infrared lamp was used to maintain a constant sample temperature of 37 °C (*see* **Note 9**). After equilibration for 50 s, ATP was added to a final concentration of 1 mM (Reproduced by permission of The American Association for the Advancement of Science ©2016 [Danquah et al., Sci. Transl. Med. 8, 366ra162, Fig. 3B])

3.3 Monitoring ATP-Induced Uptake of DNA-Staining Dyes by Transfected HEK293T Cells or by Primary Macrophages

1. Resuspend P2X7-transfected HEK293T cells in RPMI 1640 medium at 2 × 10^7 cells/ml.
2. Transfer 50 μl aliquots of cells (1 × 10^6 cells) into 5 ml polystyrol tubes.
3. To monitor DAPI uptake in real time: adjust sample volume to 300 μl with RPMI 1640 medium and sample temperature to 37 °C in a water bath and maintain sample temperature at 37 °C by using an infrared lamp while measuring (*see* **Note 9**).
4. Equilibrate at baseline for 2 min before adding 300 μl twofold concentrated ATP in warm RPMI 1640 medium containing 1 μM final concentration of DAPI and continue measuring the sample by flow cytometry.
5. Analyze stained cells by appropriate data analysis software. Representative data for DNA-staining dye uptake by flow cytometry in real time *see* Fig. 4.
6. For endpoint measurements resuspend primary macrophages in 50 μl twofold concentrated FcR-block in FACS buffer and incubate for 10 min at 4 °C.
7. Add 50 μl twofold concentrated mixture of fluorochrome-conjugated anti-CD11b and incubate further in the dark for 30 min at 4 °C.
8. Wash cells twice in 2 ml cold FACS buffer by centrifugation at 300 × *g* for 10 min at 4 °C.

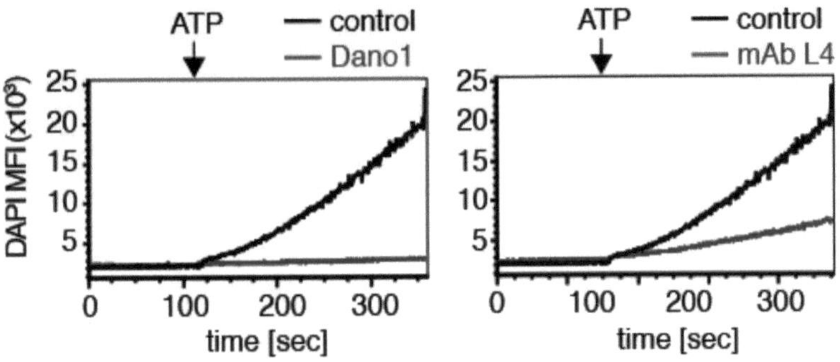

Fig. 4 Monitoring ATP-induced uptake of the DNA-staining dye DAPI by human P2X7 transfected HEK293T cells. HEK293T cells stably transfected with human P2X7 were harvested by trypsinization, washed, resuspended in a FACS tube in FACS buffer without (control) or with 100 nM of the P2X7 antagonizing nanobody Dano1 or the P2X7 antagonizing monoclonal antibody L4. Cells were incubated for 10 min at 37 °C before real time analysis by flow cytometry on a BD FACS-Canto II. An infrared lamp was used to maintain a constant sample temperature of 37 °C (*see* **Note 9**). Cells were equilibrated for 2 min before addition of 4 mM ATP and 1 μM DAPI (Reproduced by permission of The American Association for the Advancement of Science ©2016 [Danquah et al., Sci. Transl. Med. 8, 366ra162, Fig. 6B])

9. Resuspend cells were in RPMI 1640 medium containing 1 μM DAPI in the absence or presence of 1 mM ATP and incubated for 20 min at 37 °C.
10. Wash cells twice in 2 ml cold FACS buffer by centrifugation at 300 × *g* for 10 min at 4 °C.
11. Resuspend cells in 250 μl cold FACS buffer and analyze cells by flow cytometry. Representative data for end point measurement *see* Fig. 5 [13].

3.4 Monitoring ATP-Induced Externalization of Phosphatidylserine

1. Resuspend purinoreceptor-expressing primary lymphocytes from peripheral blood, spleen, lymph nodes, etc. in RPMI 1640 medium at 2 × 10^7 cells/ml.
2. Transfer 50 μl aliquots of cells (1 × 10^6 cells) into 5 ml polystyrol tubes.
3. Add 50 μl of twofold concentrated ATP in RPMI 1640 solution to cells.
4. Incubate cells for 20 min at 37 °C.
5. Wash cells twice in 2 ml cold Annexin V binding buffer by centrifugation at 300 × *g* for 10 min at 4 °C (*see* **Note 11**).
6. Resuspend cells in 100 μl cold Annexin V binding buffer containing 1 μg/ml fluorochrome-conjugated Annexin V. Optional: Additional antibody staining: resuspend the cells in 50 μl of Annexin V binding buffer containing twofold concentrated FcR-block and incubate for 10 min at 4 °C, before

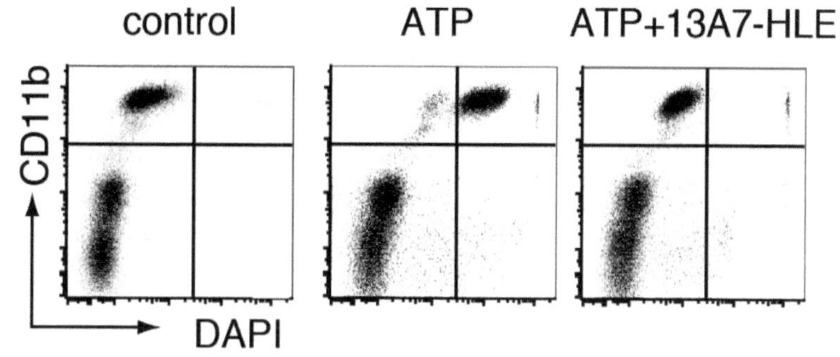

Fig. 5 Monitoring ATP-induced uptake of the DNA-staining dye DAPI by mouse peritoneal macrophages. The peritoneal cavity of a C57BL/6 mouse was flushed with 5 ml PBS to collect peritoneal macrophages. Cells were washed, resuspended in RPMI medium and stained with fluorochrome-conjugated anti-CD11b (clone M1-70) in the absence or presence of P2X7-antagonistic nanobody 13A7-HLE for 15 min at 4 °C. Cells were then further incubated for 20 min at 37 °C in RPMI containing 1 μM DAPI in the absence (control) or presence of 1 mM ATP. Cells were washed and analyzed by flow cytometry. Flow cytometric analyses were performed with a BD FACS Canto II and the FlowJo software (Reproduced by permission of The American Association for the Advancement of Science ©2016 [Danquah et al., Sci. Transl. Med. 8, 366ra162, Fig. S3A])

adding 50 μl of twofold concentrated fluorochrome-conjugated Annexin V in Annexin V binding buffer and additional antibodies (e.g., lineage markers like anti-CD4) for phenotyping the cells.

7. Incubate in the dark for 20 min at 4 °C.
8. Wash cells with 2 ml Annexin V binding buffer by centrifugation at 300 × *g* for 10 min at 4 °C.
9. Add 250 μl cold Annexin V binding buffer, add 10 μl of 20 μg/ml propidium iodide for dead cell discrimination.
10. Analyze stained cells by flow cytometry and appropriate data analysis software. *See* representative results in Figs. 6 and 7 [2, 13].

3.5 Monitoring ATP-Induced Ectodomain Shedding of Cell Surface Proteins

1. Resuspend receptor-expressing transfected or primary cells in RPMI 1640 medium at 2 × 10^7 cells/ml.
2. Transfer 50 μl aliquots of cells (1 × 10^6 cells) into 5 ml polystyrol tubes.
3. Add 50 μl twofold concentrated ATP in RPMI 1640 medium to cells (*see* **Note 12**).
4. Incubate for 20 min at 37 °C.
5. Wash cells twice in 2 ml cold FACS buffer by centrifugation at 300 × *g* for 10 min at 4 °C.

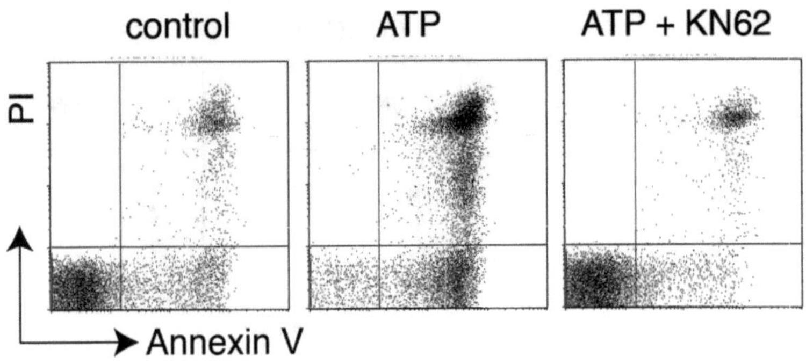

Fig. 6 Monitoring ATP-induced externalization of phosphatidylserine and uptake of the DNA staining dye propidium iodide by mouse lymph node T cells. Lymph nodes of a BALB/c mouse were gently dissected in cold (4 °C) RPMI medium and passed through a Nytex membrane (125 μm mesh, Tetko) using a syringe piston (*see* **Note 12**). Cells were washed and resuspended in FACS buffer containing magnetic beads (1 μm Dynabeads) coated with sheep anti-mouse IgG. Clustered B cells were removed by inserting the tube into a magnetic holder and carefully collecting cells remaining in suspension. Purified T cells were incubated without (control) or with 250 μM ATP for 30 min at 37 °C in the absence or presence of 10 μM of the P2X7 antagonist KN62. Cells were washed and stained with fluorochrome-conjugated Annexin-V and propidium iodide (PI) before FACS analysis (*see* **Note 11**) (Reproduced by permission of The American Association of Immunologists ©2009 [Scheuplein et al., J Immunology 182: 2898–2908, Fig. 1A])

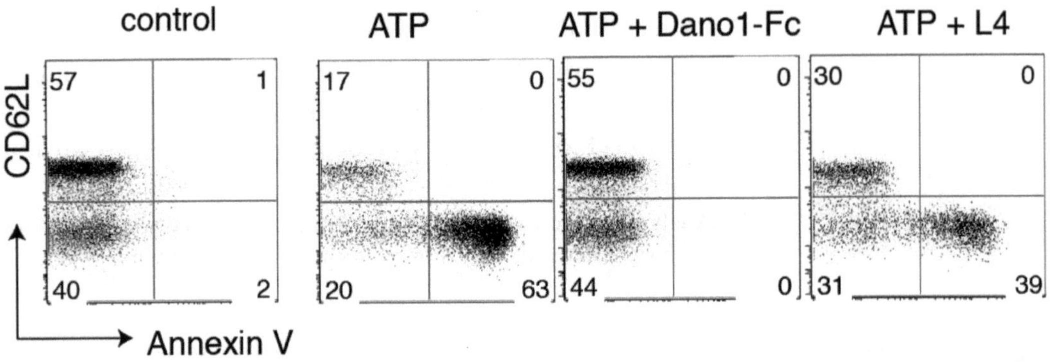

Fig. 7 Monitoring ATP-induced externalization of phosphatidylserine and ectodomain shedding of CD62L by human peripheral blood CD4 T cells. Peripheral blood mononuclear cells were purified from heparinized human blood by Ficoll density gradient centrifugation (*see* **Note 10**). Cells were washed, resuspended in RPMI medium and incubated for 20 min at 4 °C in the absence or presence of 100 nM of the P2X7 antagonizing nanobody Dano1-Fc or the monoclonal antibody L4, before addition of 4 mM ATP (final concentration). Cells were incubated for 20 min at 37 °C, washed, resuspended in FACS buffer containing 2 mM Ca^{2+} and stained with fluorochrome-conjugated Annexin V and fluorochrome-conjugated antibodies directed against CD4 (clone RM4-5) and CD62L (clone MEL-14). Flow cytometric analyses were performed with a BD FACS CantoII and the FlowJo software. Gating was performed on $CD4^+$ helper T cells. Numbers indicate percentage of cells in the respective quadrants (Reproduced by permission of The American Association for the Advancement of Science ©2016 [Danquah et al., Sci. Transl. Med. 8, 366ra162, Fig. 7C])

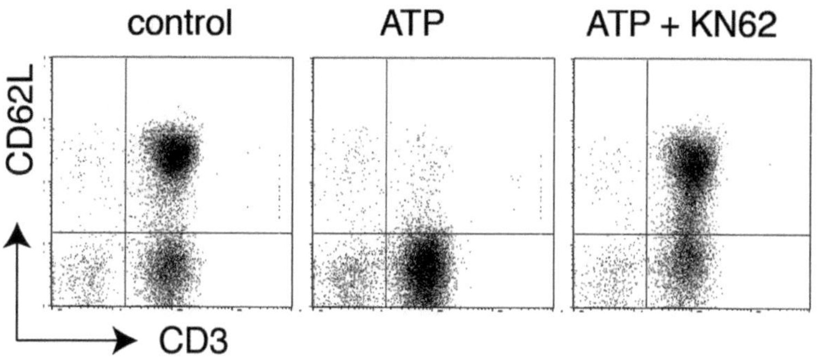

Fig. 8 Monitoring ATP-induced ectodomain shedding of CD62L by mouse lymph node T cells. T cells were prepared from lymph nodes of a BALB/c mouse as described in Fig. 6. Cells were incubated without (control) or with 250 μM ATP for 30 min at 37 °C in the absence or presence of 10 μM KN62. Cells were washed and stained with fluorochrome-conjugated antibodies directed against CD3 and CD62L before FACS analysis (Reproduced by permission of The American Association of Immunologists ©2009 [Scheuplein et al., J Immunology 182: 2898–2908, Fig. 1B])

6. Resuspend cells in 50 μl twofold concentrated FcR-block in FACS buffer and incubate for 10 min at 4 °C.
7. Prepare twofold concentrated antibody mix in FACS buffer (e.g., fluorochrome-conjugated anti-CD62L or anti-CD27) and other fluorochrome-conjugated antibodies for further phenotyping the cells. Add 50 μl cold antibody solution per tube.
8. Incubate in the dark for 30 min at 4 °C.
9. Wash cells twice in 2 ml FACS buffer by centrifugation at 300 × *g* for 10 min at 4 °C.
10. Resuspend cells in 250 μl FACS buffer, optional: add 10 μl of 20 μg/ml propidium iodide for dead cell discrimination.
11. Analyze stained cells by flow cytometry and appropriate data analysis software. *See* representative results in Figs. 7 and 8).

3.6 Monitoring ATP-Induced Assembly of the Inflammasome by Primary Monocytes

1. Resuspend human PBMC in RPMI 1640 medium at 1×10^7 cells/ml.
2. Transfer 100 μl aliquots of cells (1×10^6 cells) into 5 ml polystyrol tubes.
3. Add 100 μl twofold concentrated LPS (final concentration 1 μg/ml) in RPMI 1640 medium to the cells, incubate the cells for 2 h at 37 °C. Prepare samples without LPS as controls (*see* **Note 13**).
4. Add 200 μl of twofold concentrated ATP (final concentration 1.5 mM) in RPMI 1640 and incubate further for 30 min at 37 °C. Prepare samples without ATP as negative control and with nigericin as positive control (*see* **Note 14**).

5. Wash cell twice in 2 ml cold PBS by centrifugation at 300 × *g* for 10 min at 4 °C.
6. Resuspend cells in 100 μl cold RPMI, add 100 μl of 2% PFA, and incubate for 10 min at 4 °C.
7. Wash cell twice in 2 ml cold FACS buffer by centrifugation at 300 × *g* for 10 min at 4 °C.
8. Resuspend cells in 200 μl permeabilization buffer containing rabbit anti-ASC antibody and incubate for 90 min at 4 °C (*see* **Note 15**).
9. Wash twice in 2 ml cold FACS buffer by centrifugation at 300 × *g* for 10 min at 4 °C.
10. Resuspend cells in FACS buffer containing PE-conjugated anti-rabbit IgG and Bv421-conjugated anti-CD33 (Biolegend) and incubate in the dark for 45 min at 4 °C (*see* **Note 16**).
11. Wash cells once in 2 ml cold FACS buffer by centrifugation at 300 × *g* for 10 min at 4 °C.
12. Resuspend cells in 250 μl of cold FACS buffer.
13. Analyze stained cells by flow cytometry and appropriate data analysis software. *See* representative results for the formation of apoptosis associated speck like protein containing a caspase recruitment domain (ASC) specks formation by flow cytometry in Fig. 9 [13, 15].

3.7 Monitoring ATP-Induced Processing of Pro-IL-1β by Murine Macrophages

1. Aliquot fresh samples of heparinized blood into 50 μl samples (*see* **Note 17**).
2. Add 50 μl twofold concentrated LPS (final concentration 1 μg/ml) in RPMI 1640 medium and incubate samples for 2 h at 37 °C. Prepare parallel samples without LPS as controls (*see* **Note 18**).
3. Add 100 μl of twofold concentrated ATP (final concentration 1.5 mM) in RPMI 1640 and incubate further for 30 min at 37 °C. Prepare parallel samples without ATP as controls.
4. Wash cells twice in 2 ml cold PBS by centrifugation at 300 × *g* for 10 min at 4 °C.
5. Resuspend cells in 100 μl cold RPMI and add 100 μl of 4% PFA, incubate for 10 min at 4 °C.
6. Wash cells twice in 2 ml cold FACS buffer by centrifugation at 300 × *g* for 10 min at 4 °C.
7. Resuspend cells in 200 μl permeabilization buffer and incubate for 1 h at 4 °C.
8. Wash cells twice in 2 ml cold FACS buffer by centrifugation at 300 × *g* for 10 min at 4 °C.

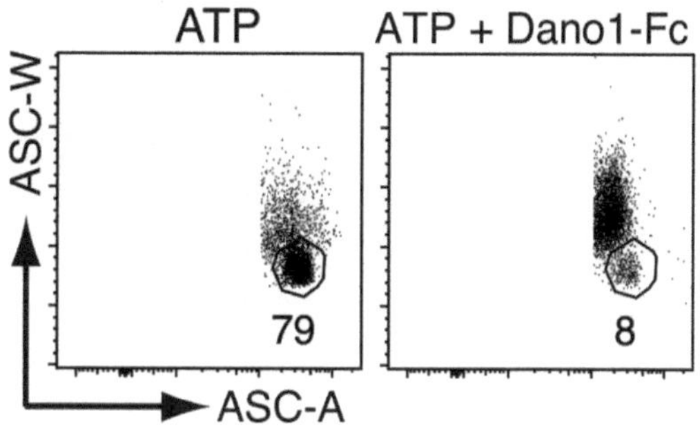

Fig. 9 Monitoring ATP-induced assembly of the inflammasome by human blood monocytes. Peripheral blood mononuclear cells were purified from human blood by Ficoll density gradient centrifugation (*see* **Note 10**), washed, and resuspended in RPMI medium. Cells were incubated for 2 h with 1 μM LPS and then further incubated for 30 min with 4 mM ATP in the absence or presence of 200 nM Dano1-Fc or control Nb-Fc. Cells were fixed with 2% paraformaldehyde, permeabilized with 0.2% saponin, and stained with rabbit anti-ASC primary antibody for 90 min at 4 °C. Cells were washed and stained for 45 min at 4 °C with PE-conjugated anti-rabbit secondary antibody and Bv421-conjugated anti-CD33. Stained cells were washed once and ASC oligomerization was measured by pulse width to pulse area profile analysis (*see* **Note 13**). Flow cytometric analyses were performed with a BD FACS Canto II and the FlowJo software. Gating was performed on CD33+ cells (Reproduced by permission of The American Association for the Advancement of Science ©2016 [Danquah et al., Sci. Transl. Med. 8, 366ra162, Fig. 7A])

9. Resuspend cells in 50 μl twofold concentrated FcR-block in FACS buffer for 10 min at 4 °C, add twofold concentrated fluorochrome-conjugated monoclonal anti-pro-IL-1β (*see* **Note 19**), anti-CD11b, anti-Ly-6C, and anti-Ly-6G, and further incubate in the dark for 30 min at 4 °C.
10. Wash cells twice in 2 ml cold FACS buffer by centrifugation at 300 × *g* for 10 min at 4 °C.
11. Resuspend cells in 250 μl of cold FACS buffer.
12. Analyze stained cells by flow cytometry and appropriate data analysis software. For representative results *see* Fig. 10 [13].

4 Notes

1. For monitoring expression of purinoreceptors use appropriate, validated mAbs, (e.g., anti-human P2X7 (L4), anti-mouse P2X7 (Hano44), anti-mouse P2X4 (RG96A246)) or validated polyclonal antibodies (e.g., anti-mouse P2X1 (CR30)). P1- and P2 purinoreceptors are expressed as transmembrane receptors that can be detected by following an extracellular staining protocol. Use primary antibody directly conjugated to

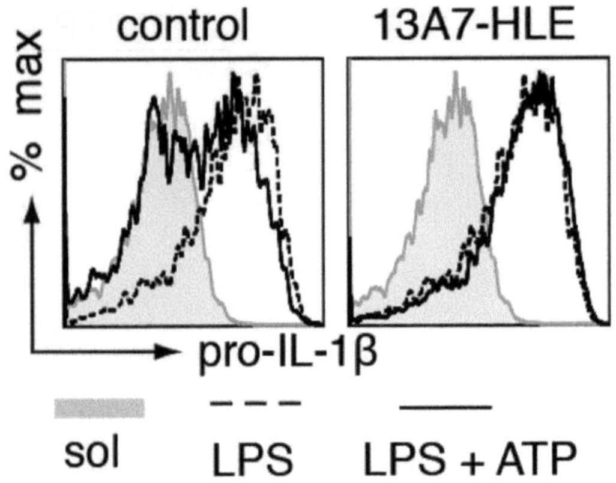

Fig. 10 Monitoring ATP-induced processing of pro-IL-1β by LPS-primed mouse monocytes. Blood was collected into a heparin-containing tube following incision of the jugular vein of a C57BL/6 mouse (*see* **Note 17**). Aliquots of heparinized blood were incubated for 2 h without (solvent) or with 1 μM LPS in the absence (control) or presence of the P2X7 antagonizing nanobody 13A7-HLE (*see* **Note 18**). ATP was added as indicated to a final concentration of 1 mM and samples were further incubated for 30 min at 37 °C. Erythrocytes were lysed with ACK lysis buffer and cells were fixed in 4% paraformaldehyde and permeabilized with 0.3% saponin, before staining with fluorochrome-conjugated antibodies directed against CD11b (M1-70), Ly-6C (HK1.4), and anti-pro-IL-1β (NJTEN3) (*see* **Note 19**). Flow cytometric analyses were performed with a BD FACS Canto II and the FlowJo software. Gating was performed on $CD11b^{+}Ly\text{-}6C^{hi}$ monocytes (Reproduced by permission of The American Association for the Advancement of Science ©2016 [Danquah et al., Sci. Transl. Med. 8, 366ra162, Fig. S3C])

fluorochromes like -AF488 or -AF647 antibody labeling kits, or detect primary antibody via a secondary fluorochrome-conjugated species-specific anti IgG antibody before analysis.

2. The P2X7 open reading frame encompasses 1583 nucleotides. It is useful to add a consensus Kozak sequence immediately upstream of the start codon.
3. Transfection of HEK293T cells using jetPEI typically results in 40–80% of cells expressing P2X7 by 24–48 h posttransfection. Transfection of lymphoma cells by jetPEI or electroporation typically results in much lower numbers of transfected cells (<1% of cells). For such cells, selection of stable transfectants by cultivation in an appropriate selection medium is recommended. The efficiency of stable transfection can be improved by linearizing the expression vector using a restriction enzyme that cuts outside of the expression cassette (e.g., in the antibiotic resistance gene used for propagating the plasmid in *E. coli* (usually ampR)).
4. The k splice variant of murine P2X7 can be activated by ADP-ribosylation, a covalent modification catalyzed by ecto-

ADP ribosyltransferase ARTC2.2 [17–19]. Using extracellular NAD^+ as substrate, ARTC2.2 catalyzes the attachment of an ADP-ribose moiety to Arg125, while nicotinamide is liberated. Even low concentrations of NAD^+ released during passaging of tissue culture cells or during preparation of primary cells from dissociated tissues are sufficient to induce ARTC2.2-mediated activation of P2X7. This can induce proteolytic cleavage of CD62L and other metalloprotease-sensitive cell surface proteins, externalization of phosphatidylserine and cell death. The ARTC2.2 blocking nanobody s + 16a can prevent ADP-ribosylation of P2X7 [3, 14]; the P2X7-blocking nanobody 13A7 and the small molecule P2X7 antagonists KN-62 and A438079 can prevent activation of ADP-ribosylated P2X7 [5, 13] (*see* **Note 5**).

5. KN-62 and A438079 are small-molecule P2X7-antagonists that prevent gating of ADP-ribosylated P2X7 at 37 °C as well as gating of P2X7 by its soluble extracellular ligand ATP [5, 13]. Nanobodies (Nbs) are highly soluble variable domains (VHH) derived from camelid heavy chain antibodies. Nb 13A7 and Nb Dano1 are potent, reversible antagonists of mouse and human P2X7, respectively. Nb 14D5 potentiates gating of mouse P2X7 by ATP and by ADP-ribosylation [13]. Nb s + 16a is a potent, reversible inhibitor of the ARTC2.2 ectoenzyme that effectively blocks ADP-ribosylation of cell surface proteins during cell preparation in response to NAD^+ released during cell preparation [3, 14]. To block ARTC2.2-catalyzed ADP-ribosylation of P2X7 during passaging of cultured cells, Nb s + 16a (0.1 μg/ml) is added to the buffers used for trypsinization and centrifugation. To block ADP-ribosylation of P2X7 during preparation of primary cells, Nb s + 16a (50 μg/mouse) is injected 30–60 min before sacrifice.
6. Injection of αCD45 permits specific labeling of intravascular leucocytes [16]. The intravenously administered antibody rapidly distributes in the vasculature and binds to vascular leukocytes. The endothelium, basement membrane and interstitial pressure impede access of the antibody to parenchymal leukocytes. A too high dose of antibody and/or too long "incubation" time in vivo can result in staining of interstitial leukocytes.
7. A careful panel design is substantial for generating valid multicolor flow cytometry data [1]. In general, select fluorochromes that match the expression levels of antigens; for example, a highly expressed lineage marker like CD4 should be detected with a dim or intermediate fluorochrome conjugate while a lowly expressed antigen should be detected with a bright fluorochrome. Select fluorochrome conjugates that do not need any or low compensation due to emission spillover when

detecting antigens on same cell. Spillover is due to the overlap of emission spectra of fluorochromes and can be compensated. For compensation controls, generate single color samples of each fluorochrome conjugate in your experiment. Compensation controls can be either stained/unstained cells or stained/unstained beads. Please note that fluorochromes are sensitive toward light. Keep your samples in the dark and on ice until direct measurement with flow cytometer. If a primary antibody without fluorochrome is used, incubation may be done in the light.

8. Fluo-4-acetylmethyl (AM) ester is a precursor of the calcium indicator Fluo-4. Cells are passively loaded with 2 μM Fluo-4-AM ester. Intracellular esterases cleave Fluo-4-AM ester to Fluo-4 within 20 min at 37 °C. Fluorescein-conjugated antibodies should be avoided when Fluo-4 or related Ca^{2+} probes are used due to the similar excitation and emission spectra.
9. Gating of the P2X7 receptor occurs at physiological temperatures. To reach and maintain 37 °C, perform all ATP incubations in a water bath. Preheat all buffers to 37 °C. In order to maintain a constant temperature of 37 °C during real time measurement in the flow cytometer, place an infrared lamp at the appropriate distance (approximately 30–40 cm) from the FACS tube during measurement. The appropriate distance should be determined previously with a FACS tube filled with FACS buffer. Preheat all buffers to 37 °C.
10. The ligand-gated P2X receptors respond to extracellular ATP. The P2X7 receptor requires relatively high concentrations of ATP (in the upper micromolar to lower millimolar range depending on the species, the cell type, and the splice variant of P2X7). Murine P2X7(a) splice variant is mainly expressed on monocytes/macrophages (EC_{50}: 500 μM), P2X7(k) splice variant is mainly expressed on T lymphocyte subclasses like Treg and NKT cells (EC_{50}: 100 μM). Allelic variants of human P2X7 exhibit different sensitivities of ATP (EC_{50}: 0.5–4 mM) [20].
11. Binding of Annexin V to phosphatidylserine is dependent on Ca^{2+}. Ensure that the staining buffer contains at least 1.5 mM Ca^{2+}.
12. Gating of P2X7 on T cells induces activation of membrane-anchored metalloproteases (ADAM17), resulting in cleavage and release of the ectodomains of CD62L, CD27, and other sensitive cell surface proteins [2, 17]. T cells of C57BL/6 mice are less sensitive to ATP-induced externalization of phosphatidylserine than T cells of BALB/c mice, owing to a lower cell surface expression of P2X7 and an allelic polymorphism (P451L) in the C-terminal cytosolic tail [19, 21].

13. The inflammasome is a large multiprotein complex composed of distinct proteins, including ASC. In many cells, there is a single complex. Staining of this complex with fluorochrome-conjugated α-ASC results in a highly concentrated fluorescent signal within the cell, resulting in a reduced pulse width and increased pulse area [15]. Detection of ASC speck depends on two signals to activate the NLRP3 inflammasome, an initial priming with LPS followed by P2X7-mediated efflux of potassium.
14. Nigericin is a microbial toxin derived from *Streptomyces hygroscopicus.* It acts as an ionophore for positively charged ions like K^+ and H^+. Nigericin induces the efflux of K^+ and is thereby a strong activator of NLRP3 inflammasome.
15. The cell membrane of living cells is not permeable to antibodies. In order to stain intracellular ASC, cells need to be fixed and permeabilized.
16. CD33 is a cell surface marker of human monocytes.
17. Blood can also be obtained by puncture of the retrobulbar venous plexus or left ventricle of the heart. Smaller volumes of blood can be taken from the tail vein. Heparin is required to prevent clotting of blood.
18. Treatment with LPS stimulates production of immature proform of IL-1β. ATP-induced formation of the inflammasome activates caspase-1 and subsequent leads to caspase-1 mediated cleavage of pro-IL-1β to mature IL-1β.
19. Monoclonal antibody NJTEN3 recognizes mouse pro-IL-1β but not processed IL-1β. The cell membrane of living cells is not permeable to antibodies. In order to stain intracellular pro-IL-1β, cells need to be fixed and permeabilized.

Acknowledgments

The authors would like to acknowledge the EU COST Program (BM1406).

References

1. Cossarizza A, Chang HD, Radbruch A, Akdis M, Andra I, Annunziato F et al (2017) Guidelines for the use of flow cytometry and cell sorting in immunological studies. Eur J Immunol 47:1584–1797
2. Scheuplein F, Schwarz N, Adriouch S, Krebs C, Bannas P, Rissiek B et al (2009) NAD+ and ATP released from injured cells induce P2X7-dependent shedding of CD62L and externalization of phosphatidylserine by murine T cells. J Immunol 182:2898–2908
3. Rissiek B, Danquah W, Haag F, Koch-Nolte F (2014) Technical Advance: A new cell preparation strategy that greatly improves the yield of vital and functional Tregs and NKT cells. J Leukoc Biol 95:543–549
4. Rissiek B, Lukowiak M, Raczkowski F, Magnus T, Mittrucker HW, Koch-Nolte F

(2018) In vivo blockade of murine ARTC2.2 during cell preparation preserves the vitality and function of liver tissue-resident memory T cells. Front Immunol 9:1580

5. Georgiev H, Ravens I, Papadogianni G, Malissen B, Forster R, Bernhardt G (2018) Blocking the ART2.2/P2X7-system is essential to avoid a detrimental bias in functional CD4 T cell studies. Eur J Immunol 48:1078–1081
6. Bartlett R, Stokes L, Sluyter R (2014) The P2X7 receptor channel: recent developments and the use of P2X7 antagonists in models of disease. Pharmacol Rev 66:638–675
7. Rissiek B, Haag F, Boyer O, Koch-Nolte F, Adriouch S (2015) P2X7 on mouse T cells: one channel, many functions. Front Immunol 6:204
8. Di Virgilio F, Dal Ben D, Sarti AC, Giuliani AL, Falzoni S (2017) The P2X7 receptor in infection and inflammation. Immunity 47:15–31
9. Moller S, Jung C, Adriouch S, Dubberke G, Seyfried F, Seman M et al (2007) Monitoring the expression of purinoceptors and nucleotide-metabolizing ecto-enzymes with antibodies directed against proteins in native conformation. Purinergic Signal 3:359–366
10. Roncador G, Engel P, Maestre L, Anderson AP, Cordell JL, Cragg MS et al (2016) The European antibody network's practical guide to finding and validating suitable antibodies for research. MAbs 8:27–36
11. Gargett CE, Wiley JS (1997) The isoquinoline derivative KN-62 a potent antagonist of the P2Z-receptor of human lymphocytes. Br J Pharmacol 120:1483–1490
12. King BF (2007) Novel P2X7 receptor antagonists ease the pain. Br J Pharmacol 151:565–567
13. Danquah W, Meyer-Schwesinger C, Rissiek B, Pinto C, Serracant-Prat A, Amadi M et al (2016) Nanobodies that block gating of the P2X7 ion channel ameliorate inflammation. Sci Transl Med 8:366ra162
14. Koch-Nolte F, Reyelt J, Schossow B, Schwarz N, Scheuplein F, Rothenburg S et al (2007) Single domain antibodies from llama effectively and specifically block T cell ecto-ADP-ribosyltransferase ART2.2 in vivo. FASEB J 21:3490–3498
15. Sester DP, Thygesen SJ, Sagulenko V, Vajjhala PR, Cridland JA, Vitak N et al (2015) A novel flow cytometric method to assess inflammasome formation. J Immunol 194:455–462
16. Anderson KG, Mayer-Barber K, Sung H, Beura L, James BR, Taylor JJ et al (2014) Intravascular staining for discrimination of vascular and tissue leukocytes. Nat Protoc 9:209–222
17. Seman M, Adriouch S, Scheuplein F, Krebs C, Freese D, Glowacki G et al (2003) NAD-induced T cell death: ADP-ribosylation of cell surface proteins by ART2 activates the cytolytic P2X7 purinoceptor. Immunity 19:571–582
18. Schwarz N, Fliegert R, Adriouch S, Seman M, Guse AH, Haag F, Koch-Nolte F (2009) Activation of the P2X7 ion channel by soluble and covalently bound ligands. Purinergic Signal 5:139–149
19. Schwarz N, Drouot L, Nicke A, Fliegert R, Boyer O, Guse AH et al (2012) Alternative splicing of the N-terminal cytosolic and transmembrane domains of P2X7 controls gating of the ion channel by ADP-ribosylation. PLoS One 7:e41269
20. Fuller SJ, Stokes L, Skarratt KK, Gu BJ, Wiley JS (2009) Genetics of the P2X7 receptor and human disease. Purinergic Signal 5:257–262
21. Adriouch S, Dox C, Welge V, Seman M, Koch-Nolte F, Haag F (2002) Cutting edge: a natural P451L mutation in the cytoplasmic domain impairs the function of the mouse P2X7 receptor. J Immunol 169:4108–4112

Chapter 9

Studying Purinoceptor Cell-Surface Expression by Protein Biotinylation

Mark T. Young

Abstract

Covalent labeling of protein with biotin (biotinylation) is a versatile technique which enables the capture and analysis of labeled protein with streptavidin-coated beads. It is particularly useful for detecting and quantifying the cell-surface expression of membrane proteins and widely used to analyze protein trafficking and the effect of mutations (particularly those which render the protein nonfunctional) on cell-surface expression. Here I describe the procedure for biotinylation and capture of cell-surface rat P2X2 receptors expressed in mammalian cells, and outline the steps in data analysis required to measure the proportion of cell-surface expressed protein of single point mutants relative to a wild-type control.

Key words Cell-surface, Protein expression, Biotinylation, P2X, Western blot, Crosslinking, Membrane protein

1 Introduction

P2X receptors are a family of plasma membrane cation channels gated by extracellular ATP which play modulatory roles in a variety of physiological processes including pain sensation and inflammation, and are thus important drug targets [1]. In recent years, several crystal structures of P2X receptors have been published, transforming our understanding of their structure–function relationship and informing a new series of structure–function experiments [2]. The cell-surface localization and trafficking of P2X receptors is highly regulated and dependent on several motifs present within their intracellular domains [3]. Clearly, when studying either the structure–function relationship or the trafficking of P2X receptors, it is of paramount importance to measure the complement of protein present at the cell surface. This is because alterations to protein trafficking as a result of loss/changes in protein-protein interactions (e.g., in heterologous expression systems), or as a result of mutations, may have a significant bearing on both the biological relevance of the system and the interpretation of data

Pablo Pelegrín (ed.), *Purinergic Signaling: Methods and Protocols*, Methods in Molecular Biology, vol. 2041,
https://doi.org/10.1007/978-1-4939-9717-6_9, © Springer Science+Business Media, LLC, part of Springer Nature 2020

showing loss of function. For example, if a single point mutant shows reduced or no function in an electrophysiological assay, it could be because the mutant protein is poorly expressed at the cell surface due to altered trafficking, rather than less sensitive to stimulation. A number of different methods have been employed in order to detect P2X receptor cell-surface expression, including (1) imaging of intact versus permeabilized cells labeled with antibodies specific to extracellular domains of the protein, or indeed engineering epitope tags for commercially available antibodies into the extracellular domains [4, 5], (2) imaging cells expressing protein constructs containing environment-specific fluorescent reporter proteins within their extracellular domain [6], and (3) biotinylation and capture of cell-surface expressed protein using streptavidin beads [7–13]. The biotin–(strept)avidin interaction [14] has a high affinity and excellent specificity, and its versatility means it has been used for labeling many different types of macromolecules [15]. Many biotin derivatives have been developed, including Sulfo-NHS-LC-biotin, which is cell-impermeant and readily labels the primary amine groups present on proteins in solution, which can be used along with streptavidin-bound sepharose beads to enable biotin capture [15]. Biotinylation also has the advantage that the protein need not be additionally labeled, and thus can be carried out on endogenously expressed protein. This protocol describes the main steps involved in carrying out a biotinylation reaction, including biotin labeling, streptavidin capture, the preparation of lysate samples for both total and cell-surface protein and Western blotting to detect protein. In addition, the protocol outlines the quantitation of the ratio of cell surface–total protein from recorded band densities.

2 Materials

Prepare all solutions using ultrapure water and analytical grade reagents. Prepare and store all reagents at room temperature, unless indicated otherwise. For the experiments presented in this protocol human embryonic kidney (HEK293T) cells transiently transfected with rat P2X2 receptor wild-type and mutant constructs bearing a C-terminal Glu-Glu (EYMPME) epitope tag were used (*see* [8, 10] for experimental methods relating to cell culture and transfection and functional analyses of the constructs). Each sample was prepared from 1× 35 mm dish of cells (plated at a density of 5×10^5 cells/dish and transfected with 1 μg of plasmid containing the desired cDNA 48 h prior to biotinylation).

2.1 Biotin Labeling

1. Phosphate-buffered saline (PBS) pH 8.0: 137 mM NaCl, 2.7 mM KCl, 10 mM Na_2HPO_4, 1.8 mM KH_2PO_4 (pH 7.4)

made up to pH 8.0 by dropwise addition of 5 M NaOH and monitored using a pH meter (*see* **Note 1**). Store on ice.

2. Biotin solution: 0.5 mg/ml EZ-link Sulfo-LC-NHS Biotin dissolved in PBS pH 8.0 (*see* **Note 2**). Store on ice.
3. 10× TG: 0.25 M Tris, 1.92 M glycine pH 8.5 (*see* **Note 3**).
4. PBS-TG: Mix 9 volumes of PBS pH 8.0 (prepared in **item 1** above) with 1 volume of 10× TG (prepared in **item 3** above). Store on ice.
5. 1.5 ml microcentrifuge tubes.
6. Benchtop microcentrifuge for 1.5 ml microcentrifuge tubes.

2.2 Cell Lysis and Protein Assay

1. Solubilization buffer: 150 mM NaCl, 20 mM Tris–HCl pH 7.4, 1 mM $MgCl_2$, 1 mM $CaCl_2$, 2% (v/v) Triton X-100, protease inhibitors (*see* **Note 4**). Store on ice.
2. Protein assay reagent or kit (e.g., Bradford assay).
3. Bovine serum albumin (BSA) (Fraction V).
4. 96-Well microtiter plate or disposable half-micro cuvettes.
5. Absorbance plate-reader or visible light spectrophotometer.
6. Rotating wheel for 1.5 ml microcentrifuge tubes.

2.3 Streptavidin Capture

1. Streptavidin agarose.
2. Solubilization buffer (*see* Subheading 2.2, **item 1**).
3. 1–200 μl Gel-loading micropipette tips.

2.4 SDS-PAGE and Western Blotting

1. 6× SDS-PAGE sample buffer: 240 mM Tris–HCl pH 6.8, 12% (w/v) SDS, 30% glycerol, 0.0018% (w/v) bromophenol blue.
2. 100 °C heating block.
3. Precast SDS-PAGE gels (*see* **Note 5**).
4. Transfer packs for Western blotting (*see* **Note 6**).
5. PVDF or nitrocellulose membranes.
6. TBS-T solution: 20 mM Tris–HCl pH 7.4, 150 mM NaCl, 0.1% (v/v) Tween 20.
7. Blocking solution: TBS-T containing 5% (w/v) nonfat milk powder (*see* **Note 7**).

2.5 Antibodies

1. Anti-Glu-Glu primary antibody.
2. Conjugated secondary antibody (*see* **Note 8**).
3. Blot detection/development system (*see* **Note 8**).

3 Methods

Carry out all procedures in Subheadings 3.1–3.3 on ice or at 4 °C. Procedures in subsequent sections should be carried out at room temperature, unless specified otherwise.

3.1 Biotin Labeling

1. Place dishes of cells on ice (*see* **Note 9**). Remove media.
2. Gently wash cells with 3× 1 ml ice-cold PBS pH 8.0, taking care not to detach cells by pipetting gently onto the side of the dish.
3. Add 1 ml biotin solution to each dish. Swirl gently and incubate on ice for 30 min.
4. Remove biotin solution. Wash cells with 2× 1 ml PBS-TG to quench the crosslinking reaction (*see* **Note 10**). Discard PBS-TG.
5. Add 1 ml PBS-TG. Gently pipet up and down to resuspend cells (*see* **Note 11**) and transfer cell suspension to a prechilled 1.5 ml microcentrifuge tube on ice.
6. Centrifuge at 4000 × *g* for 3 min in a benchtop microcentrifuge to pellet cells. Discard supernatant (*see* **Note 12**).

3.2 Cell Lysis and Protein Assay

1. Add 50 μl solubilization buffer to each sample. Mix carefully by pipetting up and down taking care to minimize bubble formation. Incubate samples on a rotating wheel for 1 h at 4 °C.
2. Centrifuge at 12,000 × *g* for 2 min to pellet cell debris. Transfer supernatant (solubilized total protein) to fresh tubes.
3. Assay samples for protein content (*see* **Note 13**). Take 20 μg total protein samples and store on ice or at −20 °C.

3.3 Streptavidin Capture

1. Resuspend streptavidin agarose beads by gentle vortex mixing. Using a P1000 pipette with a cutoff tip (*see* **Note 14**), transfer the required amount of beads into a 1.5 ml microcentrifuge tube. For eight samples take 400 μl.
2. Centrifuge the beads at 5000 rpm (no greater than 5000 × *g*) for 30 s in a benchtop microcentrifuge (*see* **Note 15**). Remove supernatant and purge beads using a gel-loading tip (push the tip through the beads to the bottom of the tube and remove all liquid).
3. Wash step: Add 500 μl solubilization buffer, cap tube and invert 20× to mix. Pellet the beads as in **step 2** above and discard supernatant. Purge the beads and repeat the wash step.
4. Resuspend beads to 50:50 (v/v) with solubilization buffer (add 200 μl for eight samples).

5. Using a cutoff P200 pipette tip (*see* **Note 14**), transfer 25 μl bead suspension per sample to a fresh 1.5 ml microcentrifuge tube. Add 500 μl solubilization buffer.
6. Add 300 μg of protein sample to the bead suspension. Incubate overnight at 4 °C on a rotating wheel to bind biotinylated protein to the beads.
7. Pellet the beads as in **step 2** above, remove supernatant and purge, taking care not to lose beads (*see* **Note 16**).
8. Wash step: Add 500 μl solubilization buffer, cap tube and invert 20× to mix. Pellet the beads as in **step 2** above and discard supernatant. Purge the beads. Repeat the wash step twice more for a total of three washes.

3.4 SDS-PAGE and Western Blotting

1. Add 25 μl 6× SDS-PAGE sample buffer to the beads and heat at 100 °C for 5 min to dissociate biotinylated protein.
2. Add 10 μl 6× SDS-PAGE sample buffer to each total protein sample and heat at 100 °C for 2 min.
3. Centrifuge samples at 12,000 × *g* for 2 min to pellet insoluble material.
4. Load entire sample on to precast polyacrylamide gels and separate proteins according to the manufacturer's instructions (*see* **Note 5**). Ensure that a positive control is loaded onto each gel to enable comparison across different experiments.
5. Transfer proteins on to PVDF or nitrocellulose membranes according to the manufacturer's instructions (*see* **Note 6**).
6. Incubate membranes in blocking solution for 1 h with gentle agitation on a rocking platform (*see* **Note** 7).
7. Incubate membranes in an appropriate dilution of primary antibody in blocking solution for 1 h with gentle agitation on a rocking platform (*see* **Note 17**).
8. Wash membranes 3× 5 min in TBS-T with gentle agitation on a rocking platform.
9. Incubate membranes in an appropriate dilution of secondary antibody in blocking solution for 1 h with gentle agitation on a rocking platform (*see* **Note 8**).
10. Wash membranes 4× 5 min in TBS-T with gentle agitation on a rocking platform.
11. Develop Western blots (if necessary) and record images (*see* **Note 8**). Example Western blots for both total (Fig. 1, top) and cell-surface (Fig. 1, bottom) protein are shown. Negative controls (−biotin) are shown on the left hand side to indicate that there is no nonspecific binding of P2X2 to the streptavidin beads. Note the differences in the band profiles between total and cell-surface protein for each sample (most clear for the

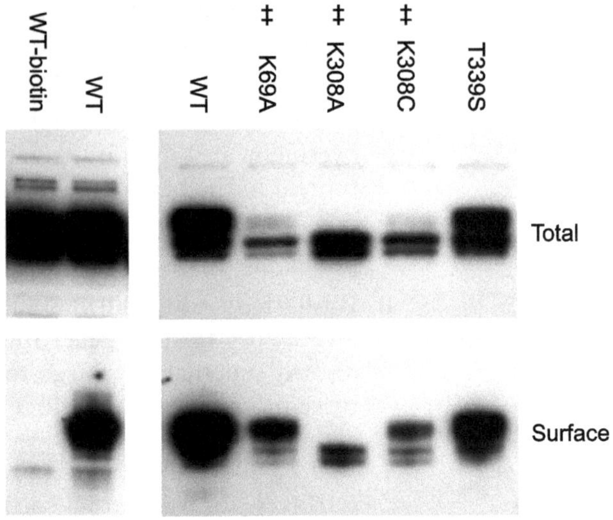

Fig. 1 Western blots (anti-EE antibody) of total and cell-surface expressed P2X2 protein following biotinylation. Top, total protein. Bottom, cell-surface protein. WT, wild-type receptor; −biotin, negative control where no biotin was included in the experiment; #, nonfunctional mutants. It is useful to note (1) the variation in total protein expression between mutants even though protein loading is the same, and (2) the difference in band profile between the total protein and cell-surface protein, showing the prevalence of higher molecular mass receptor glycoforms at the cell surface. Portions of this figure were originally published in [8]

mutants K69A and K308C); the multiple bands represent different glycoforms of the receptor and in the total protein sample the lower molecular mass (immature) glycoforms predominate whereas in the cell-surface protein sample the high molecular mass (mature) glycoforms predominate [8].

3.5 Quantitative Analysis

1. Record band intensities for both total and cell-surface protein. Choose a suitable area of the blot and record the intensity of the background. Subtract background from band intensities (*see* **Note 18**).
2. Divide the intensity of the cell-surface band by the intensity of the total protein band for each sample. The mean ratios and standard error from at least three independent experiments should be calculated and plotted as a histogram (Fig. 2). Data for the triple N-glycosylation mutant N182S/N239S/N298S (TRIP N) which does not reach the cell surface [13] is included as a negative control. Note that with this technique it is only possible to determine the relative proportion of protein expressed at the cell surface; this is because the efficiency of (1) biotinylation of cell-surface protein and (2) streptavidin capture is not known.

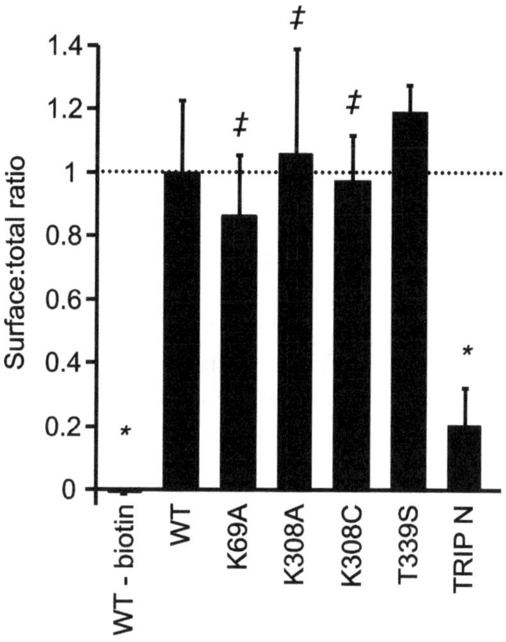

Fig. 2 Histogram showing the relative cell-surface expression of wild-type P2X2 and point mutations. Data is representative of three independent experiments; error bars indicate standard error of the mean and the mean value for wild type has been normalized to 1. TRIP N, triple N-glycosylation mutant N182S/N239S/N298S. #, nonfunctional point mutants. $*$, data significantly different from control ($p < 0.05$ as determined by one-way ANOVA)

4 Notes

1. It is important to use an alkaline pH to increase the efficiency of the biotinylation reaction. The easiest way to make PBS pH 8.0 is to use a commercially available PBS solution at pH 7.4 and alter the pH to 8.0 using NaOH and a pH meter. It is advisable to start adding 1 M NaOH and checking the pH change to avoid making the solution too alkaline.
2. EZ-link Sulfo-NHS-Biotin must be stored at −20 °C as a dry powder, and made up as required. Solutions cannot be stored and reused.
3. 10× TG is a standard solution used in the preparation of SDS-PAGE gel running buffer; it does not require pH adjustment.
4. For protease inhibitors we use 20 μl/ml of one Complete™--EDTA Protease inhibitor tablet (Sigma-Aldrich product number 000000004693132001) dissolved in 1 ml H_2O.
5. The use of precast SDS-PAGE gels (8% or 4–15% is optimal for rat P2X2) is recommended to maintain quality and

reproducibility, but good results may also be obtained with hand-cast 8% gels (including a 5% stacking gel).

6. The use of preprepared transfer packs and semidry transfer apparatus (e.g., Bio-Rad Trans-Blot Turbo) is recommended for optimum reproducibility. PVDF membranes are recommended for working with membrane proteins due to their increased hydrophobicity, but care must be taken to ensure that transfer membranes are compatible with the blot detection system (some PVDF membranes produce high background in infra-red fluorescence detection systems).
7. The optimum blocking agent (milk, BSA, casein, or proprietary formulation) is dependent on both the antibodies used and the detection system. 5% (w/v) nonfat milk powder is both cheap and readily available at supermarkets and usually works well with most antibodies.
8. The secondary antibody used depends on the detection system available. The most commonly used are conjugated to either horseradish peroxidase (HRP) (chemiluminescent detection using film or a sensitive digital camera system) or infra-red fluorescent dyes (direct detection using infra-red scanners). Infra-red fluorescence detection is preferred due to its high dynamic range and quantitative signal. If film is used for detection, it must be scanned using a system capable of scanning transparencies, and care must be taken to ensure that the signal does not exceed the dynamic range of the film.
9. Biotinylation reactions must be carried out on live cells under conditions where endocytosis is strongly inhibited, otherwise the biotin reagent may enter into the cells and label intracellular protein. The easiest way to inhibit endocytosis is to cool the sample to 4 °C [16] and this is readily achieved either by working in a cold room or placing the samples on ice.
10. The NHS group on the Sulfo-NHS-LC-Biotin cross-linker reacts with primary amines and so is quenched using TG solution. It is very important to make the biotinylation reagent up in a buffer lacking primary amines (hence the use of PBS pH 8.0 instead of a standard Tris buffer).
11. HEK293T cells are loosely adherent and may be readily detached by pipetting up and down. Other cells may need to be scraped in order to detach; alternatively lysis and solubilization may be performed in the dish but the volume of solubilization buffer must be increased to at least 100 μl in order to cover the cells.
12. At this point in the protocol the cell pellets may be frozen for later analysis at either −20 °C (for approx. 1 month) or −80 °C (for approx. 6 months).

13. The choice of protein assay reagent is a personal preference. For this protocol, the Bradford assay (using BSA as a standard) works well as long as the samples are diluted by a factor of ten in order to reduce the Triton X-100 concentration to 0.2% so that it does not interfere with the assay. From 1× 35 mm dish of cells solubilized in 50 μl buffer total protein concentrations in the range of 2–10 mg/ml are routinely obtained. In order to obtain the 300 μg protein required for the biotinylation reaction it may be necessary to increase the quantity of cells transfected.
14. In order for the beads to be able to enter the pipette tip without obstruction, it is necessary to cut off the bottom 5 mm of the tip with a clean pair of scissors.
15. It is important not to exceed 5000 rpm, as this may cause the beads to form a gel, which may then lead to problems with washing and elution.
16. Sometimes beads may enter the pipette tip (this is minimized by using gel-loading tips), or stick to the side of the tip when purging. If this happens, add more buffer to wash the tip and recentrifuge. It is also possible that centrifugation does not adequately pellet the beads. In this case place the tubes upright in a rack for 2–5 min and allow them to settle by gravity.
17. Many primary antibodies give stronger and more specific staining when incubated overnight at 4 °C instead of 1 h at room temperature. Furthermore, primary antibody may be recovered, stored at −20 °C and reused multiple times with no loss of efficiency.
18. Some image analysis software automatically adjusts the background from areas adjacent to the bands of interest (in this case it is important to check that these areas do not also contain protein signal, otherwise quantitation will not be accurate). Other software allows for lane background reduction.

References

1. North RA (2016) P2X receptors. Philos Trans R Soc Lond Ser B Biol Sci 371(1700). https://doi.org/10.1098/rstb.2015.0427
2. Pasqualetto G, Brancale A, Young MT (2018) The molecular determinants of small-molecule ligand binding at P2X receptors. Front Pharmacol 9:58. https://doi.org/10.3389/fphar.2018.00058
3. Robinson LE, Murrell-Lagnado RD (2013) The trafficking and targeting of P2X receptors. Front Cell Neurosci 7:233. https://doi.org/10.3389/fncel.2013.00233
4. Bobanovic LK, Royle SJ, Murrell-Lagnado RD (2002) P2X receptor trafficking in neurons is subunit specific. J Neurosci 22(12):4814–4824. https://doi.org/10.1523/jneurosci.22-12-04814.2002
5. Stoop R, Thomas S, Rassendren F, Kawashima E, Buell G, Surprenant A, North RA (1999) Contribution of individual subunits to the multimeric P2X(2) receptor: estimates based on methanethiosulfonate block at T336C. Mol Pharmacol 56(5):973–981
6. Xu J, Chai H, Ehinger K, Egan TM, Srinivasan R, Frick M, Khakh BS (2014)

Imaging P2X4 receptor subcellular distribution, trafficking, and regulation using P2X4-pHluorin. J Gen Physiol 144(1):81–104. https://doi.org/10.1085/jgp.201411169

7. Cao L, Broomhead HE, Young MT, North RA (2009) Polar residues in the second transmembrane domain of the rat P2X2 receptor that affect spontaneous gating, unitary conductance, and rectification. J Neurosci 29 (45):14257–14264. https://doi.org/10.1523/jneurosci.4403-09.2009
8. Young MT, Zhang YH, Cao L, Broomhead H, Jiang LH (2008) Role of the domain encompassing Arg304-Ile328 in rat P2X2 receptor conformation revealed by alterations in complex glycosylation at Asn298. Biochem J 416 (1):137–143. https://doi.org/10.1042/bj20081182
9. Roberts JA, Evans RJ (2007) Cysteine substitution mutants give structural insight and identify ATP binding and activation sites at P2X receptors. J Neurosci 27(15):4072–4082. https://doi.org/10.1523/jneurosci.2310-06.2007
10. Cao L, Young MT, Broomhead HE, Fountain SJ, North RA (2007) Thr339-to-serine substitution in rat P2X2 receptor second transmembrane domain causes constitutive opening and indicates a gating role for Lys308. J Neurosci 27(47):12916–12923. https://doi.org/10.1523/jneurosci.4036-07.2007
11. Ennion SJ, Evans RJ (2002) Conserved cysteine residues in the extracellular loop of the human P2X(1) receptor form disulfide bonds and are involved in receptor trafficking to the cell surface. Mol Pharmacol 61(2):303–311
12. Ennion S, Hagan S, Evans RJ (2000) The role of positively charged amino acids in ATP recognition by human P2X(1) receptors. J Biol Chem 275(38):29361–29367. https://doi.org/10.1074/jbc.M003637200
13. Torres GE, Egan TM, Voigt MM (1998) N-linked glycosylation is essential for the functional expression of the recombinant P2X2 receptor. Biochemistry 37(42):14845–14851. https://doi.org/10.1021/bi981209g
14. Green NM (1975) Avidin. Adv Protein Chem 29:85–133
15. Diamandis EP, Christopoulos TK (1991) The biotin-(strept)avidin system: principles and applications in biotechnology. Clin Chem 37 (5):625–636
16. Wolkers WF, Looper SA, Fontanilla RA, Tsvetkova NM, Tablin F, Crowe JH (2003) Temperature dependence of fluid phase endocytosis coincides with membrane properties of pig platelets. Biochim Biophys Acta 1612 (2):154–163

Chapter 10

Multimeric Ionotropic Purinoceptor Detection by Protein Cross-Linking

Vincent Compan and François Rassendren

Abstract

P2X receptor subunits (P2X1 to P2X7) assemble to form trimeric homomers or heteromers. Here, we describe the use of protein cross-linking to study the composition of P2X receptor complexes. This simple protocol is useful for determining the stoichiometry of P2X heteromeric receptors as well as for assessing the effect of point mutation, truncation, or concatenation on the quaternary architecture of these receptors.

Key words P2X, Cross-linking, Heteromer, Homomer, Trimer, Quaternary structure

1 Introduction

Cross-linking reagents are widely used to elucidate quaternary structure of protein complexes. These compounds create covalent link between proteins that interact or that are in close proximity to each other. These covalent bridges offer a simple method to stabilize complexes and determine molecular weight/stoichiometry of protein complexes following SDS-PAGE analysis. Different cross-linkers are commercially available and differ by their chemical reactivity, length of the spacer arm and their ability to cross the plasma membrane (*see* Fig. 1 for an overview of cross-linker features). Homobifunctional N-hydrosuccinimide (NHS) ester represents the most popular class of cross-linkers that forms bridges (cleavable or not) between lysine residues.

In the field of purinoreceptor, cross-linking studies of P2X1 and P2X3 receptors assembly provided the first biochemical evidence of the trimeric architecture of P2XR channel [1]. Trimeric architecture of P2X receptors was finally confirmed by several approaches including atomic force microscopy [2], electron microscopy [3], and the crystal structure of the truncated zebrafish zP2X4 receptor [4].

Pablo Pelegrín (ed.), *Purinergic Signaling: Methods and Protocols*, Methods in Molecular Biology, vol. 2041,
https://doi.org/10.1007/978-1-4939-9717-6_10,

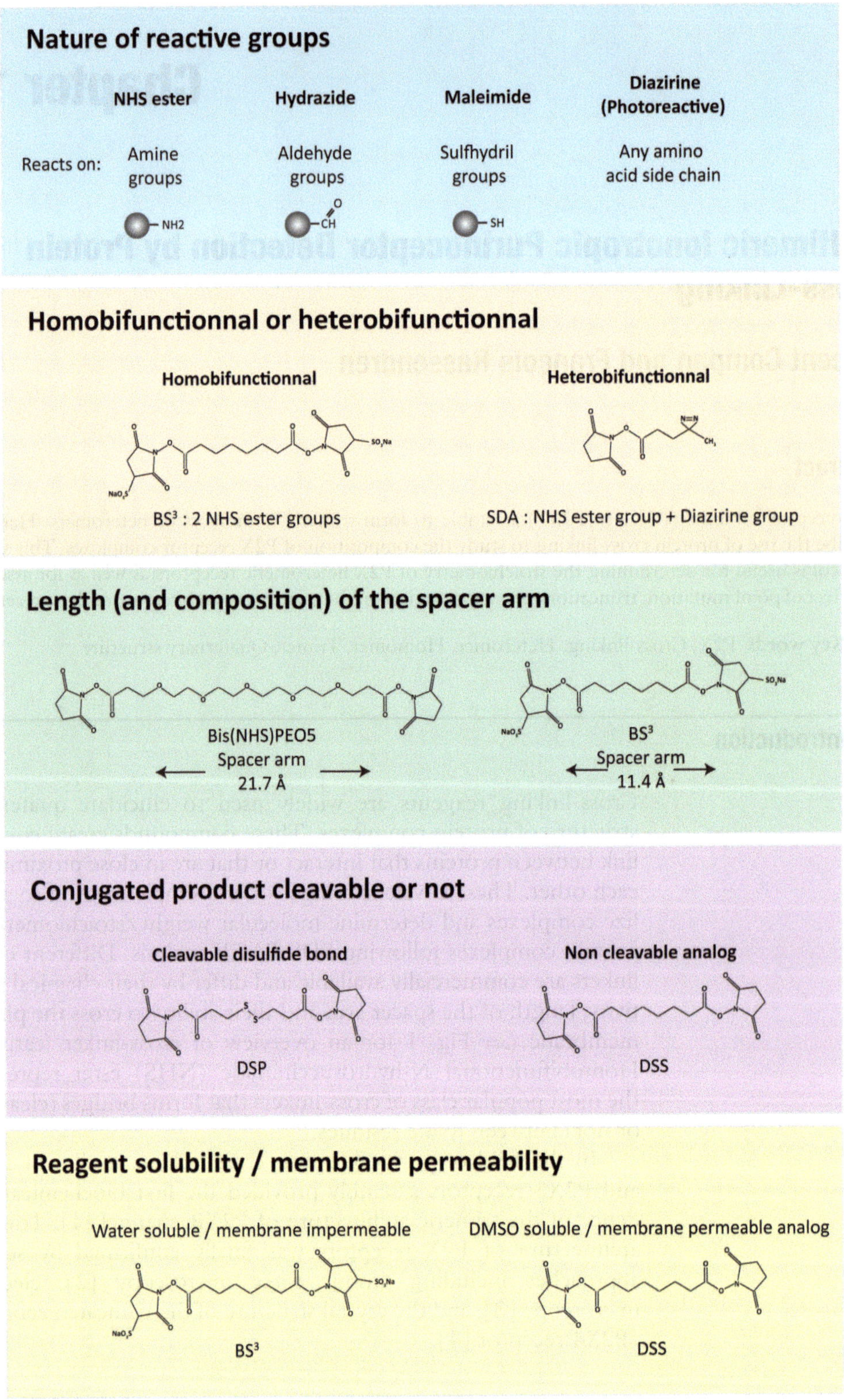

Fig. 1 Overview of the main features of protein cross-linkers. This figure illustrates the five main features of protein cross-linkers and provides examples of reagents for each category

Thanks to its simplicity and effectiveness, cross-linking strategies still represent a method of choice in structure–function studies to analyse the effects of truncation or point mutations on the quaternary structure of P2X receptors [5, 6]. We and other have also employed cross-linkers to elucidate the subunit composition of P2X receptor complexes. For instance Boumechache et al. demonstrated that in primary culture of microglia and macrophages endogenous P2X4 and P2X7 subunits interact between homotrimers rather than within heterotrimers [7]. We demonstrated that P2X5 and P2X2 subunits interact to form heteromeric receptors with alternate stoichiometries, both expressed at the plasma membrane [8].

In this chapter, we will describe a protocol that employ NHS ester based cross-linkers to study P2X receptors assembly and determine the stoichiometry of heteromeric P2X receptors.

2 Materials

2.1 Expression of the P2X Subunits in HEK293T Cell Line

1. Transfection reagent suitable to transfect HEK293T cell line (for example, we use Lipofectamine 2000, but equivalent can be also used).
2. Complete cell media appropriate to the cell line in culture (e.g., high glucose DMEM-GlutaMAX medium with 10% of fetal calf serum for HEK293T).
3. Opti-MEM culture media or equivalent media without serum.
4. Plasmids coding for the P2X subunits of interest (*see* **Note 1**).

2.2 Cross-Linking

1. Cross-linkers such as BS3, Bis(NHS)PE05, sulfo-NHS-SDA, or NHS-SDA (*see* **Note 2**).
2. Phosphate-buffered saline (PBS-CM): 1 mM $CaCl_2$ and 0.5 mM $MgCl_2$ (*see* **Note 3**).
3. Quenching solution: PBS-CM with 50 mM Tris pH 8.0.
4. UV lamp with 365 nm bulbs for diazirine based cross-linking (*see* **Note 4**).

2.3 Western Blotting

1. Cell lysis buffer: 20 mM HEPES, 100 mM NaCl, 5 mM EDTA, 1% Triton X-100, pH 7.4, supplemented with protease inhibitor cocktail.
2. Tris-glycine gels or equivalent (*see* **Note 5**).
3. Appropriate primary antibodies directed to P2X subunits and secondary antibodies.

3 Methods

3.1 HEK293T Transfection

1. Cells are plated 24 h before transfection in 6-well plate to obtain a 60–70% confluence on the day of transfection.
2. Transfection of HEK293T can be performed using Lipofectamine 2000 according to the manufacturer's instructions using 50 ng of the desired P2X cDNA expression vector (*see* **Note 6**), 100 μl Opti-MEM and 5 μl Lipofectamine 2000 per well.
3. Incubate cells with transfection mixture for 4–5 h at 37 °C and 5% CO_2, then remove transfection reagent containing media and split cells in 12-well plate with fresh culture media.
4. Cross-linking experiments can be performed the following day or 48 h after transfection.

3.2 P2X Receptor Multimer Cross-Linking Using Homobifunctional Cross-Linkers BS3 or Bis(NHS)PE05

1. Wash the cells twice with ice cold PBS-CM.
2. Incubate cells with the given cross-linker diluted in PBS-CM to a final concentration of 1–5 mM (*see* **Note 7**) for 30 min at 4 °C (*see* **Note 8**).
3. Wash cells once with ice-cold PBS-CM.
4. Incubate cells in quenching solution for 5 min at 4 °C.

3.3 P2X Receptor Multimer Cross-Linking Using Heterobifunctional Cross-Linkers Sulfo-NHS-SDA and Its Uncharged Analogue NHS-SDA

1. Cross-linking is performed as described above using 2 mM sulfo-NHS-SDA or the membrane permeant analogue NHS-SDA.
2. Incubate cells in quenching solution for 5 min at 4 °C.
3. Expose cells to UV light (365 nm) for 15 min (*see* **Note 9**) to induce diazirine group reaction to any amino acid side chain. Keep cell dishes on ice (*see* **Note 8**).

3.4 Visualization of P2X Receptor Multimer by Western Blotting

1. Scrape cells in 100 μl of cell lysis buffer per well.
2. Incubate cell lysates 30 min at 4 °C.
3. Centrifuge at 16,000 × *g* for 10 min.
4. Keep supernatant and discard pellet.
5. Determine protein concentration using BCA method or equivalent.
6. Add appropriate loading buffer (depending on the type of gels selected at **step** 7) and reducing agent such as 5% β-mercaptoethanol (*see* **Note 10**).
7. Separate protein by electrophoresis (load 10 μg of protein for each condition per lane) and transfer to nitrocellulose membranes.
8. Use appropriate primary and secondary antibodies to visualize P2X homomeric or heteromeric complexes (*see* Fig. 2).

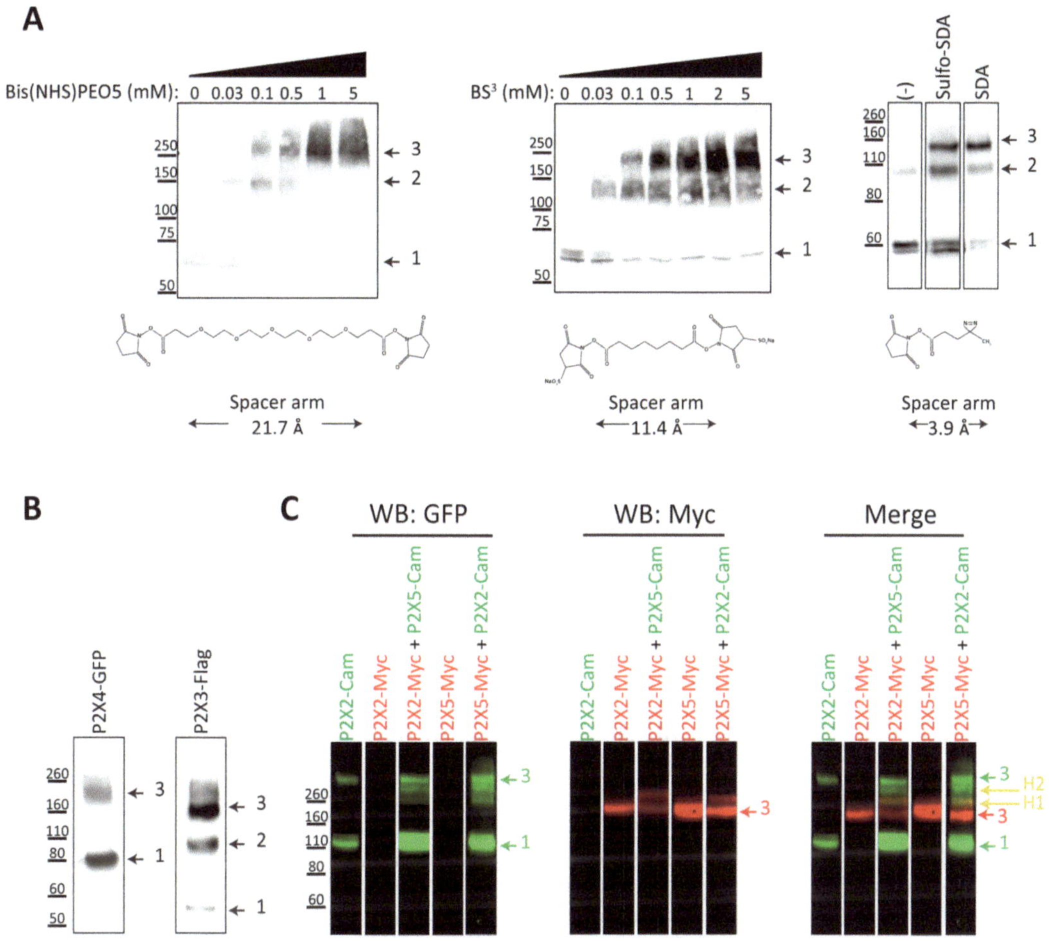

Fig. 2 Analysis of P2X receptor quaternary structure by protein cross-linking. (**a**) Trimeric architecture of homomeric P2X2 receptor is revealed by different cross-linkers. HEK293T cells expressing P2X2 subunits were incubated with increasing concentration of cross-linkers differing by their chemical reactivity (homobifunctional vs heterobifunctional), the length of their spacer arm and their membrane permeability. The structure of each cross-linker is represented below the corresponding Western blot. The numbers located at the right of each Western blot indicate the presence of monomers, dimers and trimers. (**b**) Homomeric receptors composed of P2X4-GFP or P2X3-Flag subunits are efficiently cross-linked by 5 mM BS3. (**c**) Determination of the quaternary structure of heteromeric P2X2/P2X5 receptors. HEK293T cells expressing different combination of P2X2 or/and P2X5 subunits fused to either small tag (myc tag) or YC3.1 cameleon protein (Cam) were incubated with 5 mM BS3. Complexes were separated by Western blot and revealed using sequentially GFP (pseudocolored in green) and myc antibodies (pseudocolored in red). Overlay of the two pictures revealed the presence of homotrimers and two heteromers (H1 and H2) corresponding to two different subunit stoichiometries. The presence of monomeric P2X2-Cam species likely represents intracellular pool of the subunits whose trafficking is impaired by the fusion partner

4 Notes

1. Because most P2X receptor subunits have similar molecular weight, the use of P2X subunits tagged to GFP (or any other fusion protein) might facilitate the analysis of the stoichiometry of heteromeric P2X receptors. Control experiment will have to be performed to confirm that the fusion does not alter the trimeric architecture of the corresponding homomeric P2X receptors.
2. Different cross-linkers are commercially available and differ by their (1) chemical reactivity (functional groups that bind the protein interest), (2) spacer arm composition (affecting the length and cleavability of the cross-linker), and (3) solubility (and so their membrane permeability). Homomeric and heteromeric P2X receptors are efficiently cross-linked using homobifunctional N-hydroxysuccinimide ester with spacer arm length between 11 and 22 Å (*see* Fig. 1). We also obtain good results with heterobifunctional cross-linkers that combine NHS-ester and diazirine-based photoreaction chemistries leading to the cross-link of amine-containing molecules with nearly any other functional group via long-wave UV-light activation (Fig. 1). Most of these cross-linkers have membrane permeable analogues suitable for intracellular conjugation.
3. When using N-hydroxysuccinimide ester based cross-linkers, do not use Tris-buffered solution during cross-linker incubation as Tris contains free amine group that will quench the reaction.
4. The optimal wavelength for diazirine photoactivation is around 340 nm. Avoid using UV lamps that emit light at 254 nm as this wavelength causes cell damage. Samples irradiation conditions should be optimized depending on UV lamp power and on the distance between the lamp and the sample. UV irradiation time should be less than 15 min for cross-linking of live cells. For efficient UV irradiation, use uncovered cell culture dishes and place the lamp directly above samples placed on ice.
5. For a better separation of P2X receptor complexes during electrophoresis, use precast gradient gels (such as 4–12% gradient).
6. Overexpression of P2X subunits might be toxic for cells. Minimize cytotoxicity by using low DNA amounts and low transfection reagent volume.
7. Prepare cross-linker solution immediately before use.
8. In order to minimize plasma membrane receptor internalization, use cold buffers (at 4 °C) and keep cell dishes on ice during the cross-linking step.

9. Adjust incubation time to the UV lamp power.
10. Cross-linkers with disulfide bond within the spacer arm will be cleaved using reducing agent such as β-mercaptoethanol.

References

1. Nicke A, Baumert HG, Rettinger J, Eichele A, Lambrecht G, Mutschler E, Schmalzing G (1998) P2X1 and P2X3 receptors form stable trimers: a novel structural motif of ligand-gated ion channels. EMBO J 17:3016–3028. https://doi.org/10.1093/emboj/17.11.3016
2. Barrera NP, Ormond SJ, Henderson RM, Murrell-Lagnado RD, Edwardson JM (2005) Atomic force microscopy imaging demonstrates that P2X2 receptors are trimers but that P2X6 receptor subunits do not oligomerize. J Biol Chem 280:10759–10765. https://doi.org/10.1074/jbc.M412265200
3. Mio K, Kubo Y, Ogura T, Yamamoto T, Sato C (2005) Visualization of the trimeric P2X2 receptor with a crown-capped extracellular domain. Biochem Biophys Res Commun 337:998–1005. https://doi.org/10.1016/j.bbrc.2005.09.141
4. Kawate T, Michel JC, Birdsong WT, Gouaux E (2009) Crystal structure of the ATP-gated P2X (4) ion channel in the closed state. Nature 460:592–598. https://doi.org/10.1038/nature08198
5. Ennion SJ, Evans RJ (2002) Conserved cysteine residues in the extracellular loop of the human P2X(1) receptor form disulfide bonds and are involved in receptor trafficking to the cell surface. Mol Pharmacol 61:303–311
6. Wickert LE, Blanchette JB, Waldschmidt NV, Bertics PJ, Denu JM, Denlinger LC, Lenertz LY (2013) The C-terminus of human nucleotide receptor P2X7 is critical for receptor oligomerization and N-linked glycosylation. PLoS One 8: e63789. https://doi.org/10.1371/journal.pone.0063789
7. Boumechache M, Masin M, Edwardson JM, Gorecki DC, Murrell-Lagnado R (2009) Analysis of assembly and trafficking of native P2X4 and P2X7 receptor complexes in rodent immune cells. J Biol Chem 284:13446–13454. https://doi.org/10.1074/jbc.M901255200
8. Compan V, Ulmann L, Stelmashenko O, Chemin J, Chaumont S, Rassendren F (2012) P2X2 and P2X5 subunits define a new heteromeric receptor with P2X7-like properties. J Neurosci 32:4284–4296. https://doi.org/10.1523/JNEUROSCI.6332-11.2012

Chapter 11

Multimeric Purinoceptor Detection by Bioluminescence Resonance Energy Transfer

Vincent Compan and François Rassendren

Abstract

Assays based on bioluminescence resonance energy transfer (BRET) provide a sensitive and simple method to study protein–protein interactions in live cells. Here we describe a protocol using BRET technique to investigate potential interactions between P2X subunits. This approach combined with bimolecular fluorescence complementation (BiFC) can also be employed to determine the stoichiometry of heteromeric P2X receptors.

Key words Bioluminescence, BRET, P2X, BiFC, Stoichiometry, Heteromer

1 Introduction

Bioluminescence resonance energy transfer (BRET) is a naturally occurring phenomenon observed in some marine species such as *Renilla reniformis* where a nonradiative energy transfer is naturally observed between the *Renilla* luciferase (energy donor) and the *Renilla* green fluorescent protein (energy acceptor) [1, 2]. This process has been adapted for the detection of protein–protein interactions by simply fusing one protein with the BRET donor and the other one with the BRET acceptor. Energy transfer efficiency depends on the orientation of the donor and the acceptor and is inversely proportional to the sixth power of distance between them. Different BRET donors and acceptors exist, but the most commonly used BRET donor is *Renilla* Luciferase (RLuc and its variants such as RLuc8) that catalyses the oxidation of its substrate, coelenterazine, to produce light with a peak emission around 480 nm. Emission spectra of RLuc overlaps well with the excitation spectra of YFP (and its variants such as Venus) making these fluorescent proteins suitable as BRET acceptor for RLuc. The main advantage of BRET over FRET (where donor and acceptor are fluorescent proteins) is that BRET does not require any external

Pablo Pelegrín (ed.), *Purinergic Signaling: Methods and Protocols*, Methods in Molecular Biology, vol. 2041,
https://doi.org/10.1007/978-1-4939-9717-6_11,

light excitation, improving signal to noise ratio and reducing cell damage induced by light excitation. The main limitation of BRET approach is due to the weak signals of BRET donors, although this limitation was recently partially overcame by the development of a new luciferase called NanoLuc that is much brighter than traditional RLuc BRET donor.

FRET approaches have been employed in the field of P2X receptors to analyse the architecture and size of P2X4 receptors [3] and to study cytosolic gating motions in the N- and C-terminal domains of P2X2 [4]. Interactions between P2X2 receptors and other ligand-gated ion channels such as α4β2 nicotinic channels [5], α6β4 nicotinic channels [6] or $GABA_A$ ion channels [7] were also demonstrated and characterized using FRET technique. A specific FRET approach has also been developed in biosensors able to track activation of ATP-gated P2X receptors [8].

BRET is historically more popular in the field of G protein coupled receptor and only few publications employed this approach for studying ion channels and more specifically P2X receptor. For instance, BRET was used to characterize the effect of carboxyl-terminal splicing on the physical interactions between the cytoplasmic tails of purinergic P2X2 receptors [9]. We also developed this approach to identify the interaction between P2X2 and P2X5 subunits, and to demonstrate that these two subunits can form heteromeric receptors with two different quaternary structures [10].

In this chapter, we describe a simple protocol that employ BRET approaches to study P2X subunits interactions. By combining this technique with Bimolecular Fluorescence Complementation (BiFC), BRET allows determining the stoichiometry of heteromeric P2X receptors.

2 Materials

2.1 Expression of the P2X Receptor Tagged Subunits in HEK293T

1. Transfection reagent Lipofectamine 2000 or equivalent can be used to transfect HEK293T.
2. Complete cell media appropriate to the cell line in culture (for example, high glucose DMEM-GlutaMAX medium with 10% of foetal calf serum for HEK293T).
3. Opti-MEM culture media or equivalent rich cell culture media without serum.
4. Plasmids coding for the P2X subunits tagged with either RLuc8, or YFP or hemi YFP (YC or YN) (*see* **Note 1**).

2.2 BRET Recording Using a Plate Reader

1. 96-Well plate, white, flat and micro-clear bottom, cell culture treated, sterile with lids (*see* **Note 2**) for BRET recording.

2. 96-Well plate, black, flat and micro-clear bottom, cell culture treated, sterile with lids (*see* **Note 2**) for fluorescence recording.
3. White backing tape to be stuck on the bottom of the 96-well plate the day of recording. It increases the luminescence signal by reflection.
4. A plate reader for luminescence recording equipped with two emission filters close to 480 and 535 nm.
5. HBS solution: 147 mM NaCl, 2 mM KCl, 2 mM $CaCl_2$, 1 mM $MgCl_2$, 10 mM HEPES, and 13 mM D-glucose, pH 7.4.
6. Multichannel pipettes.
7. Coelenterazine-h stock solution: 25 μM of coelenterazine-h in HBS (*see* **Note 3**).

3 Methods

3.1 General Design of the Experiment

BRET occurs when the acceptor and the donor are in close proximity (typically, within *10 nm*) that makes this approach ideal for monitoring protein–protein interactions in living cells. However, several controls have to be performed to ensure the specificity of the BRET signal, especially in overexpression system where random collisions or nonspecific oligomerization might occur. As a general rule, we strongly recommend to include the following control:

1. Use nontransfected cells to determine background signal for luminescence recording.
2. A direct fusion between the BRET donor and the acceptor might be employed as a positive control. It will result in a strong BRET signal.
3. Cells transfected with P2X-RLuc8 construct alone are necessary to calculate the net BRET signal. This will help to determine the "BRET background" due to the partial overlap *between donor and acceptor emission spectra.*
4. *BRET saturation curves must be performed* (*see* **Note 4**). In this assay, a specific BRET signal will be characterized by a hyperbolic increase of the signal whereas random collisions are characterized by a linear increase of the BRET signal (Fig. 1).

3.2 HEK293T Transfection and Plating

1. Cells are plated 24 h before transfection in 6-well plate to obtain a 60–70% confluence on the day of transfection.
2. Transfection of HEK293T can be performed using Lipofectamine 2000 according to the manufacturer's instructions, using cDNA coding for BRET donor and acceptor (*see* **Note 4**), 100 μl Opti-MEM, and 5 μl Lipofectamine 2000 per well.

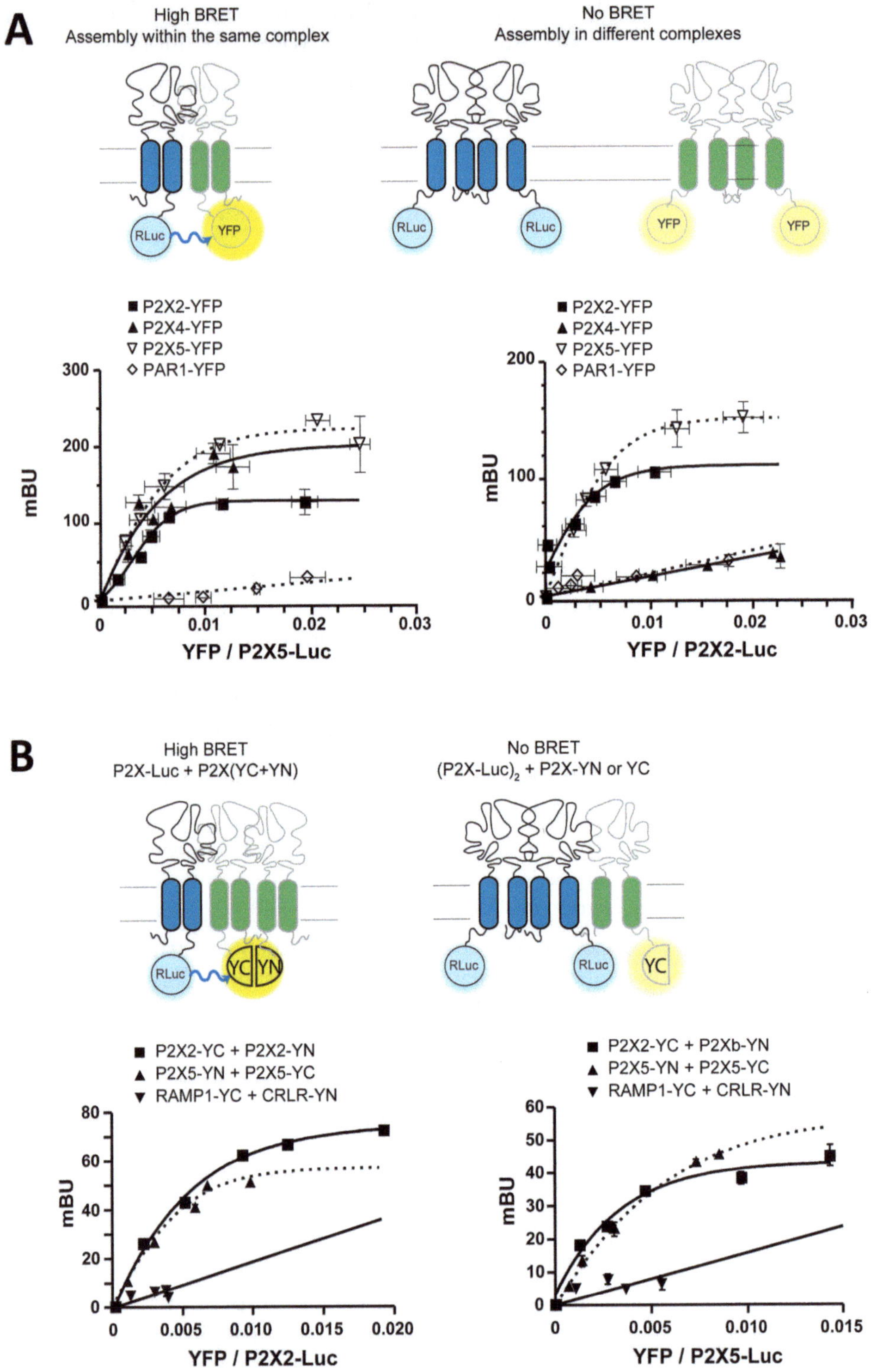

Fig. 1 Analysis of P2X receptor quaternary structure using BRET and BiFC approaches. (**a**) Top panel: cartoon illustrating how BRET assay allows to detect assembly of P2X subunits within the same complex. For simplicity, P2X receptors are only represented with two subunits. Bottom panel: BRET titration curves using

3. Incubate cells with the transfection mixture for 4–5 h at 37 °C and 5% CO_2, then remove transfection reagent containing media and split cells in two independent black and white 96-well plates (*see* **Note 5**).
4. Incubate cells for 36 h at 37 °C and 5% CO_2 (*see* **Note 6**).
5. Perform BRET experiments.

3.3 BRET and Fluorescence Recording Using a Plate Reader

1. Prewarm the plate reader at 37 °C.
2. Set up the plate reader controlling software according to record fluorescence (*see* **Note** 7).
3. Wash cells from the black plate twice with 100 μl HBS per well taking care not to detach cells.
4. Keep cells in 50 μl of HBS per well.
5. Read the YFP fluorescence ($\lambda_{ex} \approx 500$ nm, $\lambda_{em} \approx 530$ nm) to determine expression of the BRET acceptor construction.
6. Set up the plate reader controlling software to record BRET signal (*see* **Notes** 7 and **8**).
7. Prepare a stock solution of coelenterazine-h. Prepare enough volume of the stock solution in order to add 10 μl per well to read.
8. Wash cells from the white plate twice with 100 μl HBS per well taking care not to detach cells.
9. Keep cells in 40 μl of HBS per well.
10. Stick the white backing tape at the bottom of the plate.
11. Add 10 μl of the stock solution of coelenterazine-h per well to get a 5 μM final concentration in 50 μl final volume.
12. Insert the plate in the plate reader.
13. Incubate for 4 min to allow the luminescence signal to reach a steady state (*see* **Note 9**).

Fig. 1 (continued) P2X5-Luc or P2X2-Luc as the energy donors. HEK293T cells were cotransfected with a constant amount of donor and with an increasing amount of YFP fusions. BRET signals are plotted against the relative amounts of each tagged subunit. Specific, saturating, BRET signals are observed between P2X5/P2X5, P2X2/P2X2 or P2X5/P2X2 subunits, while low linear BRET is obtained between P2X2 (or P2X5) and PAR-1 or between P2X2 and P2X4. Data are expressed as mean ± SEM of at least $N = 3$ experiments. (**b**) Top panel: cartoon illustrating how combination of BRET and BiFC assays can be employed to reveal the stoichiometry of heteromeric P2X receptors. Bottom panel: BRET titration curves between P2X2-Luc (left) or P2X5-Luc (right) cotransfected with P2X2, P2X5, RAMP1, and CRL receptor fused to hemi-YFP. Specific BRET signals were observed between P2X2 and P2X5 subunits in each experimental condition, demonstrating the existence of heteromeric receptors with two different stoichiometries. Data are expressed as mean ± SEM of at least $N = 3$ experiments

14. Read the luminescence signal at ≈480 (RLuc8 filter) nm and ≈535 nm (YFP filter); determine the integration time according to the sensitivity of the plate reader (*see* **Note 7**).
15. Record the luminescence signal at 480 and 535 nm in the same experimental conditions set above. Do not forget to use cells transfected with P2X-RLuc8 construct alone to calculate net BRET signal (*see* **Note 10**).
16. Net BRET value, expressed in milliBRET units (or mBU), can be determined with the following equation:

$$\text{BRET (mBU)} = \left(\left(\frac{\text{Lum}(535\ \text{nm})}{\text{Lum}(480\ \text{nm})} \right)_{\substack{\text{P2X}-\text{RLuc8} \\ +\text{P2X}-\text{YFP}}} - \left(\frac{\text{Lum}(535\ \text{nm})}{\text{Lum}(480\ \text{nm})} \right)^{\text{P2X}-\text{RLuc8 only}} \right) \times 1000$$

17. For each condition of transfection, determine the ratio acceptor/donor by dividing the fluorescence obtained with the black plate by the luminescence measured at 480 nm in the white plate during BRET recording. Report these values (*X* axis) to the BRET value (*Y* axis) to get a saturation curve. A specific BRET signal is characterized by a hyperbolic increase of the signal whereas random collisions are characterized by a linear increase of the BRET signal (Fig. 1).

4 Notes

1. P2X subunits fused to YFP constructs are used to determine potential P2X subunit interactions. Hemi YFP constructs (YN or YC for N-term or C-term fragments of YFP respectively) are used to determine the stoichiometries of P2X heteromeric receptors by combining BRET experiments and Bimolecular Fluorescence Complementation (BiFC).
2. Other format plate (i.e., 384-well plate) can be used depending on the assay and/or luminescence signal strength. Micro-clear bottom plate allows to visualize cells during the different steps of the experiment. Black and white plates are preferentially used to record fluorescence and luminescence signal, respectively. However, depending on the fluorescence or luminescence signal strength, any type of plate can be used indifferently.
3. Coelenterazines are poorly soluble in water and must be resuspended in ethanol or methanol at 1 mM master stock solution. Keep this stock solution at −20 °C and protect it from light. For prolonged storage, resuspend coelenterazine in acidified ethanol (10 ml of ethanol and 200 μl of 3 N HCl) and store at −80 °C.

4. For BRET saturation-curve experiments, cells are cotransfected with a fixed amount of Rluc8 fusion cDNA and increasing amounts of YFP fusion cDNA (or YC and YN when combining BRET with BiFC assay). As a first try, 25 ng of P2X-RLuc8 constructs might be combined to 0, 10, 20, 50, 100, and 200 ng of P2X-YFP (or YC/YN) constructs (Fig. 1). Depending on the expression and the BRET signal, adjust the amount of cDNA.
5. Poly-L-ornithine-coated plates might be used to prevent cell detachment. Two independent plates (black and white) are used to record fluorescence and luminescence signal, respectively.
6. For BRET/BiFC experiments, cells must be incubated at 30 °C for 12 h before the reading to promote the complementation between hemi-YFP.
7. Exposure time and gain amplification of signals have to be determined to get the best signal–background ratio. Integration time must be adjusted to get the widest dynamic range and the best sensitivity for a given plate reader. For BRET recording, the integration time of the luminescence signal has to be the lowest. These values of integration time will stay approximately the same for each experiment and will just need to be adjusted depending on the transfection efficiency.
8. Detection of the luminescence signal within the filters of RLuc8 and YFP in a given well have to be recorded successively (ideally simultaneously) before recording the next well.
9. During the first 4 min following coelenterazine-h addition, luminescence signal will increase exponentially and might cause inappropriate BRET value (especially if the exposure time for each filter is $\geq$1 s). Thus, it is recommended to wait for the luminescence signal to reach a plateau.
10. For a given set of filters, plate reader, and experimental conditions, the ratio $\left(\frac{\text{Lum}(535\text{ nm})}{\text{Lum}(480\text{ nm})}\right)^{\text{P2X-RLuc8 only}}$ for the donor only will stay approximately the same during each experiment and will just have to be determined once.

References

1. Wampler JE, Hori K, Lee JW, Cormier MJ (1971) Structured bioluminescence. Two emitters during both the in vitro and the in vivo bioluminescence of the sea pansy, Renilla. Biochemistry 10:2903–2909
2. Ward WW, Cormier MJ (1979) An energy transfer protein in coelenterate bioluminescence. Characterization of the Renilla green-fluorescent protein. J Biol Chem 254:781–788
3. Young MT, Fisher JA, Fountain SJ, Ford RC, North RA, Khakh BS (2008) Molecular shape, architecture, and size of P2X4 receptors determined using fluorescence resonance energy transfer and electron microscopy. J Biol Chem 283:26241–26251. https://doi.org/10.1074/jbc.M804458200
4. Fisher JA, Girdler G, Khakh BS (2004) Time-resolved measurement of state-specific P2X2 ion channel cytosolic gating motions. J

Neurosci 24:10475–10487. https://doi.org/10.1523/JNEUROSCI.3250-04.2004

5. Khakh BS, Fisher JA, Nashmi R, Bowser DN, Lester HA (2005) An angstrom scale interaction between plasma membrane ATP-gated P2X2 and alpha4beta2 nicotinic channels measured with fluorescence resonance energy transfer and total internal reflection fluorescence microscopy. J Neurosci 25:6911–6920. https://doi.org/10.1523/JNEUROSCI.0561-05.2005
6. Limapichat W, Dougherty DA, Lester HA (2014) Subtype-specific mechanisms for functional interaction between alpha6beta4* nicotinic acetylcholine receptors and P2X receptors. Mol Pharmacol 86:263–274. https://doi.org/10.1124/mol.114.093179
7. Shrivastava AN, Triller A, Sieghart W, Sarto-Jackson I (2011) Regulation of GABA (A) receptor dynamics by interaction with purinergic P2X(2) receptors. J Biol Chem 286:14455–14468. https://doi.org/10.1074/jbc.M110.165282
8. Richler E, Chaumont S, Shigetomi E, Sagasti A, Khakh BS (2008) Tracking transmitter-gated P2X cation channel activation in vitro and in vivo. Nat Methods 5:87–93. https://doi.org/10.1038/nmeth1144
9. Koshimizu TA, Kretschmannova K, He ML, Ueno S, Tanoue A, Yanagihara N, Stojilkovic SS, Tsujimoto G (2006) Carboxyl-terminal splicing enhances physical interactions between the cytoplasmic tails of purinergic P2X receptors. Mol Pharmacol 69:1588–1598. https://doi.org/10.1124/mol.105.019802
10. Compan V, Ulmann L, Stelmashenko O, Chemin J, Chaumont S, Rassendren F (2012) P2X2 and P2X5 subunits define a new heteromeric receptor with P2X7-like properties. J Neurosci 32:4284–4296. https://doi.org/10.1523/JNEUROSCI.6332-11.2012

Chapter 12

Application of Fluorescent Purinoceptor Antagonists for Bioluminescence Resonance Energy Transfer Assays and Fluorescent Microscopy

Mark Soave, Joëlle Goulding, Robert Markus, Stephen J. Hill, and Leigh A. Stoddart

Abstract

Fluorescent antagonists offer the ability to interrogate G protein-coupled receptor pharmacology. With resonance energy transfer techniques, fluorescent antagonists can be implemented to monitor receptor–ligand interactions using assays originally designed for radiolabeled probes. The fluorescent nature of these antagonists also enables the localization and distribution of the receptors to be visualized in living cells. Here, we describe the generation of modified purinergic receptors with the NanoLuc luciferase or SNAP-tag, using the P1 adenosine A_3 receptor as an example. We also describe the procedure of characterizing a novel fluorescent purinergic antagonist using ligand-mediated bioluminescence resonance energy transfer assays and confocal microscopy.

Key words Fluorescence, Antagonist, Purinergic receptor, BRET, Confocal microscopy, Ligand binding

1 Introduction

Cell surface proteins that bind and are activated by purines can be divided into two main families, the P1 receptors that respond to adenosine and the P2 receptors that respond to nucleosides such as ATP and UTP [1–3]. The four receptors that make up the P1 family [3] and the P2Y subclass of P2 receptors [1] are all G protein-coupled receptors (GPCRs). These receptors are known to be involved in a range of both physiological and pathophysiological conditions including cardiovascular disease, neurodegenerative conditions, inflammatory conditions and cancer [1, 3]. To date, two P2Y receptors are the target for clinically used compounds, the $P2Y_2$ and $P2Y_{12}$ receptors [4]. The $P2Y_{12}$ receptor is

Mark Soave and Joëlle Goulding contributed equally to this work.

Pablo Pelegrín (ed.), *Purinergic Signaling: Methods and Protocols*, Methods in Molecular Biology, vol. 2041,
https://doi.org/10.1007/978-1-4939-9717-6_12,

the target for the antiplatelet drug clopidogrel which is widely used in the treatment and prevention of myocardial infarctions [5]. The P2Y$_2$ receptor agonist diquafosol is in use in Japan and Korea as a treatment for dry eye [6]. As GPCRs are one of the most successfully targeted class of protein for therapeutics, with over 26% of currently approved pharmaceuticals acting at GPCRs [7], there is considerable interest in developing molecules that target the P1 and P2Y receptors [4].

One of the fundamental characteristics used in drug discovery is the affinity of a candidate molecule (ligand) for its target. For GPCRs, measuring affinity is often achieved through the use of radioligand binding techniques, either by directly measuring the levels of bound labeled ligand or by measuring the ability of an unlabeled compound to block the binding of a labeled ligand. Due to the inherent safety issues associated with the use of radioisotopes and ever more stringent disposal regulations, there is an increased interest in the use of alternative ways to measure ligand binding at GPCRs. Instead of using radioisotopes, ligands can also be labeled with fluorescence molecules and the resulting fluorescent ligands can be used the same way as radioligands [8]. The binding of fluorescent ligands to a receptor of interest can be measured directly, either through the use of a high content screening platform or using a multiwell plate reader [9, 10]. Although these techniques have been successfully employed, it has been found that many fluorescent ligands display high levels of nonspecific binding which can make the signal-to-noise ratios in these assays relatively small [10, 11].

In recent years, many researchers have started to employ resonance energy transfer based techniques such as BRET (bioluminescence resonance energy transfer) or TR-FRET (time-resolved fluorescence resonance energy transfer) to overcome this issue [12]. Traditionally, both BRET and FRET are used to investigate the proximity of two proteins which are tagged with either a donor or acceptor molecule. Resonance energy is transferred from the donor molecule (either a luciferase in the case of BRET or a terbium molecule in the case of TR-FRET) to acceptor protein but only if they are within 10 nm of each other, indicating close proximity of the two tagged proteins. These techniques have been extended to use fluorescent ligands in place of the acceptor molecule. The distance constraint of energy transfer means that energy from the donor can only be transferred to fluorescent ligand that is in very close proximity, and this is taken to mean that it is bound to the receptor. Our laboratory has successfully implemented a BRET based technique (NanoBRET) that utilizes the small, bright luciferase NanoLuc (NLuc) and fluorescent ligands to monitor ligand binding at a variety of GPCRs. Due to the wide spectral emission of NLuc [13], it can be used with a variety of fluorophores

[14]. When used with fluorophores such as BODIPY 630/650 which emit in the far-red region of the spectra, we and others have found that this technique has very low levels of nonspecific binding [15, 16]. NanoBRET has been used to determine the affinity of labeled and unlabeled ligands in addition to the kinetic parameters of both labeled and unlabeled compounds.

In addition to their use in assays to determine affinity parameters, fluorescent ligands can also be used to probe the cellular localization of receptors through the use of fluorescence microscopy and to quantify receptor organization and membrane diffusion characteristics by employing them in techniques such as fluorescence correlation spectroscopy. However, fluorescence microscopy is often not as straight forward as it first seems and we have found that many fluorescent ligands can display high levels of nonspecific binding [11, 17]. This is often an inherent property of the compound and recent advances have led to fluorescent ligands that show considerably less nonspecific binding [11, 18]. Notwithstanding this, care needs to be taken to choose the correct concentration of ligand used and labeling time that reflects the properties of the fluorescent ligand. To confirm that the fluorescent ligand binds to the receptor of interest, colocalization studies can be undertaken using GPCRs tagged with intrinsically fluorescent fusion proteins (GFP, YFP, mCherry, etc.). More recently, protein tags such as SNAP-tag and CLIP-tag (from New England BioLabs) or HALO-tag (from Promega) have been utilized in colocalization studies as these tags have the advantage of allowing the choice of fluorescent dye used. Characterization of a novel fluorescent ligand can be aided by utilizing a cell line carrying a fluorescent fusion protein or tag fused to the receptor of interest. Fusion proteins or labels should be chosen to contrast to that of the fluorescent ligand being investigated and likewise tags should be concomitantly labeled with a contrasting dye.

Fluorescent ligands have been successfully used to study various aspects of purinoceptor biology. In particular, our group has used fluorescent ligands to study the adenosine A_1 and A_3 receptors [13] and the $P2Y_2$ receptor [19]. Attachment of a fluorophore to an agonist or antagonist for a given receptor can result in a molecule that can no longer bind to the receptor of interest and, as mentioned above, can display high levels of nonspecific binding which render it unsuitable for imaging techniques. Therefore, any new fluorescent ligand needs to be characterized prior to use as a tool compound and in this chapter, we will describe the steps that need to be taken to characterize a newly obtained fluorescent ligand for use in NanoBRET assays and for standard confocal imaging.

2 Materials

2.1 Expression of NanoLuc and SNAP Tagged Receptors in Immortalized Cell Lines

1. pcDNA3.1(+) vector containing the sig.NLuc-receptor or sig. SNAP-receptor fusion constructs (*see* **Note 1**).
2. Immortalized cell line, we routinely use Human Embryonic Kidney (HEK) 293 cells as they are easy to maintain and transfect.
3. Complete cell media: for HEK293 cells, use Dulbecco's Modified Eagle's Medium (DMEM) supplemented with 10% fetal calf serum (FCS) and 2 mM L-glutamine.
4. Transfection reagent. We use Fugene® HD Transfection reagent (Promega™) as we find it to result in good transfection efficiencies.
5. 100 μg/mL Geneticin®/G418 in complete media for selection of cells expressing the plasmid of interest.
6. Opti-MEM or serum free media to ensure optimum transfection efficiencies.

2.2 Ligands for the Receptor of Interest

1. 1 μM fluorescent ligand in DMSO for the receptor of interest (*see* **Note 2**). Here we have used a fluorescent antagonist selective for the adenosine A_3 receptor (A_3AR) previously synthesized in house at the University of Nottingham (AV039 [20]) to generate example data. This ligand contains the BODIPY 630/650 fluorophore.
2. 10 μM unlabeled ligands in DMSO for the receptor of interest (*see* **Note 3**).

2.3 Nano BRET Experiments

1. 1 L of HEPES Buffered Saline Solution (HBSS): 145 mM NaCl, 5 mM KCl, 2 mM Sodium Pyruvate, 1.3 mM $CaCl_2$ dihydrate, 1 mM $MgSO_4$, 10 mM HEPES, 1.5 mM $NaHCO_3$, and 10 mM D-glucose, pH 7.4.
2. 1:40 dilution of furimazine (Promega™) in HBSS. This is the optimized NanoLuc substrate for NanoBRET experiments.
3. Plate reader suitable for reading fluorescence and luminescence. We use the PHERAstar FS Plate Reader (BMG Labtech™) (*see* **Note 4**). For BRET between NanoLuc and BODIPY 630/650 as used in this example, luminescence is measured between 420 and 500 nm and fluorescence above 610 nm.
4. Sterile white 96 well plates (Greiner Bio-One™).
5. Dulbecco's modified Phosphate buffered saline (PBS): 136 mM NaCl, 1.47 mM $KHPO_4$, 8.1 mM Na_2HPO_4, and 2.68 mM KCl.
6. 5 mg/mL poly-D-lysine in PBS.
7. Data analysis software. We use GraphPad Prism but others are available.

2.4 Live Cell Fluorescence Imaging

1. Inverted confocal microscope. In this example, we use a Zeiss Laser Scanning Microscope 880 microscope, fitted with plan-apochromat 63× 1.4NA oil objective lens, 488 and 633 nm laser lines and a 488/561/633 multi beam splitter (*see* **Note 5**).
2. Lab-Tek® Chambered cover glass, 8 well (Nunc®).
3. 0.5 μM SNAP-surface® Alexa Fluor® 488 (NEB) in complete media.
4. Immersion oil Immersol 518f (Zeiss).
5. Software for image acquisition, we use the proprietary Zeiss software Zen.
6. Software for image analysis. We use a combination of Zen, FIJI (NIH) and Microsoft office PowerPoint.

3 Methods

All cell culture work should be performed under sterile conditions in a Class II biological safety cabinet.

3.1 Generation of Cell Line Expressing NanoLuc or SNAP Tagged Receptor

1. Culture HEK 293 cells to approximately 70% confluence in a T25 flask in complete media.
2. Transfection of NanoLuc or SNAP tagged receptor is performed as per manufacturers' instructions. Briefly for transfecting with Fugene in a T25 flask, prepare transfection mix with 8 μg DNA as follows: 8 μg plasmid DNA containing tagged receptor, 415 μL Opti-MEM, and 16 μL Fugene HD reagent. Briefly vortex to mix thoroughly and incubate for 5–10 min at room temperature. Add 400 μL of this mix to the T25 flask and place the flask back in the cell culture incubator.
3. 24 h after transfection replace the Opti-MEM with 8 mL complete media with the relevant selective antibiotic (selection media). For pcDNA3.1 (+) this would be with 100 μg/mL G418 in complete media.
4. Culture the cells in selection media. Nontransfected cells will die as a result of the antibiotic. Replace the media with fresh selection media when this happens (*see* **Note 6**).
5. Continue to culture the cells in selection media until no more cells die due to the antibiotic.

3.2 Saturation NanoBRET Ligand Binding

The following methodology describes how to calculate the affinity (K_D) value of a fluorescent ligand at the NLuc-tagged receptor. This assay requires a concentration range of the fluorescent ligand to bind to the receptor, and a high concentration of a competitive antagonist to define nonspecific binding.

Table 1
Example plate layout for NanoBRET saturation binding and showing final assay concentrations

	1	2	3	4	5	6	7	8	9	10	11	12
A	500 nM FL			500 nM FL + 10 μM ant								
B	250 nM FL			250 nM FL + 10 μM ant								
C	125 nM FL			125 nM FL + 10 μM ant								
D	62.5 nM FL			62.5 nM FL + 10 μM ant								
E	31.2 nM FL			31.2 nM FL + 10 μM ant								
F	15.6 nM FL			15.6 nM FL + 10 μM ant								
G	7.8 nM FL			7.8 nM FL + 10 μM ant								
H	3.9 nM FL			3.9 nM FL + 10 μM ant								

FL fluorescent ligand, *ant* antagonist

1. Add 50 μL poly-D-lysine/well of white 96-well plates for 30 min under sterile tissue culture conditions at room temperature.
2. Aspirate off the poly-D-lysine and seed 30,000 HEK293 cells expressing the NLuc-tagged receptor in 100 μL complete media per well.
3. Place the 96-well plate in a 37 °C 5% CO_2 incubator overnight.
4. Prepare serial dilution (*see* **Note 7**) of fluorescent ligand in HBSS in the presence and absence of unlabeled antagonist at the concentrations shown in Table 1 (*see* **Note 8**).
5. Aspirate off media from cells and replace with 100 μL of fluorescent ligand alone or fluorescent ligand and competing antagonist. For this and subsequent steps, the assay does not need to be done under sterile conditions.
6. Incubate the plate at 37 °C, 0% CO_2 for 2 h (*see* **Note 9**).
7. Add 10 μL of diluted (1:40) furimazine to each well. Incubate the plate at 37 °C in the dark for 5 min.
8. Place the plate on the plate reader and simultaneously read the emitted light from the NLuc (donor) and fluorescent ligand (acceptor), ensuring the correct optic module for the fluorophore conjugated on the fluorescent ligand is used (*see* **Note 10**). We routinely set the program to read for three cycles and take the second read for subsequent analysis and interpretation.
9. Calculate the BRET ratios by dividing the signal from the acceptor channel by that of the donor channel (*see* **Note 11**):

$$\text{BRET ratio} = \frac{\text{Emission from acceptor channel}}{\text{Emission from donor channel}}$$

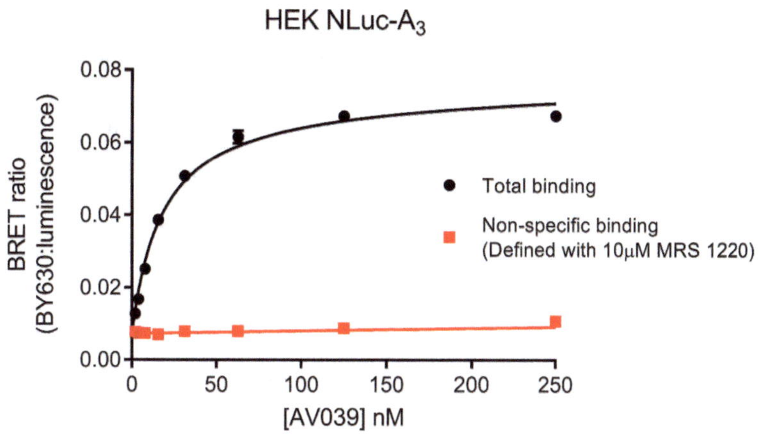

Fig. 1 Example of a saturation binding experiment showing the binding of fluorescent antagonist AV039 at the NLuc-tagged A_3AR

10. Then transfer the data into a new GraphPad Prism worksheet and analyze the data using the "one-site-total and nonspecific binding" setting. This fits the data to the following equation:

$$\text{BRET ratio} = \frac{B_{\max} * [B]}{[B] + K_D} + ((M * [B]) + C)$$

where B_{max} is the maximal response, $[B]$ is the concentration of fluorescent ligand in nM, K_D is the equilibrium dissociation constant in nM, M is the slope of the nonspecific binding component, and C is the intercept with the y axis (Fig. 1). If the results obtained appear different to the sample data provided in Fig. 1 it maybe be due to one of the following issues detailed in **Note 12**.

3.3 Competition NanoBRET Ligand Binding

The following methodology describes how to calculate IC_{50} or K_i values of unlabeled ligands at the NLuc-tagged receptor. This assay requires a fixed concentration of fluorescent ligand to bind to the receptor. The K_D of the fluorescent ligand must be known to calculate K_i values for the unlabeled ligands using the Cheng–Prusoff equation.

1. Add 50 μL poly-D-lysine/well of white 96-well plates for 30 min under sterile tissue culture conditions at room temperature.
2. Aspirate off the poly-D-lysine and seed 30,000 HEK293 cells expressing the NLuc-tagged receptor in 100 μL complete media per well.
3. Place the 96-well plate in in a 37 °C 5% CO_2 incubator overnight.
4. Prepare the unlabeled compounds at 10× the concentrations shown in Table 2.

Table 2
Example plate layout for NanoBRET competition binding final assay concentrations

	1 2 3	4 5 6	7 8 9	10 11 12
A	Total binding (FL at K_D)	Nonspecific binding (FL + 10 μM ant)	Total binding (FL at K_D)	Nonspecific binding (FL + 10 μM ant)
B	Ligand 1 × 10^{-5} M	Ligand 2 × 10^{-5} M	Ligand 3 × 10^{-5} M	Ligand 4 × 10^{-5} M
C	Ligand 1 × 10^{-6} M	Ligand 2 × 10^{-6} M	Ligand 3 × 10^{-6} M	Ligand 4 × 10^{-6} M
D	Ligand 1 × 10^{-7} M	Ligand 2 × 10^{-7} M	Ligand 3 × 10^{-7} M	Ligand 4 × 10^{-7} M
E	Ligand 1 × 10^{-8} M	Ligand 2 × 10^{-8} M	Ligand 3 × 10^{-8} M	Ligand 4 × 10^{-8} M
F	Ligand 1 × 10^{-9} M	Ligand 2 × 10^{-9} M	Ligand 3 × 10^{-9} M	Ligand 4 × 10^{-9} M
G	Ligand 1 × 10^{-10} M	Ligand 2 × 10^{-10} M	Ligand 3 × 10^{-10} M	Ligand 4 × 10^{-10} M
H	Ligand 1 × 10^{-11} M	Ligand 2 × 10^{-11} M	Ligand 3 × 10^{-11} M	Ligand 4 × 10^{-11} M

FL fluorescent ligand, *ant* antagonist

5. Prepare a high concentration solution (100 μM) of the unlabeled antagonist used in the saturation binding experiment to determine nonspecific binding (*see* **Note 13**).
6. Prepare 1 mL warm HBSS with the fluorescent ligand diluted to 10× the K_D; for example, if the K_D of the fluorescent ligand is 25 nM, prepare 1 mL HBSS with 250 nM fluorescent ligand.
7. Aspirate off media from cells and replace with 80 μL warm HBSS. For this and subsequent steps, the assay does not need to be done under sterile conditions.
8. Add 10 μL of HBSS or competing antagonist to the relevant wells, then add 10 μL of the fluorescent ligand to every well on the plate. This will dilute the competing antagonist and fluorescent ligand to a final concentration which is 10× less than the prepared solutions.
9. Add 10 μL of diluted (1:40) Furimazine to each well. Incubate the plate at 37 °C in the dark for 5 min.
10. Place the plate on the plate reader and simultaneously read the emitted light from the NLuc (donor) and fluorescent ligand (acceptor), ensuring the correct optic module for the fluorophore conjugated on the fluorescent ligand is used (*see* **Note 10**). We routinely set the program to read for three cycles and take the second read for subsequent analysis and interpretation.
11. Calculate the BRET ratios by dividing the signal from the acceptor channel by that of the donor channel:

$$\text{BRET ratio} = \frac{\text{Emission from acceptor channel}}{\text{Emission from donor channel}}$$

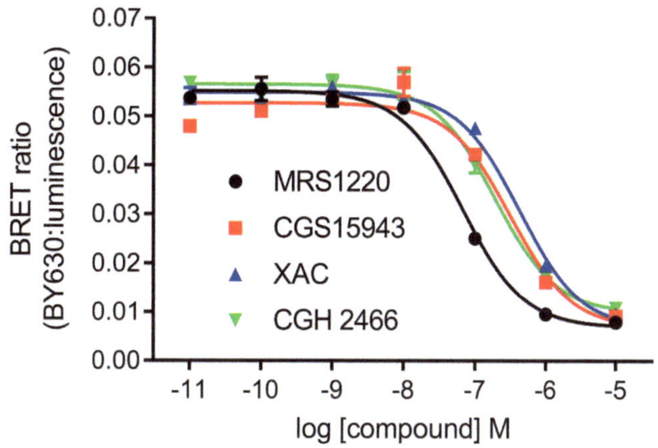

Fig. 2 Example of a competition binding experiment showing the effect of increasing concentrations of competing ligands on the binding of AV039 at the NLuc-tagged A_3AR

12. Then transfer the data into a new GraphPad Prism worksheet and fit to the nonlinear regression 'One site – Fit logIC50' setting which fits the data to the following equation:

$$\%\text{Specific binding} = \frac{100 \times [A]}{[A] + \text{IC}_{50}}$$

where $[A]$ is the concentrations of competing ligand and the IC_{50} is the molar concentration of ligand required to inhibit 50% of the specific binding of the fluorescent ligand.

13. If the K_D of the fluorescent ligand is known, then the K_i of each ligand can be calculated using the Cheng–Prusoff correction which is as follows:

$$K_i = \frac{\text{IC}_{50}}{1 + \frac{[L]}{K_D}}$$

where $[L]$ is the concentration of fluorescent ligand in nM, K_D is the K_D of the fluorescent ligand and IC_{50} is calculated as above. If the results obtained appear different to the sample data provided in Fig. 2 it maybe be due to one of the following issues detailed in **Note 11**.

3.4 Live Cell Fluorescence Imaging

1. Add 200 μL poly-D-lysine/well of a Lab-Tak 8-chambered cover glass (*see* **Note 14**) for 30 min under sterile tissue culture conditions at room temperature.
2. Aspirate off the poly-D-lysine and seed 30,000 HEK293 cells expressing the SNAP-tagged receptor in 300 μL complete media per well.

3. Place the 96-well plate in in a 37 °C 5% CO_2 incubator overnight (*see* **Note 15**).
4. Aspirate growth media from wells replace with 180 μL SNAP labeling solution for 30 min at 37 °C 5% CO_2. For this and subsequent steps, the experiment does not need to be done under sterile conditions.
5. Wash cells twice with warm HBSS before adding a final of 200 μL warm HBSS.
6. To identify nonspecific binding (*see* **Note 16**), add a high concentration of nonfluorescent antagonist for the receptor being studied (*see* **Note 17**) to the relevant wells.
7. Add fluorescent ligand to appropriate wells at 10× K_D (*see* **Note 18**) and incubate at 37 °C for 30–60 min (*see* **Note 19**).
8. Apply one drop of immersion oil to the objective, position the cover glass in the slide holder and allow to acclimatize for at least 10 min which will reduce the drift when imaging.
9. Set up the beam path for one track and two channels; channel 1 entitled "AlexaFluor 488," bandwidth range 493–629 nm, and 488 laser at 2% power; channel 2 entitled "BODIPY630/650", bandwidth range 638–737 nm, and 633 laser power 3%. Select multi beam splitter 488/561/633 and pinhole at 1 airy unit for the longer laser wavelength (633) which will give an optical slice of 0.8 μm.
10. Set up the acquisition parameters for frame size, scan speed, pixel dwell time and line averaging (*see* **Note 20**).
11. Localize cells initially using transmitted light and then if possible, dependent on color of probes used, find a bright cell using the LED or lamp light source for epifluorescence. Use this bright cell to set the detector gains/offset for image capture. If this option is not possible, bright cells can be found by live or continuous laser scanning using the acquisition parameters detailed for quick image capture. Care must be taken to turn off laser acquisition when not in use to limit bleaching of sample.
12. Define the dynamic range by setting the digital Gain and Offset for your example cell (Fig. 3, *see* **Note 21**).
13. Locate cells a new field of view and find the equatorial plane to image the receptor on the plasma membrane.
14. Collect and save four images per condition (*see* **Note 22**).

3.5 Preparing Images for Publication

1. Copy all the raw files (∗.LSM) into a new working folder, it is advisable to create a work folder for each figure.
2. Open your file within Zen, at the mid bottom of the image click "Show all" to see all the editing tools. Set brightness contrast

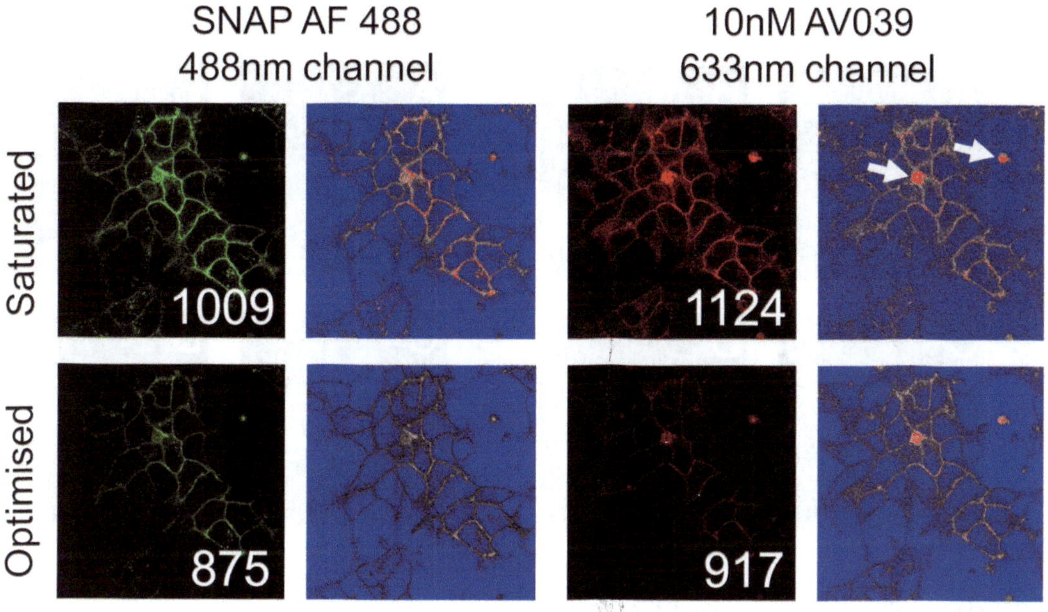

Fig. 3 Optimizing gain and offset values to obtain nonsaturated images. The dynamic range can be adjusted to optimize gain and offset which are pseudo-colored red and blue respectively. The gain (shown numerically in white) is decreased until saturated pixels are minimized for each channel (488 and 633 nm). Unless an advanced user we advise not to alter the offset. Clear saturation due to ligand aggregates (arrows) can be ignored in optimization so as not to set the gains too low. Within these panels the raw intensity is displayed, brightness and contrast can be altered for publication (as in Fig. 4) as long as all images are treated the same

for each channel (red and green), ensuring you input the same *black/white* values (numbers below the histogram left side and right side) for both experimental and matched control images (*see* **Note 23**).

3. Export separate files for each channel as a ".TIF" selecting "Full Resolution image window." (*see* **Note 24**). Each channel can be exported either as a grey scale images or in color and care should be taken in selecting which colors ensure that people who are color blind can distinguish between the images (Fig. 4).
4. Add a scale bar and export as an additional file, this will be used to copy scale bar dimensions (*see* **Note 25**).
5. Open Microsoft Office PowerPoint (*see* **Note 26**), set page properties to A4 and select "Do not compress Images" and set default target output to "*330 ppi*" within advanced settings (*see* **Note 27**).
6. Import images using the "Insert picture" command, create figure layout and at this stage use the TIF file with scale bar included to determine scale bar dimensions and create a new manually drawn scale bar using PowerPoint tools (Fig. 4). At this stage add any other additional graphics such as arrows, dashed lines etc. Do not save PowerPoint slide as TIF directly,

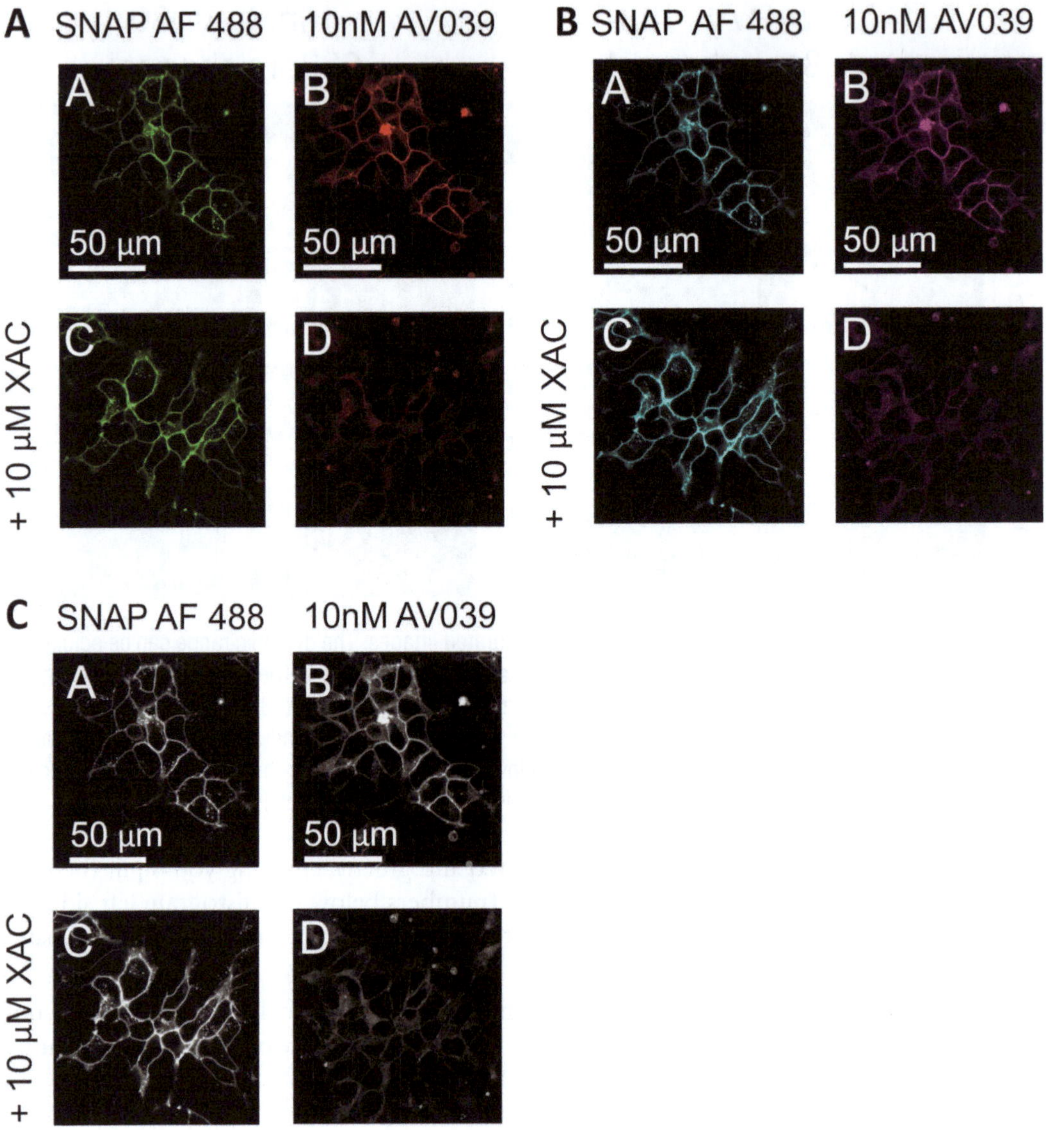

Fig. 4 Preparing images for publication. Example panels for publication which display the SNAP-tagged and ligand-bound localization of the receptor in the absence and presence of a high concentration of competing antagonist. (**a**) The first panel displays the two channels as green/red as typically described. (**b**) An alternative is cyan/magenta which may project more reliably on screen for presentation purposes but also is more suitable for color blindness. (**c**) A third alternative is greyscale which can often be more reproducible in print when hardcopy is used

as it will be downscaled to 96 ppi, despite the resolution is set to 330 ppi in advance setting.

7. Save as PDF and in the options, choose maximum resolution, for publishing. If possible, submit this file.
8. If a TIF is required by the journal, open PDF in available software (e.g., Gimp), convert to 300 ppi, crop and scale, and export as TIF.

4 Notes

1. Fluorescent ligands can either be custom synthesized or purchased commercially from companies such as Sigma-Aldrich or HelloBio. We routinely use CA200645, which is a nonselective adenosine receptor fluorescent antagonist available from HelloBio. Fluorescent ligand stocks are stored in small (5–10 μL) aliquots at −20 °C.
2. For most ligands, we solubilized them to 10 mM in DMSO unless stated in the chemical information sheet that they are soluble in water.
3. The NanoLuc and SNAP tagged constructs we use comprise of the murine $5HT_{3A}$ serotonin receptor signal sequence (sig) followed by the sequence for NanoLuc or SNAP-tag and then the receptor [22]. The start codon of the receptor is mutated from Met to Leu (ATG to CTG). This creates N-terminally NanoLuc or SNAP tagged receptor. The signal sequence is included as the N-terminus has an important role in the correct trafficking of GPCRs to the cell surface [21] and the addition of an N-terminal tag can drastically reduce receptor expression.
4. Different plate readers are available to measure NanoBRET: CLARIOstar plate reader series (BMG Labtech), Varioskan plate reader series (Thermo Scientific), and SpectraMax plate reader series (Molecular Devices).
5. Imaging with fluorescent ligands can be performed on both confocal and wide-field microscopes. The specific features required will depend on the fluorescent ligand and other fluorophores and level of resolution required. If using a wide-field microscope, deconvolution can be used post-image capture to improve the contrast and resolution of the images to allow more accurate determination of the localization of the fluorescent ligand.
6. This usually takes 2–3 weeks, but please note this timeframe can vary. Cells are then split into T75 flasks for growth and expansion.
7. For serial dilution, tips should be changed between every dilution to ensure no carry-over of a higher concentration of compound.
8. This concentration range can be altered if the fluorescent ligand is thought to be of higher or lower affinity.
9. The length of time for incubation depends on the binding kinetics of the ligand in question. We have found 2 h is usually sufficient for most fluorescent ligands to achieve equilibrium at 37 °C.

Table 3
Commonly used fluorophores and details of PHERAstar optic modules

Fluorophore	PHERAstar optic module	Wavelengths (nm) Donor, acceptor
BODIPY-FL	BRET 1 plus 1108A1	460–490, 520–550
Tetramethylrhodamine (TAMRA)	LUM 1604A1	410–490, 550–Longpass
BODIPY 630/650	LUM 1501B1	420–500, 610–Longpass

10. The optic module needed to simultaneously read the luminescence of the NLuc and the light emitted from the fluorescent ligand depends on the conjugated fluorophore on the ligand. *See* Table 3 below for a list of potential fluorophores and the module required for the PHERAstar plate reader.
11. If BRET ratios are unexpectedly high, ensure cells are still adherent on the bottom of the well and have not washed off with initial changing of serum-free media to HBSS. Also, check the luminescence values are equal across the plate. In theory, the luminescence should be similar; however, it could be that the physiochemical properties of the ligand are affecting the luminescence output. Although a decrease in luminescence may be due to efficient transfer of energy to the acceptor molecule. Additionally, one can turn the plate around and read the plate 'upside down' to ascertain if there's any discrepancies occurring from the way in which the substrate is added to the plate, or if the plate reader has a systematic error in reading certain positions in the plate. Ensure the luminescence values are not saturating the detector. This happens when the light output is too great for the detector's threshold and results in exactly the same luminescence reading for every affected well. Either reduce the gain settings on the plate reader or seed the cells at a lower density.
12. If results are not as expected, check the ligand concentrations are right for the expected affinities. For saturation binding experiments, *see* the graphs provided below as a guide (Fig. 5). It is possible the fluorescent ligand is sticking to the plastic of the microcentrifuge tubes or the plastic of the 96-well plate rather than staying in solution which would result in the graph obtained in Fig. 5c. Try adding 0.1% or 0.2% bovine serum albumin (BSA) to the HBSS when diluting the ligands and repeat the experiment. Ensure the incubation times and temperatures are suitable for the ligands in question, be careful of the conditions if using labile ligands such as adenosine and always prepare these fresh on the day of experiment. Be aware if competing for fluorescent ligand binding with agonists as the

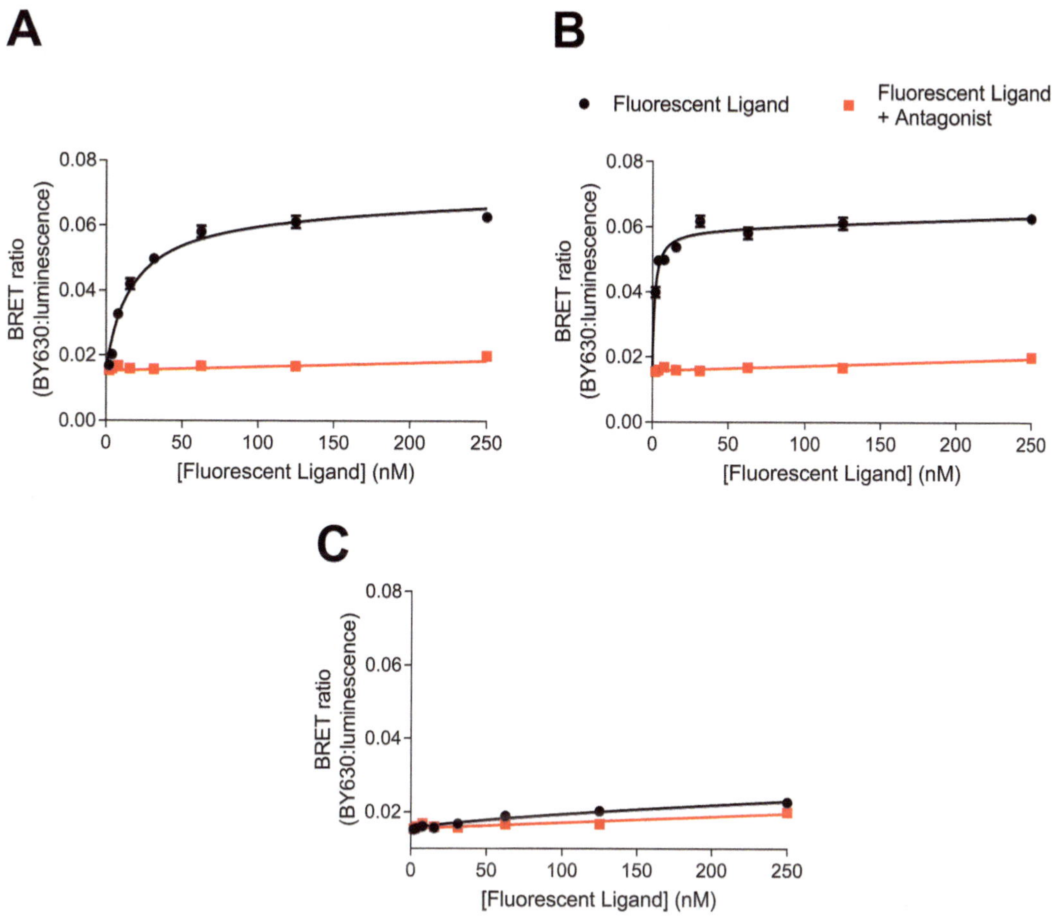

Fig. 5 Simulated data of saturation binding with an uncharacterized fluorescent ligand. (**a**) A good concentration range for the affinity of the ligand in question. The resulting affinity estimate will be accurate. (**b**) The concentration range is too high for the affinity of the ligand, so the estimate of the affinity will be inaccurate. Try again with a lower concentration range. (**c**) There is no evidence of specific binding over these ligand concentrations

receptors will undergo internalization which can affect the resulting BRET ratios.

13. The high concentration of antagonist in row A acts to define nonspecific binding and can be used to determine if a competing ligand being tested is fully competitive (reaches the same level of binding inhibition) or not.
14. Whilst there are various slide setups on which cells can be plated we recommend using the 8-well Nunc Lab Tek chambered cover glass slides. The cover glass slides have eight separate wells atop a cover glass 1–1.5 thickness allowing multiple conditions within the same experimental slide. Glass thickness is important for optimizing image quality and should be 1–1.5 for confocal microscopy.

15. Dependent on the cell type and use of fusion protein or tag it may be necessary to employ 2-day plating to allow cells either time to settle or to express the required protein/tag at a sufficient level. If this longer plating time is employed seeding density should be reduced accordingly.
16. The level of nonspecific binding can be dependent on the chemistry of the fluorescent ligand for example its hydrophobicity. Observing high levels of labeling but significant nonspecific binding it may be pertinent to reduce the concentration of fluorescent antagonist and the laser power can be increased as required to compensate for this.
17. It is possible to determine the required concentration of unlabeled antagonist from competition NanoBRET binding assays. Commonly a high concentration (10 μM) of selective antagonist is used.
18. K_D parameters can either be obtained from the literature, contained within supplier documents or could be derived in-house for example using the NanoBRET binding assay. Use of ligand at 1–10× its K_D should give a sufficient proportion of receptor binding for visualization whilst also being at a displaceable concentration in order to study nonspecific binding.
19. Incubation time depends on the kinetics of the fluorescent ligand. Dependent on research query, using a 30–60 min incubation time is likely to be sufficient for purpose.
20. A slower scan speed will increase the information gained per pixel however the resultant longer dwell time of the laser at each pixel will increase the potential for cell damage through phototoxicity. A compromise is therefore met between speed of acquisition, phototoxicity and resolution (Fig. 6). By choosing line averaging a number of scans are made at a faster speed and less time spent per pixel per scan which reduces photodamage. The 100-STACK set up (Table 4) defines parameters which can be selected to collect a time series of 100 sequential images. These images are then combined in a post-acquisition step (e.g., in Fiji), in effect averaging the intensity of each pixel and reducing pixel dwell time and photodamage.
21. Adjust digital gain on brightest cell to minimize red pixels ensuring to scan through a few *Z*-positions in case original focus was not at the brightest point. The digital gain can amplify the signal however higher settings will also amplify noise. Adjust offset if needed to minimize blue pixels. The digital offset affects contrast and should be kept close to zero (we recommend keeping at zero unless an advanced user), decreasing too far may result in faint data being lost.

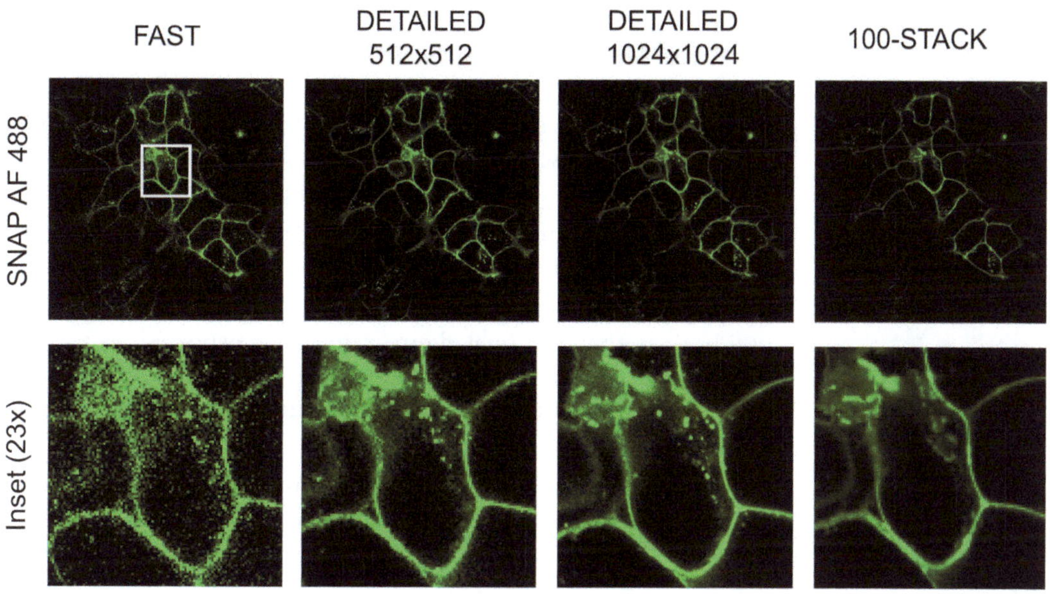

Fig. 6 Images of SNAP tagged receptor using different acquisition parameters. Adjusting acquisition parameters (frame size, scan speed, pixel dwell time and line average) allows image capture of either a fast example picture or of a more detailed image which could be used for publication. The 100-stack image is formed post-acquisition using image analysis software from 100 sequential images which individually would have poor resolution but combined are comparable to the detailed 1024 × 1024

Table 4
Acquisition parameters for fast, detailed and 100-stack scans

Acquisition parameter	Fast scan	Detailed scan	Detailed scan	100-Stack
Frame size	512 × 512	1024 × 1024	512 × 512	512 × 512
Scan speed	8	6	7	9
Pixel dwell time (μs)	1.52	2.06	2.06	1.03
Line averaging	1	8	8	1

22. Always check publication specifications for the intended journal. Note that a 512 × 512 pixel image is suitable for publication when scaled at 1.7 in. (4.3 cm). This means that in a composite figure you aim not to enlarge a 512 × 512 image more than 1.7 in. This will ensure that the image will be 300 pixels per inch (ppi), which is the standard requirement for publication figures.
23. Alternatively, open ∗.LSM image in Fiji, a pop up window "Bio-Format Import options" window will open, at the bottom left unclick "Autoscale," and click OK. To set black/white values (brightness contrast), click Ctrl + Shift + C, in this window click "SET," type in the minimum and maximum

values, make a note of these numbers and click ok. Do the same for all the other images including controls.

24. If using Fiji, covert image to RGB TIF (command: Image\Type \RGB color), save the new resulting image as a .Tif. These files will then be used for the figure.

25. The scale bar is added separately as this allows ease of editing if figure specifications change without needing to return to the original raw files and exporting again.

26. Figure assembly is easier done in PowerPoint, but the drawback is that the final figure cannot be exported as a JPEG without accruing artifacts. Other software such as Inkscape or Adobe Illustrator will keep TIF images intact when exporting the final figure.

27. Also check the register of your computer to ensure that it exports figures at 300 ppi.

References

1. Abbracchio MP, Burnstock G, Boeynaems J-M, Barnard EA, Boyer JL, Kennedy C, Knight GE, Fumagalli M, Gachet C, Jacobson KA, Weisman GA (2006) International Union of Pharmacology. LVIII: Update on the P2Y G protein-coupled nucleotide receptors: from molecular mechanisms and pathophysiology to therapy. Pharmacol Rev 58:281–341
2. Coddou C, Yan Z, Obsil T, Huidobro-Toro JP, Stojilkovic SS (2011) Activation and regulation of purinergic P2X receptor channels. Pharmacol Rev 63:641–683
3. Fredholm BB, IJzerman AP, Jacobson KA, Linden J, Müller CE (2011) International Union of Basic and Clinical Pharmacology. LXXXI. Nomenclature and classification of adenosine receptors, an update. Pharmacol Rev 63:1–34
4. Burnstock G (2017) Purinergic signalling: therapeutic developments. Front Pharmacol 8:661
5. Sarafoff N, Byrne RA, Sibbing D (2012) Clinical use of clopidogrel. Curr Pharm Des 18:5224–5239
6. Lau OC, Samarawickrama C, Skalicky SE (2014) P2Y2 receptor agonists for the treatment of dry eye disease: a review. Clin Ophthalmol 8:327–334
7. Garland SL (2013) Are GPCRs still a source of new targets? J Biomol Screen 18:947–966
8. Stoddart LA, Kilpatrick LE, Briddon SJ, Hill SJ (2015) Probing the pharmacology of G protein-coupled receptors with fluorescent ligands. Neuropharmacology 98:48–57
9. Stoddart LA, Vernall AJ, Denman JL, Briddon SJ, Kellam B, Hill SJ (2012) Fragment screening at adenosine-A(3) receptors in living cells using a fluorescence-based binding assay. Chem Biol 19:1105–1115
10. Arruda MA, Stoddart LA, Gherbi K, Briddon SJ, Kellam B, Hill SJ (2017) A non-imaging high throughput approach to chemical library screening at the unmodified adenosine-A3 receptor in living cells. Front Pharmacol 8:908
11. Vernall AJ, Stoddart LA, Briddon SJ, Ng HW, Laughton CA, Doughty SW, Hill SJ, Kellam B (2013) Conversion of a non-selective adenosine receptor antagonist into A3-selective high affinity fluorescent probes using peptide-based linkers. Org Biomol Chem 11:5673–5682
12. Stoddart LA, White CW, Nguyen K, Hill SJ, Pfleger KD (2016) Fluorescence- and bioluminescence-based approaches to study GPCR ligand binding. Br J Pharmacol 173:3028–3037
13. Stoddart LA, Johnstone EK, Wheal AJ, Goulding J, Robers MB, Machleidt T, Wood KV, Hill SJ, Pfleger KD (2015) Application of BRET to monitor ligand binding to GPCRs. Nat Methods 12:661–663
14. Stoddart LA, Kilpatrick LE, Hill SJ (2018) NanoBRET approaches to study ligand binding to GPCRs and RTKs. Trends Pharmacol Sci 39:136–147
15. Soave M, Stoddart LA, Brown A, Woolard J, Hill SJ (2016) Use of a new proximity assay (NanoBRET) to investigate the ligand-binding characteristics of three fluorescent ligands to

the human beta1-adrenoceptor expressed in HEK-293 cells. Pharmacol Res 4:e00250

16. Christiansen E, Hudson BD, Hansen AH, Milligan G, Ulven T (2016) Development and characterization of a potent free fatty acid receptor 1 (FFA1) fluorescent tracer. J Med Chem 59:4849–4858
17. Rose RH, Briddon SJ, Hill SJ (2012) A novel fluorescent histamine H(1) receptor antagonist demonstrates the advantage of using fluorescence correlation spectroscopy to study the binding of lipophilic ligands. Br J Pharmacol 165:1789–1800
18. Stoddart LA, Vernall AJ, Bouzo-Lorenzo M, Bosma R, Kooistra AJ, de Graaf C, Vischer HF, Leurs R, Briddon SJ, Kellam B, Hill SJ (2018) Development of novel fluorescent histamine H1-receptor antagonists to study ligand-binding kinetics in living cells. Sci Rep 8:1572
19. Conroy S, Kindon ND, Glenn J, Stoddart LA, Lewis RJ, Hill SJ, Kellam B, Stocks MJ (2018) Synthesis and evaluation of the first fluorescent antagonists of the human P2Y2 receptor based on AR-C118925. J Med Chem 61:3089–3113
20. Vernall AJ, Stoddart LA, Briddon SJ, Hill SJ, Kellam B (2012) Highly potent and selective fluorescent antagonists of the human adenosine A_3 receptor based on the 1,2,4-triazolo (4,3-a)quinoxalin-1-one scaffold. J Med Chem 55:1771–1782
21. Dong C, Filipeanu CM, Duvernay MT, Wu G (2007) Regulation of G protein-coupled receptor export trafficking. Biochim Biophys Acta 1768:853–870
22. Keppler A, Pick H, Arrivoli C, Vogel H, Johnsson K (2004) Labeling of fusion proteins with synthetic fluorophores in live cells. Proc Natl Acad Sci U S A 101:9955–9959

Chapter 13

Detection of Extracellular ATP in the Tumor Microenvironment, Using the pmeLUC Biosensor

Elena De Marchi, Elisa Orioli, Anna Pegoraro, Elena Adinolfi, and Francesco Di Virgilio

Abstract

ATP is one of the main components of the tumor microenvironment, where it affects cell growth, tumor progression and antitumor immune response. The development of the pmeLUC probe, a luciferase engineered to be expressed on the outer facet of the plasma membrane, allowed real-time measurement of extracellular ATP in vitro and in vivo systems, among which the tumor microenvironment. Here we describe the experimental procedures to measure extracellular ATP levels in the tumor microenvironment of three different cancer models generated by the implant of pmeLUC-expressing tumor cells into the appropriate mice strain: ACN human neuroblastoma (*nude/nude* mice host), WEHI-3B murine leukemia (BALB/c host), and B16F10 murine melanoma (C57Bl/6 host). The procedure to obtain stable expression of pmeLUC in different cell types and methods for the measurement of extracellular ATP with pmeLUC in vitro are also described.

Key words Extracellular ATP, Plasma membrane luciferase, Luciferin, Luminescence, Luminometry

1 Introduction

ATP is the main intracellular energy intermediate produced by phosphorylation, fermentation and cellular respiration involved in virtually all cellular responses [1]. ATP is also passively (e.g., following cell injury or death) or actively (via exocytosis or channel-mediated transport) released into the extracellular milieu, and as such is the essential component of an ubiquitous and highly plastic signaling system (purinergic signaling), with an established role in neurotransmission, inflammation and cancer [2, 3]. Measurement of the ATP concentration has been a crucial issue for many years, with several methods developed for in vitro and in vivo applications. In vitro systems include techniques based on aptamer [4], micro-electrode biosensor [5], the engineered sensor Perceval [6], FRET probes [7], HPLC-based methods [8], and fluorescent dyes [9]. These applications are widely used and most of them have a

Pablo Pelegrín (ed.), *Purinergic Signaling: Methods and Protocols*, Methods in Molecular Biology, vol. 2041, https://doi.org/10.1007/978-1-4939-9717-6_13,

high sensitivity, but the limit of fluorimetric or colorimetric kits is the cell harvesting and lysis required to extract the intracellular ATP content and the consequent inaccurate measurement of ATP levels [10]. Moreover, most of these techniques are not amenable to in vivo studies, and their kinetics are not compatible with those of biological systems. So far, the best and most widely used approach used to measure ATP both in vitro and in vivo is to take advantage of the bioluminescent luciferin-luciferase reaction. Firefly luciferase-based probes [10–12] have high sensitivity and affinity and can be also localized to subcellular organelles, a very useful feature since ATP levels vary widely in different intracellular or pericellular compartments. Luciferase, derived from the firefly *Photinus pyralis*, is an enzyme that in the presence of magnesium catalyzes the ATP-dependent oxidation of the substrate D-luciferin with emission of light. The reaction is described below [13]:

$$\text{ATP} + \text{luciferin} + O_2 \rightarrow \text{oxyluciferin} + \text{AMP} + \text{PP}_i + CO_2 + \text{light}$$

The intensity of light emission is proportional to the amount of ATP present and can be easily measured with a luminometer (for further details refer to [10, 11]). Luciferase has been engineered to be targeted to the mitochondrial matrix (mtLUC) [14, 15], the cytosol (cytLUC) [8], the subplasma membrane (pmLUC) [16, 17], the outer side of the plasma membrane (pmeLUC) [12, 18], or the nucleus (nucLUC) [16].

The pmeLUC biosensor, developed by Di Virgilio and collaborators, was engineered by appending to the luciferase cDNA a N-terminal leading sequence and a C-terminal glycosyl phosphatidylinositol (GPI) anchor of the folate receptor [12]. Thanks to these modifications, pmeLUC is targeted to and expressed on the outer face of the plasma membrane, thus allowing detection of ATP levels in the extracellular environment [10]. This probe can be used to engineer all cell types amenable to transfection, to allow analysis of the extracellular ATP concentration in a multiplicity of in vitro and in vivo settings. We have used HEK293T cells stably transfected with pmeLUC (HEK293T-pmeLUC) to monitor the extracellular ATP concentration in many different disease models: tumors, graft-vs-host reactions, contact dermatitis, tissue transplantation, type 2 diabetes. To this aim, pmeLUC-transfected HEK293T cells are inoculated at tumor or inflammation sites to report the extracellular ATP concentration. Alternatively, cancer cells themselves can be engineered with pmeLUC to function as direct sensors of the ATP concentration in the tumor microenvironment (TME) [19–22]. Virtually all cancer cell lines can be stably transfected with pmeLUC and used to start tumors in different mouse strains. Here we will describe three experimental procedures developed in our laboratory.

2 Materials

2.1 Cells

The pmeLUC probe can be transfected virtually in any cell type with an appropriate transfection method. We describe here the protocols used for human ACN neuroblastoma, mouse WEHI-3B leukemia and mouse B16F10 melanoma cells, that differ for transfection technique and host mice strain.

2.2 Cell Culture

1. Complete culture medium for ACN and B16F10 cells: RPMI 1640 medium supplemented with 10% fetal bovine serum (FBS), 100 U/ml penicillin and 100 μg/ml streptomycin, and nonessential amino acids, pH 7.4.
2. Complete culture medium for WEHI-3B cells: Iscove modified Dulbecco's medium supplemented with 10% FBS, 100 U/ml penicillin and 100 μg/ml streptomycin, pH 7.4.

2.3 Transfection of ACN Cells

1. Complete cell culture medium for ACN cells.
2. ACN cells.
3. TransIT-2020 transfection reagent or equivalent reagent of ACN cells.
4. Opti-MEM or equivalent reduced serum medium.
5. 1.5 ml sterile centrifuge tubes.
6. 10 cm Petri dishes for tissue culture.
7. 10 μg pmeLUC encoding plasmid in 60 μl of sterile water.
8. Selection solution: 0.8 mg/ml G-418 in culture media.

2.4 Transfection of WEHI-3B Cells

1. Complete cell culture medium for WEHI-3B.
2. WEHI-3B cells.
3. Electroporation buffer T, R and electrolytic buffer E for WEHI-3B cells.
4. Electroporation cuvettes.
5. Microporator MPT-100 and microporator pipette.
6. Power supply.
7. 10 ml conical tubes.
8. 6-Well plates.
9. 3 μg pmeLUC plasmid in 60 μl of sterile water.
10. 0.4 mg/ml G-418 solution in culture media.

2.5 Transfection of B16F10 Cells

1. Complete cell culture medium for B16F10.
2. B16F10 cells.
3. Lipofectamine LTX and PLUS reagent or equivalent reagent to transfect B16F10 cells.

4. Opti-MEM or equivalent reduced serum medium.
5. 1.5 ml sterile centrifuge tubes.
6. 6-Well plates.
7. 3 μg pmeLUC plasmid in 60 μl of sterile water.
8. 0.4 mg/ml G-418 solution in culture media.

2.6 In Vitro Extracellular ATP Measurement for ACN Cells

1. 6-Well plates.
2. RPMI 1640 medium.
3. 15 mg/ml D-luciferin sodium salt stock solution dissolved in sterile PBS. Protect from light and store aliquots at −80 °C.
4. 100 mM adenosine-5′-triphosphate disodium salt (ATP) dissolved in 0.1 M Tris-base solution at pH 7.5. Store in 0.5 ml aliquots at −80 °C.
5. Total body luminometer (PerkinElmer Caliper IVIS 100 System™ or similar).
6. Living Image® software.

2.7 In Vitro Extracellular ATP Measurement for WEHI-3B Cells

1. 96-Well plates.
2. Iscove modified Dulbecco's medium.
3. 15 mg/ml D-luciferin stock solution dissolved in sterile PBS. Protect from light and store aliquots at −80 °C.
4. 100 mM adenosine-5′-triphosphate disodium salt (ATP) dissolved in 0.1 M Tris-base solution at pH 7.5. Store in 0.5 ml aliquots at −80 °C.
5. Total body luminometer (PerkinElmer Caliper IVIS 100 System™ or similar).
6. Living Image® software.

2.8 In Vitro Extracellular ATP Measurement for B16F10 Cells

1. 24-Well plates.
2. RPMI 1640 medium.
3. D-Luciferin solution: 15 mg/ml D-luciferin sodium salt stock solution dissolved in sterile PBS. Protect from light and store aliquots at −80 °C.
4. 100 mM adenosine-5′-triphosphate disodium salt (ATP) dissolved in 0.1 M Tris-base solution at pH 7.5. Store in 0.5 ml aliquots at −80 °C.
5. Total body luminometer (PerkinElmer Caliper IVIS 100 System™ or similar).
6. Living Image® software.

2.9 In Vivo Extracellular ATP Measurement

1. 15 mg/ml D-luciferin sodium salt stock solution in sterile PBS. Protect from light and store aliquots at −80 °C.
2. Phosphate buffered saline (PBS): 137 mM NaCl, 4.3 mM Na_2HPO_4, 1.4 mM KH_2PO_4, 2.7 mM KCl, pH 7.4.
3. Isofluorane.
4. Syringe and 29-G needle.
5. ACN cells stably expressing pmeLUC (ACN-pmeLUC).
6. B16F10 cells stably expressing pmeLUC (B16F10-pmeLUC).
7. WEHI-3B cells stably expressing pmeLUC (WEHI-3B-pmeLUC).
8. *nude/nude* mice 4–6 weeks old.
9. C57Bl/6 mice 4–6 weeks old.
10. BALB/c 4–6 weeks old.
11. Total body luminometer (PerkinElmer Caliper IVIS 100 System™ or similar).
12. Apparatus for inhalation anesthesia.
13. Living Image® software.

3 Methods

3.1 ACN Cell Transfection

1. Plate 2 × 10^6 ACN cells into 100 × 20 mm cell culture Petri dishes (grow to 80% confluence) 24 h before transfection.
2. The day of transfection, mix gently 10 μg of pmeLUC plasmid DNA by pipetting in 250 μl of Opti-MEM medium and add to 30 μl of room temperature prewarmed and vortexed TransIT-2020 Transfection Reagent.
3. Incubate the mixture for 30 min at room temperature.
4. Add drop by drop the reaction mix (250 μl Opti-MEM Medium, 10 μg pmeLUC DNA, 30 μl TransIT-2020 transfection reagent) to ACN cells.
5. Incubate for 16 h at 37 °C in a 5% CO_2 humidified incubator.
6. Remove the reaction mix and wash with 2 ml of sterile PBS.
7. Add 2 ml of complete cell medium and incubate for 24 h in a 5% CO_2 humidified incubator at 37 °C.

3.2 WEHI-3B Cell Transfection

1. Suspend 6 × 10^6 WEHI-3B cells (*see* **Note 1**) in 10–12 μl of the appropriate mix of electroporation T and R buffers and mix with 3 μg of plasmid's pmeLUC DNA (*see* **Note 2**).
2. Add 3 ml of Electrolytic Buffer E to electroporation cuvette.
3. Set the power supply according to the following protocol:

Voltage (1–2500 V)	1250 V
Pulse width (1–100)	40 ms
Pulse number (1–10)	1

4. Pipet gently DNA and cell suspension to mix completely.
5. Collect DNA/cell suspension with the microporator pipette and transfer the mix to the electroporation cuvette.
6. Shock one time at the desired voltage.
7. After electroporation, transfer the cell suspension to 6-well plate.
8. The day after remove the reaction mix by washing with sterile PBS and add Iscove-modified Dulbecco' s complete culture medium.
9. Culture the cells for 24 h at 37 °C in a 5% CO_2 humidified incubator.

3.3 B16F10 Cell Transfection

1. Seed B16F10 cells to 70% confluence (60,000 cells/well) 24 h before transfection.
2. The day of transfection, dilute 3 μg pmeLUC DNA in 500 μl Opti-MEM medium, then add 0.33 μl PLUS Reagent per μg of DNA used.
3. Incubate for 5 min at room temperature.
4. Add 1 μl Lipofectamine LTX per μg DNA.
5. Incubate this reaction mix for 30 min in the dark at room temperature.
6. Add drop by drop the reaction mix to each well.
7. Incubate overnight at 37 °C in a 5% CO_2 humidified incubator.
8. The day after remove the reaction mix from each well, add 2 ml of RPMI 1640 complete culture medium.
9. Incubate for 24 h in a 5% CO_2 humidified incubator at 37 °C.

3.4 Selection of Stably Transfected ACN Cell Clones

1. Supplement ACN cell cultures with G-418 (0.8 mg/ml) selection solution and culture for 72 h.
2. Supplement with 0.4 mg/ml G-418 and culture 48 h to get rid of nontransfected cells.
3. Keep ACN-pmeLUC cells under selection with 0.2 mg/ml G-418.
4. Dilute ACN-pmeLUC polyclone to a single cell per well in a 96-well plate. Place the cell plate in a 5% CO_2 humidified incubator at 37 °C.
5. Check cell clone growth with a phase contrast microscope at 20× magnification every 5 days for about 30–40 days.

6. Transfer positive colonies to a 6-well plate in RPMI medium supplemented with 0.2 mg/ml G-418. It is useful to split each successfully growing cell clone into two separate wells, one to be used for luminescence analysis and the other to be kept under sterile conditions for expansion.

3.5 Selection of Stably Transfected B16F10 and WEHI-3B Cell Clones

1. Supplement cell cultures with 0.4 mg/ml G-418 and culture for 2 weeks to allow successful selection of pmeLUC-transfected cells.
2. Dilute B16F10 or WEHI-3B-pmeLUC polyclone to a single cell per well in a 96-well plate. Place the cell plate in a 5% CO_2 humidified incubator at 37 °C.
3. Check clone growth with a phase contrast microscope at 20× magnification after 10–14 days. It is useful to split each successfully growing cell clone into two separate wells, one to be used for luminescence analysis and the other to be kept under sterile conditions for expansion.

3.6 In Vitro Validation of ACN Transfected Cells

1. After expansion of the colony in the 6-well plate, add 100 μl/well of D-luciferin solution and start luminescence acquisition in IVIS luminometer. Set an acquisition time of 5 min and a medium binning. After basal acquisition, add 1 mM ATP and repeat the acquisition, setting 5 min acquisition time and medium binning.
2. Select a region of interest (ROI) on each well. Keep the same area of ROI and analyze the light intensity as photons per seconds using Living Image® software.
3. Identify the clones with the highest expression of pmeLUC by assessing the highest increase of luminescence following ATP stimulation (*see* **Notes 3, 4** and **5**).

3.7 In Vitro Validation of B16F10 and WEHI-3B Transfected Cells

1. After colony growth in the 96-well plate, add 25 μl of D-luciferin solution to each well and start luminescence acquisition in IVIS luminometer. Set an acquisition time of 5 min and a large and medium binning for B16F10-pmeLUC and WEHI-3B-pmeLUC cells, respectively. After basal acquisition add 500 μM ATP and repeat the acquisition setting 5 min time and the appropriate binning.
2. Select manually a region of interest (ROI) on each well. Record and analyze the intensity of luminescence as photons per seconds using Living Image® (PerkinElmer) software.
3. Identify clones with the highest expression of pmeLUC by assessing the highest increase of luminescence following ATP addition (*see* **Notes 3, 4** and **5**).

3.8 In Vivo Analysis of Extracellular ATP in in a Model of Human Neuroblastoma

The pmeLUC biosensor expressed by cancer cells can be used to directly analyze extracellular ATP levels in the TME in different experimental tumor mouse models (*see* **Note 6**).

1. Anesthetize mice with 2.5% isoflurane and inject subcutaneously 3×10^6 ACN-pmeLUC cells suspended in 100 μl of sterile PBS into the right limb of *nude/nude* mice.
2. After 7 days from inoculum anesthetize mice with 2.5% isofluorane and inject intra peritoneum (i.p.) 150 mg/kg D-luciferin solution allowing 15 min for biodistribution.
3. Place mice into the total body IVIS luminometer to capture luminescence from dorsal view.
4. Set the acquisition time at 3 min and a medium binning.
5. Luminescence emission is measurable from day 7 until the end of the experimental procedure (typically days 18–28).
6. Quantify photon emission using the Living Image® software, selecting a ROI in the tumor area and recording luminescence intensity as photons/second/cm^2/steradian (p/s/cm^2/sr) (Fig. 1).

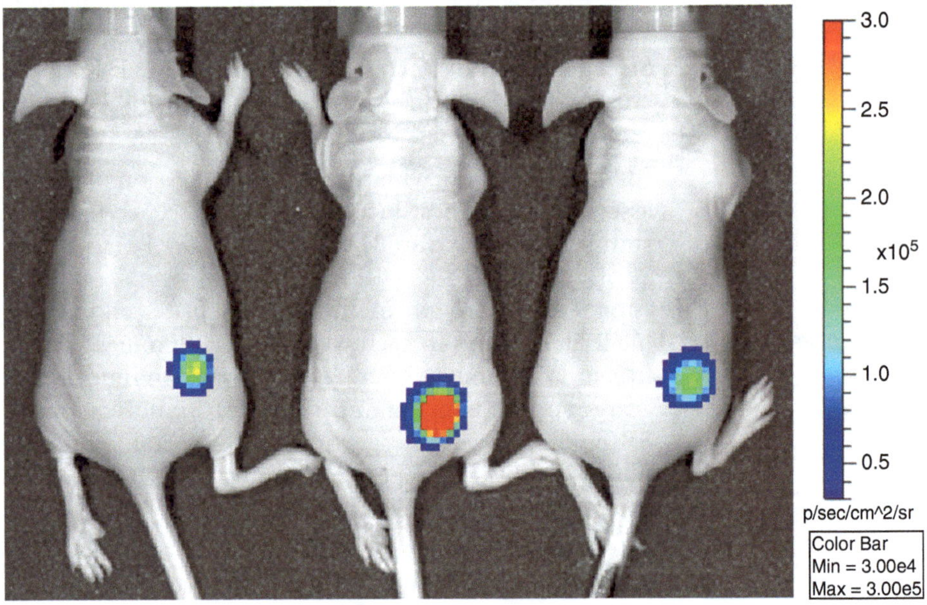

Fig. 1 Measurement of ATP-dependent bioluminescence (p/s/cm^2/sr) in *nude/nude* mice inoculated with ACN human neuroblastoma cells stably transfected with pmeLUC (ACN-pmeLUC cells) at post-inoculum day 9. Slightly different bioluminescence emission depends on the given tumor size in the different animals due to variable growth kinetics in the individual hosts

3.9 In Vivo Analysis of Extracellular ATP in in a Melanoma Mouse Model

The pmeLUC biosensor expressed by cancer cells can be used to directly analyze extracellular ATP levels in the TME in additional experimental tumor mouse models (*see* **Note 5**).

1. Anesthetize mice with 2.5% isofluorane and inject subcutaneously into the right limb of C57Bl/6 mice 2.5×10^5 B16F10-pmeLUC cells in 100 μl of sterile PBS.
2. Five days after inoculum anesthetize mice with 2.5% isoflurane and inject i.p. 150 mg/kg D-luciferin solution allowing 15 min for biodistribution.
3. Place mice into the total body IVIS luminometer capturing luminescence from dorsal view.
4. Set acquisition time at 5 min and a medium binning.
5. Luminescence emission is measureable from day 5 until the end of the experimental procedure (typically days 12–15). Often luminescence (i.e., extracellular ATP) is detectable before the appearance of a palpable tumor mass.
6. Quantify photon emission using the Living Image® software, selecting a ROI in the tumor area and recording luminescence intensity as $p/s/cm^2/s$ (Fig. 2) (*see* **Note 6**).

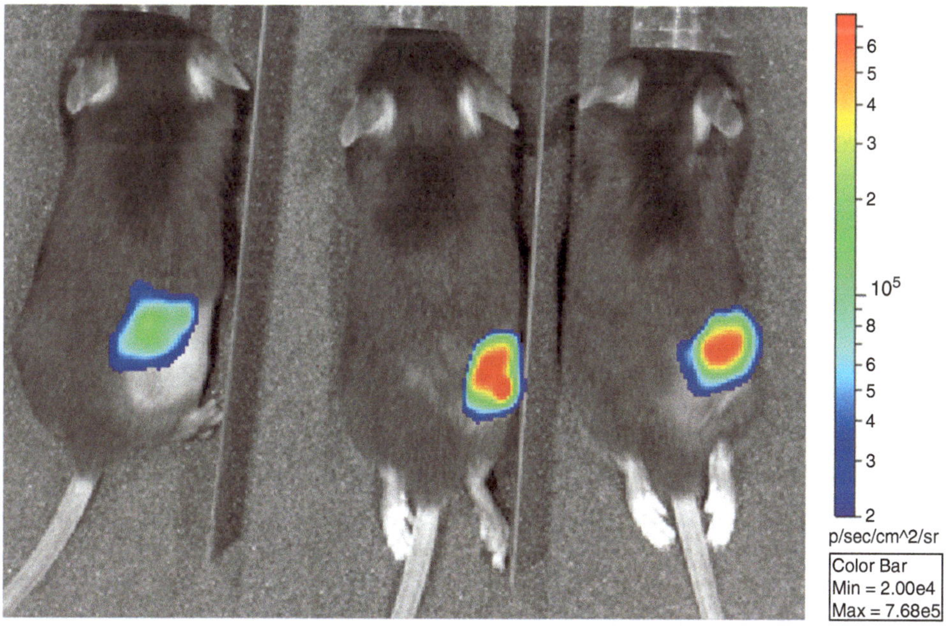

Fig. 2 ATP-dependent bioluminescence emission ($p/s/cm^2/sr$) from C57Bl/6 mice inoculated with B16F10 mouse melanoma cells stably transfected with pmeLUC (B16F10-pmeLUC) at post-inoculum day 14. Slightly different bioluminescence emission depends on the given tumor size in the different animals due to variable growth kinetics in the individual hosts

3.10 In Vivo Analysis of Extracellular ATP in in a Leukemia Mouse Model

1. Anesthetize mice with 2.5% isofluorane and inject subcutaneously into the right limb of BALB/c mice 2×10^6 WEHI-3B-pmeLUC cells in 100 μl of sterile PBS.
2. Five days from inoculum anesthetize mice with 2.5% isoflurane and inject i.p. 150 mg/kg D-luciferin solution allowing 15 min for biodistribution.
3. Place mice into the total body IVIS luminometer capturing luminescence from dorsal view.
4. Set acquisition time at 5 min and a medium binning.
5. Luminescence emission is measureable from day 5 until the end of the experimental procedure (typically days 12–15). Often luminescence (i.e., extracellular ATP) is detectable before the appearance of a palpable tumor mass.
6. Quantify photon emission using the Living Image® software, selecting a ROI in the tumor area and recording luminescence intensity as p/s/cm^2/s (Fig. 3) (*see* **Note 6**).

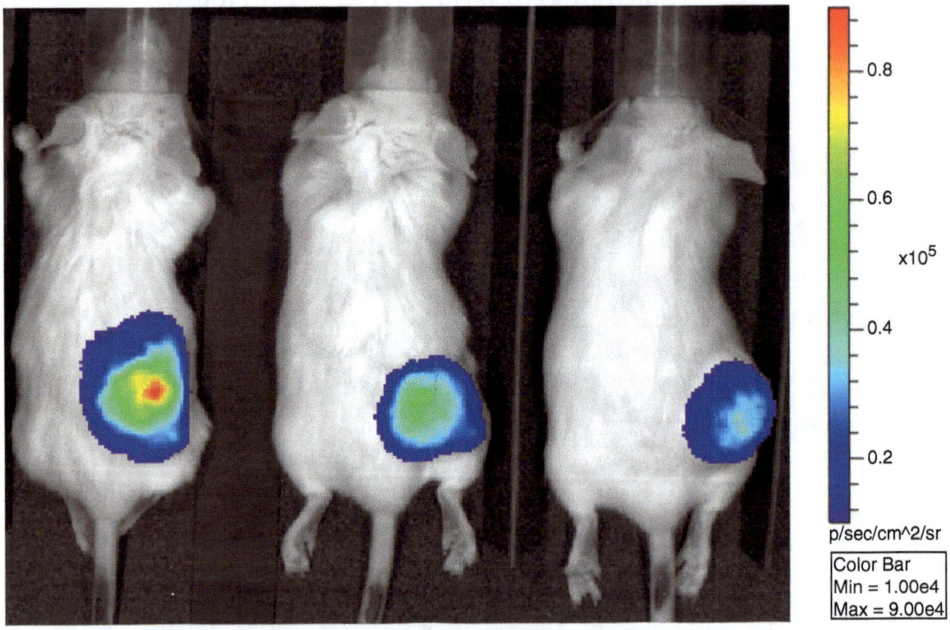

Fig. 3 ATP-dependent bioluminescence emission (p/s/cm^2/sr) from Balb/c mice inoculated with WEHI-3B murine leukemia cells stably transfected with pmeLUC (WHEI-3B-pmeLUC) cells at post-inoculum day 14. Slightly different bioluminescence emission depends on the given tumor size in the different animals due to variable growth kinetics in the individual hosts

4 Notes

1. WEHI-3B cells are mouse BALB/c myelomonocytes that show many characteristics of human myelomonocytic leukemia. The best transfection procedure for WEHI-3B cells, that grow both in adhesion and in suspension, consists in the use of high-voltage electric shocks to introduce DNA into cells. Due to the toxicity of the procedure, usually transfection by electroporation requires a high number of cells to obtain a reasonable number of transfectants.
2. In the WEHI-3B transfection procedure, the use of a mixture of electroporation R and T buffers is due to the presence of two different cell populations, growing in adhesion or in suspension, as described above. R buffer is optimized for adherent cells, while T buffer is generally used for primary T and B cells, PBMCs, monocytes, and bone marrow-derived cells. In our hands, a 1:1 ratio of the two buffer proved optimal for WEHI-3B transfection with the pmeLUC construct.
3. In some cell lines plasma membrane expression of the pmeLUC construct is not optimal, thus to enhance it cells can be incubated overnight with 1 mM DTT (dithiothreitol), or kept at 21 °C for 2 h before luminescence recording.
4. Expression and localization of pmeLUC can be checked by immunofluorescence staining with anti c-myc antibodies (the pmeLUC construct includes a c-myc tag). To this aim, we routinely used anti c-myc-SC-40 monoclonal antibody.
5. To check for expression of pmeLUC probe on the outer surface of the plasma membrane, luminescence emission is assessed before and after addition of ATP to the cell culture. Only clones showing a significant increase in luminescence emission are selected for experiments. It is also useful to treat pmeLUC-transfected cells with the cell-impermeant ATP-hydrolyzing enzyme apyrase. Luminescence decrease upon apyrase addition confirms correct localization of pmeLUC on plasma membrane.
6. Unless *nude/nude* mice are used, we suggest to shave the area of tumor cell injection and luminescence acquisition, as hairs (especially black hairs) absorb a substantial fraction of emitted luminescence.

Acknowledgments

F.D.V. is supported by grants from the Italian Association for Cancer Research (n. IG 13025 and IG 18581), the Ministry of Health of Italy (n. RF-2011-02348435), the Ministry of Education of Italy (n. 20178YTNWC). E.A. is supported by grant from the Italian Association for Cancer Research (n. IG 16812). Both F.D.V. and E.A. are supported by institutional funds from the University of Ferrara. This work was funded with support from the European H2020 office through the COST Action BM1406 "Ion channels and Immune response."

References

1. Nath S, Villadsen J (2015) Oxidative phosphorylation revisited. Biotechnol Bioeng 112:429–437
2. Di Virgilio F, Sarti AC, Falzoni S, De Marchi E, Adinolfi E (2018) Extracellular ATP and P2 purinergic signalling in the tumour microenvironment. Nat Rev Cancer 18:601–618
3. Burnstock G (2007) Physiology and pathophysiology of purinergic neurotransmission. Physiol Rev 87:659–797
4. Ozalp VC, Nielsen LJ, Olsen LF (2010) An aptamer-based nanobiosensor for real-time measurements of ATP dynamics. Chembiochem 11:2538–2541
5. Llaudet E, Hatz S, Droniou M, Dale N (2005) Microelectrode biosensor for real-time measurement of ATP in biological tissue. Anal Chem 77:3267–3273
6. Berg J, Hung YP, Yellen G (2009) A genetically encoded fluorescent reporter of ATP:ADP ratio. Nat Methods 6:161–166
7. Dalton CM, Szabadkai G, Carroll J (2014) Measurement of ATP in single oocytes: impact of maturation and cumulus cells on levels and consumption. J Cell Physiol 229:353–361
8. Manfredi G, Yang L, Gajewski CD, Mattiazzi M (2002) Measurements of ATP in mammalian cells. Methods 26:317–326
9. Lee DH, Kim SY, Hong JI (2004) A fluorescent pyrophosphate sensor with high selectivity over ATP in water. Angew Chem Int Ed Engl 43:4777–4780
10. Morciano G, Sarti AC, Marchi S, Missiroli S, Falzoni S, Raffaghello L, Pistoia V, Giorgi C, Di Virgilio F, Pinton P (2017) Use of luciferase probes to measure ATP in living cells and animals. Nat Protoc 12:1542–1562
11. Di Virgilio F, Pinton P, Falzoni S (2016) Assessing extracellular ATP as danger signal in vivo: the pmeLuc system. Methods Mol Biol 1417:115–129
12. Pellegatti P, Falzoni S, Pinton P, Rizzuto R, Di Virgilio F (2005) A novel recombinant plasma membrane-targeted luciferase reveals a new pathway for ATP secretion. Mol Biol Cell 16:3659–3665
13. DeLuca M, McElroy WD (1974) Kinetics of the firefly luciferase catalyzed reactions. Biochemistry 13:921–925
14. Jouaville LS, Pinton P, Bastianutto C, Rutter GA, Rizzuto R (1999) Regulation of mitochondrial ATP synthesis by calcium: evidence for a long-term metabolic priming. Proc Natl Acad Sci U S A 96:13807–13812
15. Patergnani S, Baldassari F, De Marchi E, Karkucinska-Wieckowska A, Wieckowski MR, Pinton P (2014) Methods to monitor and compare mitochondrial and glycolytic ATP production. Methods Enzymol 542:313–332
16. Michels AA, Nguyen VT, Konings AW, Kampinga HH, Bensaude O (1995) Thermostability of a nuclear-targeted luciferase expressed in mammalian cells. Destabilizing influence of the intranuclear microenvironment. Eur J Biochem 234:382–389
17. Gajewski CD, Yang L, Schon EA, Manfredi G (2003) New insights into the bioenergetics of mitochondrial disorders using intracellular ATP reporters. Mol Biol Cell 14:3628–3635
18. Pellegatti P, Raffaghello L, Bianchi G, Piccardi F, Pistoia V, Di Virgilio F (2008) Increased level of extracellular ATP at tumor sites: in vivo imaging with plasma membrane luciferase. PLoS One 3:e2599
19. Michaud M, Martins I, Sukkurwala AQ, Adjemian S, Ma Y, Pellegatti P et al (2011) Autophagy-dependent anticancer immune responses induced by chemotherapeutic agents in mice. Science 334:1573–1577

20. Bianchi G, Vuerich M, Pellegatti P, Marimpietri D, Emionite L, Marigo I, Bronte V, Di Virgilio F, Pistoia V, Raffaghello L (2014) ATP/P2X7 axis modulates myeloid-derived suppressor cell functions in neuroblastoma microenvironment. Cell Death Dis 5: e1135
21. Pietrocola F, Pol J, Vacchelli E, Rao S, Enot DP, Baracco EE et al (2016) Caloric restriction Mimetics enhance anticancer immunosurveillance. Cancer Cell 30:147–160
22. Lecciso M, Ocadlikova D, Sangaletti S, Trabanelli S, De Marchi E, Orioli E et al (2017) ATP release from chemotherapy-treated dying leukemia cells elicits an immune suppressive effect by increasing regulatory T cells and tolerogenic dendritic cells. Front Immunol 8:1918

Chapter 14

Using Amperometric, Enzyme-Based Biosensors for Performing Longitudinal Measurements of Extracellular Adenosine 5-Triphosphate in the Mouse

Edward Beamer and Tobias Engel

Abstract

Adenosine 5-triphosphate (ATP) functions in the central nervous system as an extracellular signaling molecule. While much progress has been made in understanding the circumstances under which it is released, from in vitro preparations, in vivo has proven more challenging. Microdialysis followed by high-performance liquid chromatography has been employed to demonstrate a spike in extracellular concentrations under some pathological conditions including seizures, but this method lacks the sensitivity to detect extracellular ATP at concentrations present under normal physiological conditions. An alternative approach, the use of amperometric, enzyme-based microelectrode biosensors for measuring extracellular ATP in vivo have been employed in the rabbit. Here, we describe a protocol for measuring ATP concentrations using these biosensors in the mouse, simultaneously with electroencephalogram recordings. This approach is ideal for investigating the relationship between ATP release and seizures.

Key words Microelectrode biosensors, Amperometric detection, Enzyme-based, Adenosine-5-triphosphate, In vivo electrochemistry

1 Introduction

1.1 Adenosine 5-Triphosphate Signaling in the Central Nervous System

The purine nucleoside, adenosine-5′-triphosphate (ATP), discovered by Karl Lohmann in 1929 [1], is ubiquitously present in all cell types throughout all organisms. That it also functions as an extracellular signaling molecule was first proposed in the 1950s [2], while Burnstock [3] first suggested that ATP functions as a neurotransmitter. In the central nervous system (CNS), extracellular ATP (eATP) activates a diverse set of purinergic P2 receptors, effecting processes such as neuroinflammation and cell proliferation, differentiation, and death [4]. eATP also serves as a substrate for ectoenzymes [5] producing metabolites, which mediate different, often opposing, physiological effects [4]. Elevations in eATP have been

Pablo Pelegrín (ed.), *Purinergic Signaling: Methods and Protocols*, Methods in Molecular Biology, vol. 2041,
https://doi.org/10.1007/978-1-4939-9717-6_14,

proposed following various acute brain insults, including traumatic brain injury (TBI) [6], cerebral ischemia/stroke [7, 8], and seizures [9–12].

For a number of reasons, the direct measurement of eATP, particularly in vivo, is extremely challenging. Firstly, enzymatically degraded in the extracellular milieu by a host of ectoenzymes, eATP has an extremely short half in CNS tissue, in the order of seconds [13]. eATP concentrations are maintained by constant cellular release and extracellular degradation [11]. When using methods, such as microdialysis, in which eATP is removed from the pericellular environment, the eATP pool continues to deplete after the source of resupply has been removed. Using these methods, ATP concentrations drop below the threshold for accurate measurement [12]. Pharmacological approaches to the study of ATPergic signaling have suffered from poor specificity of available drugs [14], and the confounding effects of ATP metabolites, particularly adenosine [15].

1.2 Methods for Measuring ATP

ATP measurements have largely been carried out in in vitro preparations. The principal methods rely on calculating the rate of ATP-dependent enzymatic reactions or the binding of ATP to purinergic receptors and the subsequent membrane depolarization. ATP-dependent enzymatic reactions can produce electroactive molecules or bioluminescence proportional to the ATP concentration and can be detected by electrode or camera, respectively. P2 receptor-based methods for ATP measurement, meanwhile, involve the use of patch-clamp recordings, either in whole "sniffer" cells transfected with purinergic receptors [16], or through isolated membrane "sniffer" patches [17]. Enzyme-based bioluminescence assays and electrochemical biosensors have been the primary tools for the measurement of eATP in ex vivo and cell culture-based studies.

Measuring ATP in vivo has largely relied on microdialysis followed by high-performance liquid chromatography. While this method has been sufficient for detecting increases in ATP following TBI [6] and stroke [7, 8], it is likely to greatly underestimate the extracellular concentration in the tissue. The half-life of ATP is extremely short, so it is likely that much of the ATP in the dialysate will have been degraded before being measured. This also means that physiological concentrations of ATP fall below the threshold for detection using this method. An alternative approach is to use enzyme-based microelectrode biosensors to detect ATP concentrations in situ. Gourine et al. [18] employed this approach in rabbits to demonstrate an elevation in hypothalamic extracellular ATP concentrations in response to systemic inflammation. While enzyme-based biosensors are not optimal for measuring absolute concentrations of analytes, they have the advantage of a much higher temporal resolution (1 measurement per second), when compared to alternative methods, such as microdialysis (1 measurement per minute). This makes the approach particularly well equipped for investigating fast dynamics in extracellular ATP concentrations. The possibility to record ATP concentrations with a high temporal resolution simultaneously with EEG allows for the elucidation of the temporal relationship between seizures, ATP release and ATP degradation.

1.3 Amperometric Microelectrode Biosensors

Amperometric microelectrode biosensors utilize an enzyme or cascade of enzymes, held in close apposition with a charged electrode. Enzymes are selected in such a way that an electroactive molecule is produced in proportion to the analyte of interest. This electroactive molecule, typically hydrogen peroxide (H_2O_2), is oxidized on the electrode surface giving a signal. For ATP, this electrochemical detection relies upon the enzymes glycerol kinase and glycerol-3-phosphate kinase [19], held in a gel matrix covering the electrode (Fig. 1). If glycerol is present at a concentration of >0.05 mM, the rate of the reaction is entirely dependent on the concentration of ATP (Fig. 2). When using these biosensors to make recordings in vitro, glycerol must be added to the recording solution, while in vivo, the biosensor tip should be coated in glycerol prior to

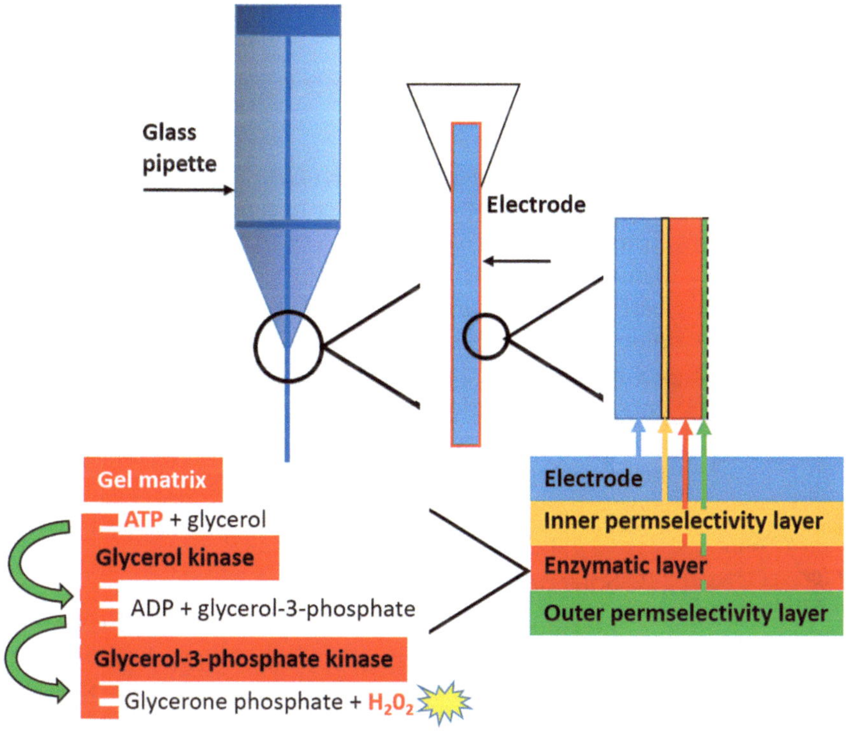

Fig. 1 Amperometric microelectrode biosensor for detection of ATP in vivo. Enzymes in gel matrix catalyze breakdown of ATP and glycerol to glycerone phosphate and H_2O_2. Permselective layers limit interference from intermediate substrates on the enzyme cascade and electroactive molecules, such as reactive oxygen species

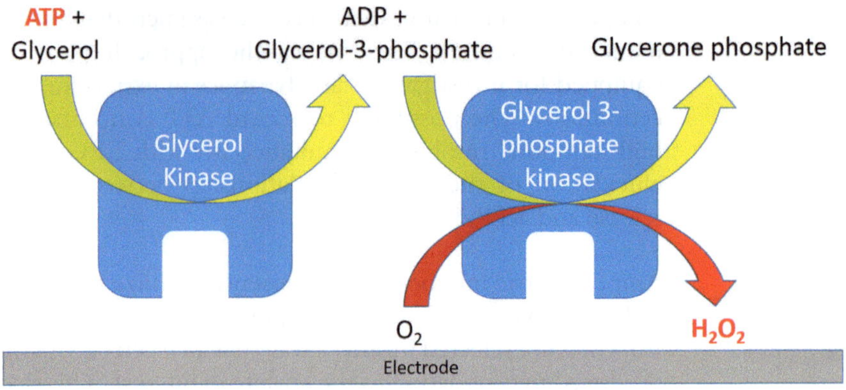

Fig. 2 Enzyme cascade. Provided glycerol is present in sufficient concentrations (>0.05 mM), enzyme cascade produces H_2O_2 proportional to concentration of extracellular ATP. H_2O_2 oxidizes on electrode surface generating a signal

insertion (*see* **Note 1**). Because multienzyme biosensors are also sensitive to intermediate substrates and other electroactive molecules, it is necessary to take null recordings, using similar biosensors, but lacking the first enzyme in the cascade. This signal can be subtracted from the signal from the analyte of interest. In the case of ATP measurements, these "null sensors" are coated with a gel matrix with glycerol-3-phosphate kinase, but no glycerol kinase.

2 Materials

2.1 Hardware

In principle, the hardware and software used for taking in vivo measurements of ATP concentrations in mice could be sourced from a number of suppliers or manufactured in house. While some of the details of the protocol are general to all hardware, here we report on the use of a detection system produced by pinnacle technology (Lawrence, KS, USA), with the enzymes and gel matrix applied to the electrode tip by Sarissa Biomedical (Coventry, UK). Hardware and software for in vivo measurements are available as a full set, including everything necessary for performing measurements. This hardware is listed below.

2.2 List of Hardware Necessary for Measuring ATP In Vivo

1. Powered USB hub.
2. Soldering iron.
3. Preamplifier for electroencephalogram (EEG) and biosensor measurements (Pinnacle Technology, Lawrence, KS, USA) with two channels for EEG recording (gain × 100, high pass filters: 1 Hz).

4. Data controlling/acquisition system for recording EEG and biosensor measurements (Pinnacle Technology, Lawrence, KS, USA).
5. Mouse commutator/Swivel.
6. EEG/EMG test source (Pinnacle Technology, Lawrence, KS, USA).
7. Stereotaxic frame suitable for mouse surgery.
8. USB cable.
9. Test load for biosensor (Pinnacle Technology, Lawrence, KS, USA).
10. 23-G needle.
11. 0.1 in. EEG screws with wire leads for EEG recording.
12. EEG/EMG/biosensor mouse headmount (Pinnacle Technology, Lawrence, KS, USA).
13. Intracerebral guide cannula with stylet for microdialysis.
14. In vivo biosensor (Pinnacle Technology, Lawrence, KS, USA) coated with gel matrix with glycerol kinase and glycerol 3-phoshate kinase (ATP sensor) and glycerol 3-phosphate kinase ("null" sensor) (*see* **Note 2**).

2.3 Software

Software suitable for recording EEG and biosensor measurements is necessary. Sirenia acquisition (Pinnacle Technology, Lawrence, KS, USA) is recommended. Data analysis can be performed using the same software, or can be exported for analysis in a dedicated analysis package such as MATLAB (Mathworks, MA, USA).

2.4 Drugs and Reagents

1. Isoflurane.
2. Analgesic cream (5% eutectic mixture of local anesthetics).
3. Glycerol.
4. Calibration solution (phosphate-buffered saline (PBS) with glycerol): 137 mM NaCl, 2.7 mM KCl, 10 mM Na_2HPO_4, 1.8 mM KH_2PO_4, 2 mM glycerol, pH 7.4.
5. Adenosine 5′triphosphate stock solution: 1 mM adenosine 5′triphosphate dissolved in sterile PBS. Stock should be made fresh daily and discarded following experiment.
6. Rehydration buffer: 10 mM NaPi buffer, pH 7.4, 100 mM NaCl, 1 mM $MgCl_2$, 2 mM glycerol.

3 Methods

3.1 Overview

Mice are anaesthetized and implanted with EEG electrodes and a cannula for the insertion of the biosensor. A headmount is held in place on top of the skull, which serves the purpose of holding the preamplifier in place during recording and transmitting the EEG signal. A microelectrode biosensor is passed through the cannula into the anatomical region of interest and is connected to the preamplifier. The preamplifier is connected to an amplifier above the cage, via a swivel. The amplifier is connected to a laptop of desktop PC via a USB cable and biosensor recordings and EEG can be recorded simultaneously using dedicated software. The biosensor is calibrated before and after insertion, using known concentrations of ATP in solution.

3.2 Care of Sensors

Sensors should be stored between 2 and 8 °C and used within the date indicated on the box (*see* **Note 3**). Prior to use, sensors should be rehydrated for 10 min in rehydration buffer. Once rehydrated, it is important that biosensors do not dry out. Following rehydration, sensors can be stored, immersed in the rehydration buffer between 2 and 8 °C for up to 4 days. Biosensors can be used two or three times, but they will lose sensitivity through enzyme degradation, physical damage, and fouling of the sensor surface. Care should be taken that sensors retain sufficient sensitivity before being used.

3.3 Setup and Testing of Hardware and Software

The data conditioning/acquisition system (amplifier) is attached to a metal frame, which is mounted on two stands and placed either side of the cage. The connecting cable is used to connect the amplifier to the mouse commutator (swivel), which hangs down from the metal frame toward the cage. The height of the metal plate can be adjusted in such a way that the swivel hangs into the cage at an appropriate height to record from the mouse. The amplifier is connected to a laptop or PC via USB, through the USB hub, which ensures stable operation. The swivel is attached to the preamplifier, in such a way that white markings on each are on the same side.

While different software can be used for data analysis, the use of Sirenia Acquisition (Pinnacle Technology, KS, USA) is recommended for recording simultaneous EEG and ATP measurements. Prior to the taking of measurements, the hardware and software should be tested with a test load, which is a 10 Ω resistor, producing a 60 nA current when the channel voltage is set to 0.6 V. When the software is running and this test load is attached to the preamplifier, a 60 nA signal should be seen on the software screen. In order to test the hardware and software for EEG recording, a test source, which applies sinusoidal waveforms can be used. This piece of hardware has a 6-pin socket, into which the 6-pin plug in the preamplifier can be connected. If the hardware and software are set up and working correctly, pressing the button should lead to sinusoidal waves of fluctuating wavelength appearing on the EEG channel on the software screen.

3.4 Surgery

All surgery should be performed under sterile conditions. The mouse is fully anaesthetized and placed in a stereotaxic frame suitable for surgery with mice. The mouse is placed on the nose bar and ear bars are coated in analgesic cream (5% EMLA) prior to insertion. When the mouse is held in place, an incision is made through the skin lengthways from between the ears to the nape of the neck. Analgesic cream is applied to the skull surface using a sterile cotton bud, starting from the center, spiraling outward to avoid contamination of the site of surgery. Holes for EEG electrodes and the screw are made in the skull using a 26-G needle and electrodes screwed in deep enough only that they remain in place. Placement of electrodes should be planned prior to beginning surgery and are dependent on the anatomical region from which ATP measurements are to be taken (*see* **Note 4**).

1. Anaesthetize mice with isoflurane, place in a stereotaxic frame with analgesic cream applied to the ear bars prior to insertion.
2. Cut skin to expose skull, lengthways from between the ears to the nape of the neck.
3. Use 26-G needle to make holes in the skull for EEG electrodes.
4. Screw in EEG electrodes deep enough that they hold in place, but no deeper.
5. Use stereotaxic frame to line up cannula at correct coordinates for biosensor to measure from anatomical region of interest.
6. Make hole in the skull at correct location with 26-G needle.
7. Push cannula into the hole at correct depth for recording.
8. Apply dental cement and allow to harden (approximately 5–10 min).
9. Place headmount on top of cement (*see* **Note 5**).
10. Put EEG wires through EEG sensors on headmount and solder in place (*see* **Note 6**).
11. Apply a second coating of dental cement to hold headmount and cannula in place.
12. Inject 0.05 mL buprenorphine to prevent pain windup.
13. Place stylet tip into cannula, remove isoflurane and allow mouse to recover with oxygen.
14. Place mouse in incubator and allow to recover until it is moving comfortably.
15. Allow mouse to recover for 5–7 days prior to recording, providing soft chow and checking weight and welfare twice daily.

3.5 Calibration of Biosensors

Because enzyme-based biosensors continually lose sensitivity, calibration should be performed both before and after a period of measurement. Data should be calibrated according to a line drawn between these points (*see* **Note 7**). Prior to making a calibration recording, a 1 mM ATP stock solution must be made by dissolving ATP in calibration buffer. In order to calibrate the biosensor, it should be connected to the preamplifier and hardware as described in previous sections. The preamplifier is equipped with a two-pin plug, which connects to two sockets on the biosensor. The water heater is turned on and the jacketed beaker is filled with 20 mL buffer solution and allowed to reach a temperature of 37 °C. The stir bar should be placed in the solution and the motorized stirrer should be set to such a speed as it turns at the maximum speed possible without causing a vortex in the solution. The biosensor is introduced using the calibration holder, ensuring that the biosensor extends below the surface of the buffer.

1. Fill jacketed beaker with 20 mL calibration solution.
2. Switch on heater and wait until solution is stable at 37 °C.
3. Turn on electric stirrer and increase speed to maximum possible without causing vortex.
4. Connect the biosensor to the hardware in use.
5. Lower biosensor into calibration solution.
6. Initiate file recording.
7. Allow time for the biosensor to asymptote to a stable baseline. This can take up to 1 h.
8. When the biosensor has reached a stable baseline, add injections of the ATP stock solution into the buffer, using the software to annotate the points of injection. Add three injections of 1 mM ATP stock to the calibration buffer, in order to increase the concentration of ATP in the buffer, stepwise from 0 μM, to 5 μM, to 10 μM and finally to 20 μM. At each step, allow the recording to asymptote before adding another injection. Typically, this will take around 5 min. In detail, into 20 mL calibration buffer allow signal to asymptote in calibration buffer (buffer ATP concentration 0 μM); then inject 100 μL ATP stock solution (buffer ATP concentration: 5 μM); then allow signal to asymptote; then inject further 100 μL ATP stock solution (buffer ATP concentration: 10 μM); allow signal to asymptote; and finally inject 200 μL ATP stock solution (buffer ATP concentration 20 μM).

3.6 Recording

The mouse is anaesthetized prior to connection of the head mount to the preamplifier and insertion of the biosensor through the guide cannula. The biosensor is dipped in glycerol and carefully inserted through the guide cannula and connected to the preamplifier via the two-pin plug. The mouse is placed in the cage with access to food and water and the preamplifier is connected to the recording hardware via the swivel. The mouse is left for a minimum of 1 h to allow recovery from anesthesia, allow biosensor to asymptote and for run-down of ATP associated with tissue damage from injection to metabolize. During this time, the software should be initialized in order to ensure that EEG and biosensor measurements are being taken. After a minimum of 1 h, recording can be commenced and the experimental protocol started. After the experiment is finished, recording can be stopped and the mouse removed from the preamplifier.

1. Anesthetize mouse with isoflurane.
2. Connect headmount to preamplifier via 6-pin plug.
3. Dip tip of biosensor in glycerol.
4. Carefully insert biosensor through cannula into recording site.
5. Connect biosensor to preamplifier via 2-pin plug.
6. Connect preamplifier to swivel and place mouse in cage with access to food and water.
7. Initiate software and ensure that biosensor and EEG recording is working.
8. Allow mouse to recover with biosensor and preamplifier in place for a minimum of 1 h prior to initiating recording.
9. Perform assay (e.g., injection of chemoconvulsant).
10. Following completion of experiment, stop recording and remove mouse from preamplifier.

3.7 Null Measurements

The current from ATP biosensors includes not only ATP but also intermediate products on the enzymatic pathway and electroactive molecules. For this reason, it is important to employ a null sensor in order to measure the amount of signal, which is independent of ATP. Null sensors are identical to ATP sensors, but without the first enzyme in the cascade (glycerol kinase).

Ideally, ATP measurements should be taken simultaneously with null measurements from the same anatomical region and the null recording subtracted from the ATP recording. This, however, is unlikely to be possible in the mouse. One approach is to take simultaneous measurements from equivalent anatomical regions in different hemispheres [18]. The validity of this, however, is dependent on the assumption that both hemispheres will show similar dynamics of ATP release and metabolism/seizure activity. This may be the case for studies where slow changes are expected, such as circadian fluctuations, but is unlikely to be appropriate for studies where fast, dynamic changes in ATP concentrations are expected, such as with seizures. An alternative approach is to take multiple null measurements from different mice in order to characterize the relationship between EEG and null measurements. It should be borne in mind that microelectrode biosensors, while extremely good for investigating changes in the concentration of an analyte, are not optimal for measuring absolute concentrations. While our experience is that the activity recorded on the null sensor is negligible, it represents an important control measure.

3.8 Post Measurement Procedures

When the experiment is complete, mice should be anaesthetized, sacrificed, their brains removed and snap frozen. 15 μm thick brain sections should be cut on a cryostat and stained with hematoxylin and eosin, in order to confirm the anatomical region from which recordings were taken.

4 Notes

1. In order for the production of H_2O_2 to be proportional to the concentration at the biosensor, glycerol must be present at a saturating concentration. For in vivo measurements, the tip of the electrode can be dipped in glycerol solution prior to insertion, ensuring that the entire recording surface is coated.
2. The microelectrode biosensor is optimized for in vivo use and is manufactured as a single probe that includes a built in Ag/AgCl reference electrode and a Pt-Ir working electrode. The electrode has a sensing cavity 176 μM in diameter and 1 mm in length. This cavity is protected with an epoxy tip in order to prevent damage on insertion through the guide cannula.
3. Sensors should be stored between 2 and 8 °C and rehydrated prior to use. While they can be used for up to 3 or 4 recordings, they should be discarded if they become insufficiently sensitive.
4. Placement of electrodes should be planned prior to beginning surgery and are dependent on the anatomical region from which ATP measurements are to be taken. There is limited space on top of the mouse skull for the electrodes, headmount, and cannula, so careful planning of recording sites is necessary.
5. It is important to ensure that both the preamplifier and cannula can be inserted simultaneously. Prior to applying the second round of dental cement, it is useful to attach the preamplifier and cannula stylet to make sure that they both fit. If they do not, the position of the headmount can be adjusted.
6. Be careful not to hold soldering iron in place for too long as it can damage the hardware and injure the mouse.
7. Calibration should be performed both before and after a period of measurement and the gradient between these lines used to calibrate data throughout the experiment, in order to take into account the continual depletion of the signal, resulting from degradation of enzymes and fouling of the biosensor.

Acknowledgments

This work was supported by funding from the Health Research Board (HRA-POR-2015-1243 to T.E.); Science Foundation Ireland (13/SIRG/2098 and 17/CDA/4708 to T.E.), from the H2020 Marie Skłowdowksa-Curie Actions Individual Fellowship (753527 to E.B.) and from the European Union's Horizon 2020 research and innovation program under the Marie Sklowdowska-Curie grant agreement (No. 766124 to T.E.).

References

1. Langen P, Hucho F (2008) Karl Lohmann and the discovery of ATP. Angew Chem Int Ed Engl 47:1824–1827
2. Holton FA, Holton P (1954) The capillary dilator substances in dry powders of spinal roots; a possible role of adenosine triphosphate in chemical transmission from nerve endings. J Physiol 126:124–140
3. Burnstock G (1972) Purinergic nerves. Pharmacol Rev 24:509–581
4. Burnstock G, Krugel U, Abbracchio MP, Illes P (2011) Purinergic signalling: from normal behaviour to pathological brain function. Prog Neurobiol 95:229–274
5. Dombrowski KE, Ke Y, Brewer KA, Kapp JA (1998) Ecto-ATPase: an activation marker necessary for effector cell function. Immunol Rev 161:111–118
6. Franke H, Grummich B, Hartig W, Grosche J, Regenthal R, Edwards RH, Illes P, Krugel U (2006) Changes in purinergic signaling after cerebral injury — involvement of glutamatergic mechanisms? Int J Dev Neurosci 24:123–132
7. Melani A, Turchi D, Vannucchi MG, Cipriani S, Gianfriddo M, Pedata F (2005) ATP extracellular concentrations are increased in the rat striatum during in vivo ischemia. Neurochem Int 47:442–448
8. Frenguelli BG, Wigmore G, Llaudet E, Dale N (2007) Temporal and mechanistic dissociation of ATP and adenosine release during ischaemia in the mammalian hippocampus. J Neurochem 101:1400–1413
9. Wu PH, Phillis JW (1978) Distribution and release of adenosine triphosphate in rat brain. Neurochem Res 3:563–571
10. Frenguelli BG, Wall MJ (2016) Combined electrophysiological and biosensor approaches to study purinergic regulation of epileptiform activity in cortical tissue. J Neurosci Methods 260:202–214
11. Lietsche J, Imran I, Klein J (2016) Extracellular levels of ATP and acetylcholine during lithium-pilocarpine induced status epilepticus in rats. Neurosci Lett 611:69–73
12. Dona F, Conceicao IM, Ulrich H, Ribeiro EB, Freitas TA, Nencioni AL, Da Silva Fernandes MJ (2016) Variations of ATP and its metabolites in the hippocampus of rats subjected to pilocarpine-induced temporal lobe epilepsy. Purinergic Signal 12:295–302
13. Picher M, Burch LH, Boucher RC (2004) Metabolism of P2 receptor agonists in human airways: implications for mucociliary clearance and cystic fibrosis. J Biol Chem 279:20234–20241
14. Beamer E, Goloncser F, Horvath G, Beko K, Otrokocsi L, Kovanyi B, Sperlagh B (2016) Purinergic mechanisms in neuroinflammation: an update from molecules to behavior. Neuropharmacology 104:94–104
15. Masino SA, Kawamura M Jr, Ruskin DN (2014) Adenosine receptors and epilepsy: current evidence and future potential. Int Rev Neurobiol 119:233–255
16. Pangrsic T, Potokar M, Stenovec M, Kreft M, Fabbretti E, Nistri A, Pryazhnikov E, Khiroug L, Giniatullin R, Zorec R (2007) Exocytotic release of ATP from cultured astrocytes. J Biol Chem 282:28749–28758
17. Brown P, Dale N (2002) Spike-independent release of ATP from Xenopus spinal neurons evoked by activation of glutamate receptors. J Physiol 540:851–860
18. Gourine AV, Dale N, Llaudet E, Poputnikov DM, Spyer KM, Gourine VN (2007) Release of ATP in the central nervous system during systemic inflammation: real-time measurement in the hypothalamus of conscious rabbits. J Physiol 585:305–316
19. Dale N, Frenguelli BG (2009) Release of adenosine and ATP during ischemia and epilepsy. Curr Neuropharmacol 7:160–179

Chapter 15

Fluorescent Labeling and Quantification of Vesicular ATP Release Using Live Cell Imaging

Kirstan A. Vessey, Tracy Ho, Andrew I. Jobling, Anna Y. Wang, and Erica L. Fletcher

Abstract

Adenosine triphosphate (ATP) is actively transported into vesicles for purinergic neurotransmission by the vesicular nucleotide transporter, VNUT, encoded by the gene, solute carrier 17, member 9 (*SLC17A9*). In this chapter, methods are described for fluorescent labeling of VNUT positive cells and quantification of vesicular ATP release using live cell imaging. Directions for preparation of viable dissociated neurons and cellular labeling with an antibody against VNUT and for ATP containing synaptic vesicles with fluorescent ATP markers, quinacrine or MANT-ATP, are detailed. Using confocal microscope live cell imaging, cells positive for VNUT can be observed colocalized with fluorescent ATP vesicular markers, which occur as discrete puncta near the cell membrane. Vesicular release, stimulated with a depolarizing, high potassium physiological saline solution induces ATP marker fluorescence reduction at the cell membrane and this can be quantified over time to assess ATP release. Pretreatment with the voltage gated calcium channel blocker, cadmium, blocks depolarization-induced membrane fluorescence changes, suggesting that VNUT-positive neurons release ATP via calcium-dependent exocytosis. This technique may be applied for quantifying vesicular ATP release across the peripheral and central nervous system and is useful for unveiling the intricacies of purinergic neurotransmission.

Key words Retina, VNUT, Purine, P2X7, Adenosine, Purinergic, Solute carrier 17, member 9 (*SLC17A9*), Dopamine, Dopaminergic

1 Introduction

One of the primary neurotransmitters of the purinergic system, extracellular adenosine 3′-triphosphate (eATP) mediates a range of physiological responses on binding to cell surface purinergic receptors [1, 2]. Release of eATP has been suggested to occur via a range of mechanisms: conductive mechanisms following physiological and mechanical stimuli, which involves passive, osmotic diffusion through ATP permeable channels [3] and also standard vesicular release [4–8]. Despite a long history of recognition of ATP as a neurotransmitter spanning back nearly 50 years [4], it

Pablo Pelegrín (ed.), *Purinergic Signaling: Methods and Protocols*, Methods in Molecular Biology, vol. 2041,
https://doi.org/10.1007/978-1-4939-9717-6_15,

was not until 2008 that the first and only so far identified vesicular transporter for active accumulation of ATP within synaptic vesicles was determined [9]. Vesicular nucleotide transporter (VNUT) protein, encoded by the solute carrier 17, member 9 (*SLC17A9*) gene, was found to be a novel member of an anion transporter family. It was found to actively drive accumulation of nucleotides (i.e., ATP, ADP, and UTP) into secretory vesicles and to be required for exocytosis of ATP from various tissue and cell types [9]. In addition to neurons, VNUT has been found in astrocytes, endothelial cells, and microglia, suggesting that vesicular release of eATP is not limited to traditional neurotransmission [10–17].

Vesicular release of ATP can be quantified using a number of different techniques including postsynaptic electrophysiology [4, 18], ATP quantification assays of extracellular media following stimulation of exocytosis [19–22], and fluorescent tagging of ATP vesicles combined with subsequent live cell imaging of exocytosis [19, 20, 22, 23]. Fluorescent tagging of ATP containing vesicles using quinacrine and MANT-ATP is useful for live cell imaging, allowing changes in fluorescence at the cell membrane to be used as a measure vesicle exocytosis in a range of physiological systems [19, 20, 22, 23]. Quinacrine has a high affinity for ATP [24] and has been shown to be taken up by purinergic vesicles in a variety of cells [19, 20, 22, 23]. As quinacrine has a fluorescent excitation 417 nm/emission 496 nm spectrum, imaging of vesicles within cells and vesicular release from cells can be tracked using common wavelengths found on standard fluorescent and also confocal microscopes. MANT-ATP is a fluorescent analogue of ATP, which has also been shown to label ATP stores and has a fluorescent spectrum similar to quinacrine [22, 23]. Fluorophore labeling of ATP containing vesicles has been applied and found to be useful for measuring exocytosis from nonneural cells such as endothelial cells [19], pancreatic acini [22], and microglia [23], as well as neuronal cells in the central nervous system [25] and periphery [26].

In the retina, VNUT has been found to be expressed in dopaminergic amacrine cells [20] and may also be expressed by photoreceptors, bipolar, amacrine cells, Müller cells, and astrocytes [27]. Furthermore, ATP release has been detected from retinal neurons including cholinergic, GABAergic [8, 28] and dopaminergic neurons [20]. Recently, we have developed methods to track VNUT protein expression and vesicular release of ATP from retinal neurons using live cell imaging [20]. The technique involves dissociating the retinal neurons using enzymatic digest, labeling VNUT-positive cells with a fluorescent antibody while colabeling ATP containing vesicles with fluorescent ATP markers (quinacrine or MANT-ATP) and imaging vesicular dynamics using confocal microscopy. Analysis of the live cell images over time to detect changes in fluorescent ATP markers at the cell membrane in response to depolarizing stimuli can be used as a measure of ATP release. Using this technique, we have shown that dopaminergic neurons in the retina also release ATP via calcium-dependent exocytosis [20]. While this method has been used with retinal samples, it is robust enough to be applied to numerous other tissue types. Overall, this technique is useful for unveiling the role purinergic neurotransmission plays in intercellular signaling and could be applied to the quantification of vesicular ATP release across the central and peripheral nervous system.

2 Materials

2.1 Source of Neural Cells

Retinal or other neural cells can be sourced from rodents or cell cultures as desired. For our experiments we have used both 6–8-week-old, adult C57BL/6J mice (http://jaxmice.jax.org/strain/013636.html) and Dark Agouti rats of either sex, sourced from Animal Resources Centre, WA, Australia. All experimental procedures using animals were performed in accordance with the Association for Research in Vision and Ophthalmology (ARVO) Statement for the Use of Animals and The University of Melbourne Animal Ethics Committee (Ethics #: 1112260). All rats and mice were housed and maintained in plastic cages with ad libitum access to food and water under a 12 h:12 h light–dark cycle.

2.2 Solutions

1. Physiological saline solution (PSS): 137 mM NaCl, 2.5 mM KCl, 10 mM HEPES, 28 mM D-glucose, 1.25 mM NaH_2PO_4, 1 mM $MgCl_2$, and 2 mM $CaCl_2$, pH 7.4.
2. PSS + BSA buffer: PSS containing 0.5% bovine serum albumin.
3. High potassium physiological saline solution (K^+PSS): 29.5 mM NaCl, 110 mM KCl, 10 mM HEPES, 28 mM D-glucose, 1.25 mM NaH_2PO_4, 1 mM $MgCl_2$, and 2 mM $CaCl_2$, pH 7.4 (*see* **Note 1**).
4. 400 μM of cadmium (Cd^{++}).
5. Ketamine.
6. Xylazine.
7. Sodium pentobarbital.

2.3 Neural Dissociation and Vesicular/Cellular Labeling

1. Postnatal Neural Dissociation Kit: MACs Miltenyi Biotec, Cat#130-094-802, preprepare reagents as directed by the manufacturer and as detailed in **items 2–6**.
2. Resuspend Enzyme A with 1 mL storage buffer. Aliquot and store at −20 °C.
3. Aliquot Enzyme P and store at −20 °C.
4. Add 3.38 μL beta-mercaptoethanol to 2.5 mL buffer Z. This solution is stable for 1 month.

5. Enzyme mix 1 = 12 μL Enzyme P + 2.34 mL Buffer Z. Solution Z should be prewarmed to room temperature.
6. Enzyme mix 2 = 18 μL Enzyme A + 36 μL Buffer Y. This should be kept on ice until use.
7. 1:100 dilution of guinea pig anti-VNUT antibody reactive with mouse and rat, we use the antibody from Merck Millipore, Cat#ABN83.
8. 1:500 dilution of secondary VNUT goat anti-guinea pig Alexa 594 antibody.
9. 50 μM *N*-methylanthraniloyl-adenosine 5′ triphosphate trisodium salt (MANT-ATP).
10. 10 μM quinacrine.
11. 50 μM fluorescent false neurotransmitter, FFN102, for labeling dopaminergic vesicles (optional [20]).
12. 400 μM cadmium chloride.
13. 0.4% trypan blue solution.

2.4 Laboratory Equipment and Consumables

1. Two pairs of Dumont Style #5 tweezers.
2. Small Aesculap straight spring microscissors.
3. Dissecting microscope with illumination from above.
4. Fire-polished Pasteur pipettes, diameter to 0.8 mm and edges rounded.
5. 75 μm Corning® cell strainer.
6. Centrifuge.
7. Automated cell counter.
8. 35 mm glass coverslip-bottomed petri dish.

2.5 Specialist Equipment

Confocal microscope: A laser scanning confocal microscope is required for viewing and imaging via 10× air, 20× air, 40× oil, and 63× oil objectives. For this study, we used an inverted Leica SP5, confocal laser scanning microscope from Leica microsystems.

3 Methods

3.1 Collect Retinal Samples

1. Euthanize animal. A single mouse or rat can be used for each experiment. Mice are deeply anesthetized by an intraperitoneal injection of ketamine (130 mg/kg) and xylazine (27 mg/kg) and sacrificed by cervical dislocation. Rats are deeply anesthetized by intraperitoneal injection of ketamine (130 mg/kg) and xylazine (27 mg/kg) and sacrificed using intraperitoneal injection of sodium pentobarbital (120 mg/kg).

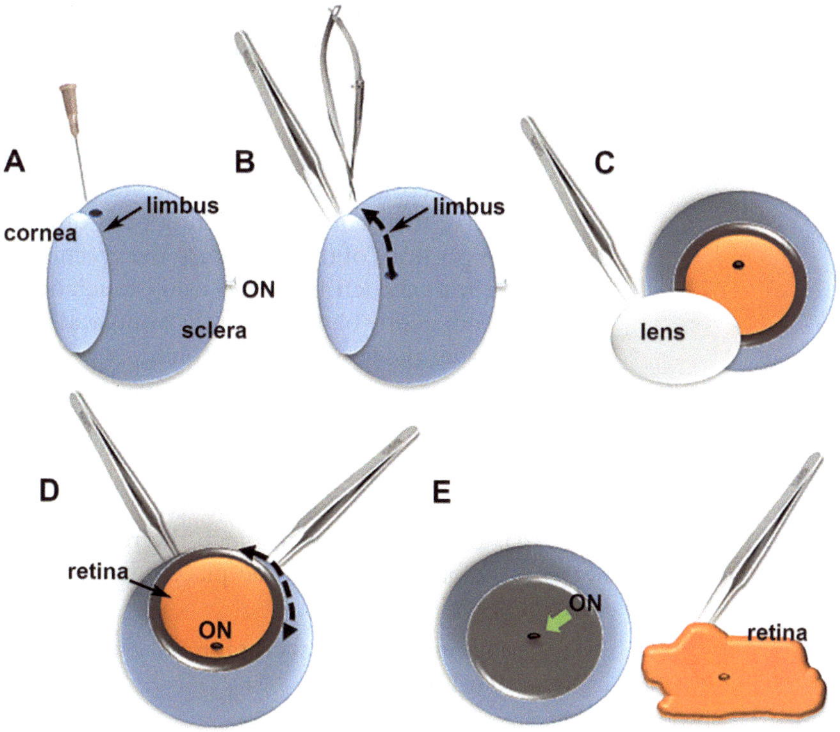

Fig. 1 Dissecting the eye. (**a**) Clean the eye of extraocular tissue, lay the eye under the dissecting microscope and puncture the globe through the sclera posterior to the limbus using a 26 G needle. (**b**) Using fine tweezers, grasp the cornea to steady the sample and use fine spring microscissors to cut around the sclera posterior to the limbus. (**c**) Use fine tweezers to grasp the sclera and steady the sample on its side. Use closed fine tweezers in the other hand to gently roll the lens in the eyecup and then roll the lens out toward you. (**d**) Use closed fine tweezers to loosen the retina away from the eyecup, slowly blunt dissecting down toward the optic nerve (ON) at the back. (**e**) Use fine tweezers to pinch the optic nerve and release the retina from the eyecup. Optimally, the retina should come out as a sheet

2. Collect eyes. In mice, eyes are proptosed using forceps and pinched at the optic nerve to remove from the socket. In rats, scissors are required to cut away the conjunctiva and cut the optic nerve at the posterior of the eye.
3. Dissect retinal tissue from the eye. This technique is the same for mice and rats and is illustrated in schematic in Fig. 1. The technique described is for right-handed people.
4. Lay the eye on laboratory tissue wet with PSS under the dissecting microscope and trim away the extraocular muscles and tissues.
5. Orient the eye on its side, cornea facing the left and optic nerve stump facing the right. Puncture the globe through the sclera using a 26 G needle tip just posterior to the limbus, at the top (Fig. 1a).
6. Insert the small spring scissor tip into the hole created and make a small cut to extend the hole.

7. Grip the cornea with the fine tweezers (left hand) and continue extending the original cut around the sclera, posterior to the limbus, using the fine scissors in the right hand, cutting toward the microscope. Rotate the eye with the tweezers and continue cutting as required (Fig. 1b).
8. Once the entire cornea is cut around, lay the eyecup upright on the optic nerve in a standard dissecting petri dish and remove the anterior portions of the eye. Grasp the eyecup by the sclera at the back with the left hand and using touching tweezers in the right hand, gently blunt dissect the zonula adherens around the lens causing the lens to move around.
9. Tip the eyecup upright toward you and roll the lens out and away (Fig. 1c).
10. Run the touching tweezers between the sclera/choroid/RPE and retina to gently loosen the retina from the eyecup (Fig. 1d). Use the forceps to pinch the retina from the optic nerve at the back of the eye and place the sheet of retina into ice-chilled PSS (Fig. 1e).
11. Place two mouse retinae or one rat retinae into 5 mL of PSS on ice in a 50 mL Falcon tube.

3.2 Neural Dissociation Preparation

1. After dissection, let retinae settle to the base of the Falcon tube and gently remove PSS, then dissociate retinal cells using the Postnatal Neural Dissociation Kit.
2. Add 2.30 mL of room temperature enzyme mix 1 and agitate to remove tissue from bottom of tube.
3. Incubate for 15 min at 37 °C preferably in a tube rotator at 15 rpm.
4. Add 35 μL enzyme mix 2.
5. Dissociate tissue by using a 5 mL pipette set to 2 mL. Pipette up and down ten times gently and slowly. Avoid bubbles by keeping the tip of the pipette below the surface of the liquid.
6. Incubate for 10 min at 37 °C using tube rotator at 15 rpm.
7. Add 17 μL enzyme mix 2.
8. Dissociate as in **step 5**.
9. Dissociate further using fire polished Pasteur pipettes. Pipette up and down 35 times, slowly and avoid bubbles.
10. Apply dissociated preparation to a 75 μm strainer and strain cells into a 50 mL Falcon tube.
11. Add 20 mL of PSS + BSA to stop the enzyme digest.
12. Centrifuge at 130 × *g* for 10 min at 4 °C.
13. Remove supernatant and resuspend in 0.3 mL of PSS + BSA.

3.3 Fluorescently Label VNUT Positive Neurons and ATP Containing Vesicles

1. Incubate cells with 30 μL of the guinea pig anti-VNUT antibody for 15 min at 4 °C (*see* **Note 2**).
2. Wash cells with 20 mL of PSS and centrifuge at 130 × *g* for 10 min at 4 °C.
3. Remove the supernatant and resuspend cells in 4 mL of PSS + BSA.
4. Add 8 μL VNUT secondary antibody and 50 μM MANT-ATP or 10 μM quinacrine and incubate for 15 min at 4 °C. Additionally, dopamine-containing vesicles, which are found in VNUT positive neurons [20], can be colabeled using 50 μM Fluorescent false neurotransmitter.
5. Wash cells with 20 mL of PSS and centrifuge at 130 × *g* for 10 min at 4 °C.
6. Remove the supernatant and resuspend cells in 4 mL of PSS. Keep cells on ice until use.

3.4 Check Cell Viability/Number Using Countess Automated Cell Counter (See Note 3)

1. Add 10 μL of trypan blue Solution with 10 μL of cell sample to a 1 mL tube and mix gently by pipetting a couple of times.
2. Add 10 μL of mixed sample to cell counter slide.
3. Insert slide into an automated cell counter.
4. Zoom in, focus and count. In the retina at least 5 × 10^7 cells/mL are expected and viability should be around 85–95%.

3.5 Live Cell Imaging Using Confocal Microscopy

For viewing and imaging apply 200 μL of cell sample to a glass coverslip-bottomed petri dish and cover the sample with a small glass cover slip. Place the petri dish on an inverted confocal laser scanning microscope. We used an SP5 Leica microsystems, with the temperature-controlled chamber set at 37 °C. Cells need to be viewed and imaged quickly. Cells can be viewed initially with a ×40 oil objective and should look like the example shown in Fig. 2a when imaged using the Differential Interference Contrast or brightfield/transmitted light setting. Once the cells are brought into focus, use the red filter set to find cells positive for VNUT labeling and check subsequently for quinacrine/MANT-ATP using the green filter set (Fig. 2b). In the retina, all VNUT positive cells are also positive for quinacrine/MANT-ATP. VNUT positive cells are sparse and difficult to find as because they colocalize exclusively with dopaminergic neurons in the retina when using the suggested VNUT antibody [20], and so they represent less than 1% of the total retinal population.

Once a cell is visualized, live cell imaging should be undertaken with the ×63 microscope objective, with ×7 times zoom. The fluorescent dyes for the ATP-containing and dopamine-containing vesicles excite with the 488 and 380 nm lasers, respectively [23, 29], while VNUT-positive labeling will fluoresce with

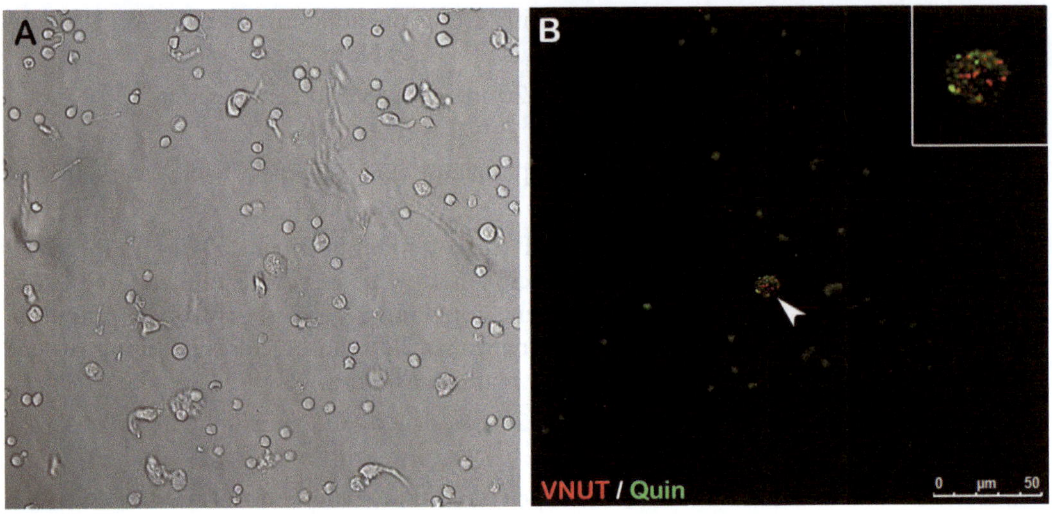

Fig. 2 Low magnification view of dissociated retinal neuron preparation labeled with VNUT and quinacrine. (**a**) Differential interference contrast image of dissociated retinal preparation collected using a ×40 objective, at 1× zoom. (**b**) Confocal image of dissociated retinal preparation collected using a ×40 objective, at 1× zoom. The white arrow indicates a VNUT positive labeled neuron (red puncta) that is also positive for the ATP vesicle label quinacrine (green puncta), which is magnified in the top right inset. VNUT positive neurons were quite scarce, representing less than 1% of the total retinal population. Scale, 50 μm

594 nm laser excitation. Scan speed should be optimized for image quality. Still shots were collected at 100 Hz and each channel was captured independently to ensure no bleed through. In Fig. 3, examples of still shots collected of VNUT-positive cells (red) that are also positive for the dopamine marker, FFN-102 (blue) and the ATP markers, quinacrine (green, Fig. 3a–d) or MANT-ATP (green, Fig. 3e–h) are presented.

For the purposes of collecting live cell images over time, red (VNUT) and green channel (Quinacrine or MANT-ATP) data can be collected concurrently at 1400 Hz. To assess ATP release from labeled cells, neuronal preparations in PSS should be imaged over a 1 min epoch in response to:

1. No treatment, to assess for fluorescence bleaching and basal vesicle release.
2. Slowly add 200 μL of PSS to assess for fluorescence bleaching and basal vesicle release in response to changes in volume.
3. Slowly add 200 μL of K^+PSS for neuronal depolarization and assess vesicle release.
4. Slowly add 200 μL of K^+PSS to cells that have been preincubated with 400 μM of cadmium to depolarize neuronal preparations where voltage-gated calcium channel-dependent synaptic transmission is blocked by cadmium.

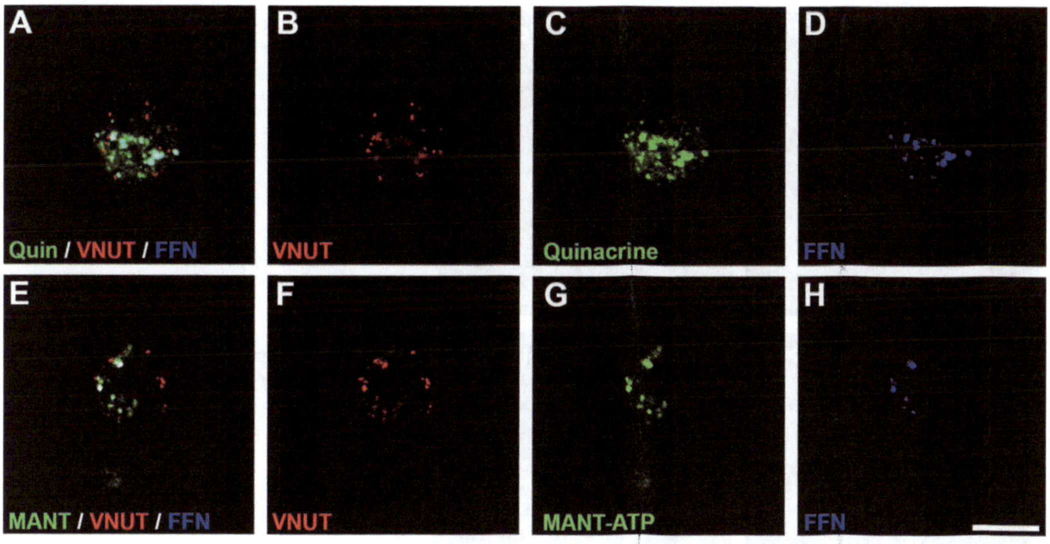

Fig. 3 High magnification image of VNUT positive neurons colocalized with ATP and dopamine-vesicular labels. (**a**) Colocalized image of VNUT-positive neuron (red) showing coexpression of purinergic vesicles labeled with quinacrine (green) and dopaminergic vesicles labeled with False Fluorescent Neurotransmitter 102 (FFN102; blue). Separate images of each channel presented in (**a**) are presented, showing (**b**) VNUT-positive labeled vesicles; (**c**) quinacrine-labeled ATP-containing vesicle; and (**d**) FFN102-labeled, dopamine containing vesicles. (**e**) Colocalized image of VNUT-positive neuron (red) showing coexpression of purinergic vesicles labeled with MANT-ATP (green) and dopaminergic vesicles labeled with FFN102 (blue). Separate images of each channel presented in (**e**) are presented, showing (**f**) VNUT-positive labeled vesicles; (**g**) MANT-ATP-labeled ATP-containing vesicle; and (**d**) FFN102-labeled, dopamine containing vesicles. Scale, 7.5 μm

Treatments were administered within the first 15–20 s of imaging and for the purposes of analysis, slight differences in the start time can be normalized by considering only the 10 s epoch before treatment as the start time. In Fig. 4, examples of a cell treated with PSS (Fig. 4a–c) and a cell treated with K^+PSS (Fig. 4d–f) are presented. Treatments were administered during imaging using a custom made microinjection system (*see* **Note 4**). The needle-tip was positioned next to the coverslip over the cells and fixed in place at the edge of the petri dish using blue tack.

3.6 Image Analysis

Live cell imaging data was exported as .avi files. Changes in fluorescence intensity of quinacrine or MANT-ATP puncta near the cell membrane were assessed as a marker of ATP release over time using a customized Image J 1.43 Freeware script (NIH, USA). Specifically, the script was written such that the .avi file was opened directly in ImageJ, the user was able to select a region of interest (ROI) around the edge of the cell surface which included many puncta, defined using the freehand selection tool. This ROI was added to the ROI Manager. The color image was split into red, green, and blue channels. The ROI was reapplied to the green channel. The "mean" fluorescence level was measured for the first frame and each

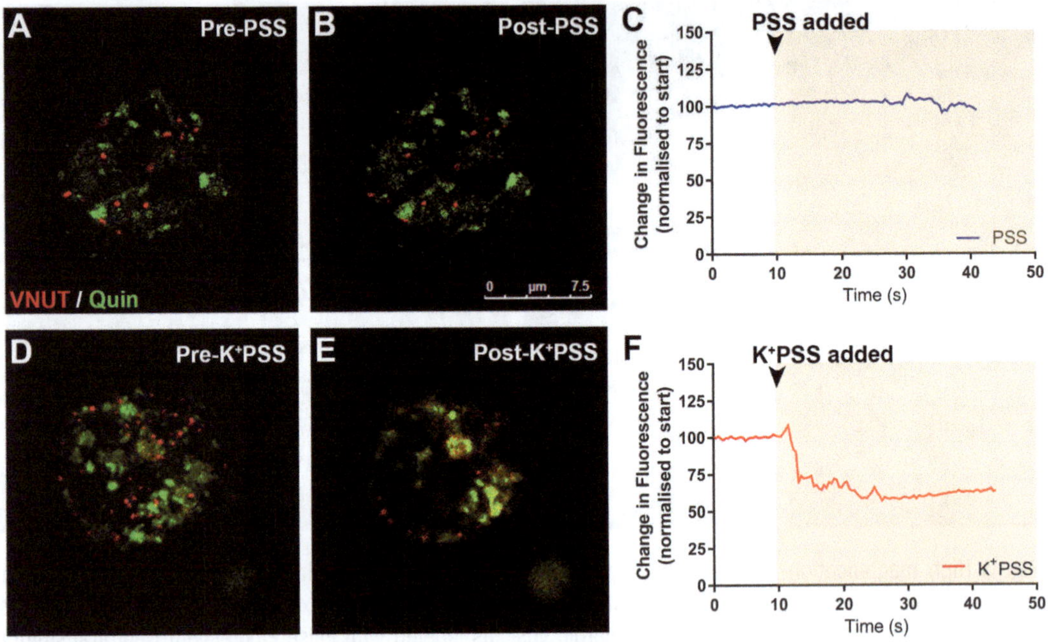

Fig. 4 Time lapse image analysis of ATP-marker fluorescence intensity near the cell membrane can be used as an indicator of ATP release under various treatment conditions. (**a–c**) Example of a VNUT-positive cell (red) showing similar levels of quinacrine vesicle fluorescence (green) before (**a**) and after addition of a control physiological saline solution, PSS (**b**). (**c**) Quinacrine fluorescence at the cell membrane can be quantified over time and normalized to mean fluorescence in the 10 s before treatment. Normalized mean intensity near the membrane (blue line) was not altered in response to addition of PSS over the recording interval (orange highlighted region). (**d–f**) Example of a VNUT-positive cell (red) showing changes in levels of quinacrine vesicle fluorescence (green) between (**d**) before and (**e**) after treatment to induce depolarizing-conditions, where 55 mM K^+PSS was added. (**f**) Quinacrine fluorescence at the cell membrane can be quantified over time and normalized to mean fluorescence in the 10 s before treatment. Normalized mean intensity near the membrane (red line) was reduced in response to addition of K^+PSS over the recording interval indicating vesicular release of ATP (orange highlighted region). Scale, 7.5 μm

subsequent frame and added to the "Results" table. These data were then copied and pasted into excel and normalized to average intensity at baseline over the first 10 s prior to treatment. In Fig. 4, examples of the analysis of a cell treated with PSS, which showed no mean change in fluorescence intensity (Fig. 4c), and a cell treated with K^+PSS, which showed a reduction of fluorescence at the cell membrane (Fig. 4f), are presented. For the purposes of statistical analysis, multiple cells should be recorded across multiple experiments ($>n = 6$ mice/rats) with a range of treatments each day. Subsequently, results can be expressed as the Mean ± Standard Error of the Mean (SEM) for each cell. One-way ANOVA was used to assess the effect of treatment conditions and differences between treatment groups were considered significant when $p < 0.05$.

4 Notes

1. This K^+PSS solution is designed to be added in equal parts to the PSS so that the final concentration of potassium is 55 mM. The concentration increase in potassium of this solution has been osmotically equalized by an equivalent reduction in sodium.
2. The VNUT antibody can be preconjugated to a secondary fluorophore using a commercial kit (e.g., DyLight® 594 Conjugation Kit, from Abcam, Melbourne, Australia, Cat#ab201801). This will reduce a couple of steps in this protocol and potentially preserve retinal cell number and viability.
3. Other cell counting and viability techniques can be used. Cells can be labeled with Trypan blue solution then 10 μL of cells can be loaded onto a hemocytometer slide and examined immediately under a microscope at low magnification (×10). The number of blue stained cells and the number of total cells can be counted per area (usually 1 mm^2).

   The percentage of viable cells = [1.00 − (Number of blue cells ÷ Number of total cells)] × 100 × 2 for the dilution with trypan blue.

   To calculate the number of viable cells per mL: Number of viable cells × 10^4 × 1.1 = cells/mL.
4. The custom-made perfusion needle was prepared by combining a 30 G diabetic needle to thin tubing (0.35 mm) using superglue. The other end of the tubing was fitted with a 26 G needle attached to a syringe. The tubing and syringe were backfill loaded with the solution to be perfused. Similar commercial needle and syringe products are available (e.g., IO-Kit from World Precision Instruments).

References

1. Abbracchio MP, Burnstock G, Verkhratsky A, Zimmermann H (2009) Purinergic signalling in the nervous system: an overview. Trends Neurosci 32(1):19–29. https://doi.org/10.1016/j.tins.2008.10.001
2. Burnstock G (2007) Physiology and pathophysiology of purinergic neurotransmission. Physiol Rev 87(2):659–797. https://doi.org/10.1152/physrev.00043.2006
3. Taruno A (2018) ATP release channels. Int J Mol Sci 19(3). https://doi.org/10.3390/ijms19030808
4. Burnstock G, Satchell DG, Smythe A (1972) A comparison of the excitatory and inhibitory effects of non-adrenergic, non-cholinergic nerve stimulation and exogenously applied ATP on a variety of smooth muscle preparations from different vertebrate species. Br J Pharmacol 46(2):234–242
5. Lazarowski ER (2012) Vesicular and conductive mechanisms of nucleotide release. Purinergic Signal 8(3):359–373. https://doi.org/10.1007/s11302-012-9304-9
6. Loiola EC, Ventura AL (2011) Release of ATP from avian Muller glia cells in culture.

Neurochem Int 58(3):414–422. https://doi.org/10.1016/j.neuint.2010.12.019
7. Reigada D, Mitchell CH (2005) Release of ATP from retinal pigment epithelial cells involves both CFTR and vesicular transport. Am J Physiol Cell Physiol 288(1): C132–C140. https://doi.org/10.1152/ajpcell.00201.2004
8. Santos PF, Caramelo OL, Carvalho AP, Duarte CB (1999) Characterization of ATP release from cultures enriched in cholinergic amacrine-like neurons. J Neurobiol 41 (3):340–348
9. Sawada K, Echigo N, Juge N, Miyaji T, Otsuka M, Omote H, Yamamoto A, Moriyama Y (2008) Identification of a vesicular nucleotide transporter. Proc Natl Acad Sci U S A 105 (15):5683–5686. https://doi.org/10.1073/pnas.0800141105
10. Geisler JC, Corbin KL, Li Q, Feranchak AP, Nunemaker CS, Li C (2013) Vesicular nucleotide transporter-mediated ATP release regulates insulin secretion. Endocrinology 154 (2):675–684. https://doi.org/10.1210/en.2012-1818
11. Haanes KA, Kowal JM, Arpino G, Lange SC, Moriyama Y, Pedersen PA, Novak I (2014) Role of vesicular nucleotide transporter VNUT (SLC17A9) in release of ATP from AR42J cells and mouse pancreatic acinar cells. Purinergic Signal 10(3):431–440. https://doi.org/10.1007/s11302-014-9406-7
12. Harada Y, Hiasa M (2014) Immunological identification of vesicular nucleotide transporter in intestinal L cells. Biol Pharm Bull 37 (7):1090–1095
13. Iwatsuki K, Ichikawa R, Hiasa M, Moriyama Y, Torii K, Uneyama H (2009) Identification of the vesicular nucleotide transporter (VNUT) in taste cells. Biochem Biophys Res Commun 388 (1):1–5. https://doi.org/10.1016/j.bbrc.2009.07.069
14. Larsson M, Sawada K, Morland C, Hiasa M, Ormel L, Moriyama Y, Gundersen V (2012) Functional and anatomical identification of a vesicular transporter mediating neuronal ATP release. Cereb Cortex 22(5):1203–1214. https://doi.org/10.1093/cercor/bhr203
15. Oya M, Kitaguchi T, Yanagihara Y, Numano R, Kakeyama M, Ikematsu K, Tsuboi T (2013) Vesicular nucleotide transporter is involved in ATP storage of secretory lysosomes in astrocytes. Biochem Biophys Res Commun 438 (1):145–151. https://doi.org/10.1016/j.bbrc.2013.07.043
16. Sathe MN, Woo K, Kresge C, Bugde A, Luby-Phelps K, Lewis MA, Feranchak AP (2011) Regulation of purinergic signaling in biliary epithelial cells by exocytosis of SLC17A9-dependent ATP-enriched vesicles. J Biol Chem 286(28):25363–25376. https://doi.org/10.1074/jbc.M111.232868
17. Sesma JI, Kreda SM, Okada SF, van Heusden C, Moussa L, Jones LC, O'Neal WK, Togawa N, Hiasa M, Moriyama Y, Lazarowski ER (2013) Vesicular nucleotide transporter regulates the nucleotide content in airway epithelial mucin granules. Am J Physiol Cell Physiol 304(10):C976–C984. https://doi.org/10.1152/ajpcell.00371.2012
18. Ziogas J, Vessey K (2001) Angiotensin-induced enhancement of excitatory junction potentials evoked by periarteriolar nerve stimulation and vasoconstriction in rat mesenteric arteries are both mediated by the angiotensin AT1 receptor. Pharmacology 63(2):103–111. https://doi.org/10.1159/000056120
19. Bodin P, Burnstock G (2001) Evidence that release of adenosine triphosphate from endothelial cells during increased shear stress is vesicular. J Cardiovasc Pharmacol 38 (6):900–908
20. Ho T, Jobling AI, Greferath U, Chuang T, Ramesh A, Fletcher EL, Vessey KA (2015) Vesicular expression and release of ATP from dopaminergic neurons of the mouse retina and midbrain. Front Cell Neurosci 9:389. https://doi.org/10.3389/fncel.2015.00389
21. Mitchell CH, Carre DA, McGlinn AM, Stone RA, Civan MM (1998) A release mechanism for stored ATP in ocular ciliary epithelial cells. Proc Natl Acad Sci U S A 95(12):7174–7178
22. Sorensen CE, Novak I (2001) Visualization of ATP release in pancreatic acini in response to cholinergic stimulus. Use of fluorescent probes and confocal microscopy. J Biol Chem 276 (35):32925–32932. https://doi.org/10.1074/jbc.M103313200
23. Dou Y, Wu HJ, Li HQ, Qin S, Wang YE, Li J, Lou HF, Chen Z, Li XM, Luo QM, Duan S (2012) Microglial migration mediated by ATP-induced ATP release from lysosomes. Cell Res 22(6):1022–1033. https://doi.org/10.1038/cr.2012.10
24. Irvin JL, Irvin EM (1954) The interaction of quinacrine with adenine nucleotides. J Biol Chem 210(1):45–56
25. Menendez-Mendez A, Diaz-Hernandez JI, Ortega F, Gualix J, Gomez-Villafuertes R, Miras-Portugal MT (2017) Specific temporal distribution and subcellular localization of a functional vesicular nucleotide transporter (VNUT) in cerebellar granule neurons. Front Pharmacol 8:951. https://doi.org/10.3389/fphar.2017.00951

26. Alund M, Olson L (1979) Depolarization-induced decreases in fluroescence intensity of gastro-intestinal quinacrine-binding nerves. Brain Res 166(1):121–137

27. Moriyama S, Hiasa M (2016) Expression of vesicular nucleotide transporter in the mouse retina. Biol Pharm Bull 39(4):564–569. https://doi.org/10.1248/bpb.b15-00872

28. Neal M, Cunningham J (1994) Modulation by endogenous ATP of the light-evoked release of ACh from retinal cholinergic neurones. Br J Pharmacol 113(4):1085–1087

29. Rodriguez PC, Pereira DB, Borgkvist A, Wong MY, Barnard C, Sonders MS, Zhang H, Sames D, Sulzer D (2013) Fluorescent dopamine tracer resolves individual dopaminergic synapses and their activity in the brain. Proc Natl Acad Sci U S A 110(3):870–875. https://doi.org/10.1073/pnas.1213569110

Chapter 16

Using FRET-Based Fluorescent Sensors to Monitor Cytosolic and Membrane-Proximal Extracellular ATP Levels

Klaus E. Kaschubowski, Axel E. Kraft, Viacheslav O. Nikolaev, and Friedrich Haag

Abstract

The assessment of local concentrations of extracellular ATP (eATP) at the site of receptor binding remains a challenge in the field of purinergic signaling. In many cases, biosensors exploiting the principle of Förster resonance energy transfer (FRET) have provided useful tools to visualize local concentrations of metabolites. A series of FRET-based biosensors based on the epsilon subunits of bacterial ATP synthases have been described for the visualisation of ATP. These sensors carry ATP-sensing units with different affinities for ATP, permitting imaging of ATP under the widely different concentration conditions found in subcellular locations such as the cytoplasm and the membrane-proximal extracellular space.

Key words FRET, Microscopy, FACS, Live-cell imaging, Biosensor, Extracellular ATP

1 Introduction

The purine nucleotide ATP is abundant within cells, where it plays an important role as the prime currency of energy metabolism. Although under steady-state conditions ATP is present only in low nanomolar concentrations outside of cells (reviewed in [1]), extracellular ATP (eATP) provides the raw material for a system of cellular communication known as purinergic signaling. ATP is released into the extracellular space either from dead or damaged cells or as the result of regulated secretion for signaling purposes. Once outside of cells, eATP either acts directly on purinergic P2X or P2Y receptors or is metabolized by the combined action of the ectoenzymes CD39 and CD73 to adenosine, which acts on adenosinergic P1 receptors [2].

A challenge in the field of purinergic signaling is to assess whether the ATP concentration achieved locally at the site of its interaction with the receptor is sufficient to trigger a given signaling pathway. A sensitive and widely used method to measure

Pablo Pelegrín (ed.), *Purinergic Signaling: Methods and Protocols*, Methods in Molecular Biology, vol. 2041,
https://doi.org/10.1007/978-1-4939-9717-6_16,

extracellular ATP concentrations is the luciferase reaction, where light is produced during the ATP-dependent cleavage of luciferin. Since luciferase is a soluble protein, this method is most commonly used to measure bulk ATP concentrations in solutions such as cell supernatants or lysates. This limitation has been overcome by genetically engineering luciferase to be expressed on the cell surface. However, although this method is highly sensitive and quantifiable, it is not well suited for live-cell imaging because of the low amount of light generated by the luciferase reaction.

A commonly used approach to visualize local concentrations of metabolites for imaging purposes is to exploit the mechanism of Förster resonance energy transfer (FRET) [3]. This effect describes the transfer of energy from a donor to an acceptor fluorophore when the two are in very close proximity. Genetically encoded FRET-based biosensors have been widely used to visualize the concentrations of various metabolites, such as cyclic AMP (cAMP) and calcium [4, 5]. Such sensors are typically constructed by fusing suitable fluorophores to the N- and C-termini of a receptor protein that binds the metabolite to be measured with high specificity and that undergoes a major conformational change in response to binding. Binding of the metabolite then causes motion of the fluorophores relative to each other, thereby changing their capacity to transfer energy.

Following this principle, a FRET-based biosensor for ATP has been constructed based on the epsilon subunit of the ATP-synthase from *Bacillus* species, which noncovalently binds ATP without metabolizing it [6]. Subunits from different species of Bacillus vary widely in their affinity to ATP, permitting the construction of sensors suited for the detection of ATP in different concentration ranges. Thus, the subunit from B. subtilis (Bs) has optimal sensitivity in the millimolar range, whereas the subunit from *Bacillus PS3* species detects changes in the micromolar range. Mutation of two arginine residues critical for binding ATP to lysine (RRKK variants) provides a good negative control yielding the background FRET ratio in the absence of any ATP binding. This sensor, called ATEAM (ATP indicator based on epsilon subunit for analytical measurements), carries monomeric superenhanced cyan fluorescent protein lacking the 11 C-terminal amino acids (mseCFPΔ11) and circularly permutated monomeric Venus (cp173-mVenus, a close relative of the yellow fluorescent protein YFP) as donor and acceptor fluorophores, respectively [6]. By adding an N-terminal signal sequence and a C-terminal signal sequence for GPI-anchoring into the cell membrane, we adapted the originally cytoplasmic sensors for expression at the cell surface (*see* Fig. 1).

2 Materials

2.1 Sensor Components (See Fig. 1)

1. Sensors for detection of cytoplasmic ATP: Plasmids Bs.cyt and Bs_RRKK.cyt.
2. Sensors for detection of membrane-proximal eATP: Plasmids PS3.GPI and PS3_RRKK.GPI.

2.2 Imaging System

1. Inverted fluorescence microscope.
2. Fluorescence light source emitting at 440 nm. The emission wavelength should be close to the maximum spectral absorbance of the donor fluorophore (in our case CFP with an absorbance maximum at 436 nm).
3. Beam-splitter including a filter cube consisting of a dichroic mirror and two emission filters for the donor and acceptor fluorophores. For the CFP/Venus donor-acceptor pair we routinely use the DV2 cube 05-EM containing a 505dcxr dichroic mirror with D480/30m (CFP channel) and D535/40m (YFP channel) emission filters.
4. CMOS Camera (QIMAGING optiMOS, or similar).
5. Arduino digital input/output board.
6. Microscopy software (Micro-Manager1.4.5 together with ImageJ, or similar) (*see* **Note 1**).

2.3 Live-Cell Measurement

1. Imaging buffer (ECS): 15 mM HEPES pH 7.4, 140 mM NaCl, 5 mM KCl, 10 mM D-glucose, 0.1% BSA, 1 mM $CaCl_2$, and 1 mM $MgCl_2$. Adjust pH to 7.4 with 1 M NaOH.
2. Glass coverslips with a diameter of 25 mm.

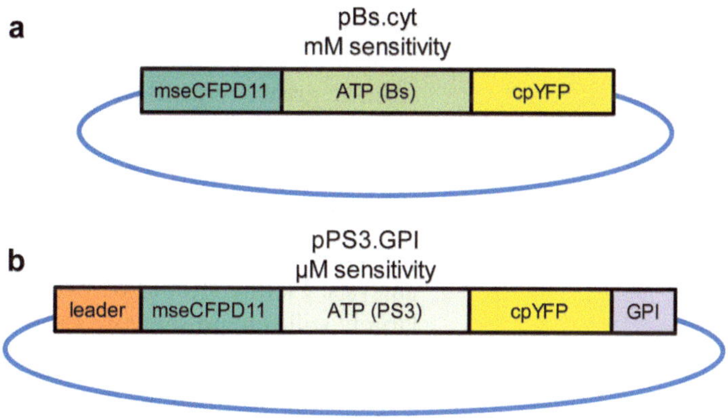

Fig. 1 Composition of FRET-based ATP sensors used for measuring cytosolic (**a**) and membrane-proximal extracellular ATP (**b**). The sensor in (**b**) contains the mouse Ig-kappa leader sequence and the human folate receptor GPI signal sequence

3. 5 mg/mL solution of bovine serum albumin (BSA).
4. 0,1 mg/mL poly-L-lysine (70 kDa) solution.
5. Adherent cells expressing the FRET sensor.
6. Cell chamber for microscopy (e.g., Attofluor™ Cell Chamber from Invitrogen).

2.4 Calculation of Spectral Bleedthrough

1. Six-well plate with adherent cells (HEK293 cell line) plated on autoclaved glass cover slides in Dulbecco's Modified Eagle Medium (DMEM) supplemented with 10% fetal calf serum, L-glutamine, and antibiotics.
2. Plasmid coding for the donor fluorescent protein (e.g., pECFP-N1).
3. Lipofectamine 2000 transfection reagent or equivalent.

2.5 Flow Cytometry Unit and Analysis Software

1. Flow cytometry (FACS) unit equipped with a violet laser and channels for the detection of the fluorophores BV421 (emission filter 450-40) and BV510 (emission filter 525-50).
2. Flow cytometry analysis software package FlowJo (Treestar, Ashland, OR, USA). For determination of FRET ratios we used version 9.9.4.

3 Methods

3.1 Preparation of Cells

1. When analyzing suspension cells pretreat coverslips with poly-L-lysine by adding a spot of 5 μL BSA on the coverslip (use a rectangular coverslip to smear the BSA evenly across the coverslip, creating a thin, continuous film), let dry for 10–20 min and then spot 5 μL of poly-L-lysine onto the BSA-coated coverslips (using a rectangular coverslip to create a thin film as described above for BSA). Let dry for 10–20 min (if desired, small measuring chambers can be created on coverslips by using silicone grease (e.g., Baysilone) and O-rings).
2. For cells growing in suspension, prepare a cell suspension at a density of approximately 4×10^6/mL. Keep the cells cold until they are ready for use.
3. For adherent cells, plate the cells on coverslips the day before analysis. Plating should be done at a density such that the cells largely still remain separate on the day of analysis.
4. Place coverslips into the cell chamber and add 20 μL of the cell suspension. Incubate for 20 min at room temperature to allow for attachment of the cells.
5. Wash cells with 300 μL warm (37 °C) ECS buffer to remove dead and nonadherent cells (*see* **Note 2**).
6. Add 300 μL warm (37 °C) ECS buffer before beginning measurement.

3.2 Determination of the Spectral Bleedthrough Factor

Due to overlap in the emission spectra of CFP and Venus a certain amount of light emitted directly from CFP is recorded in the Venus channel in the absence of any energy transfer. This effect is termed bleedthrough. Its magnitude is determined by determining the FRET ratio of cells that express only CFP. The mean FRET ratio of cells expressing only CFP and no YFP yields the correction factor *b*.

1. Transfect HEK293 cells with a plasmid encoding solely for the CFP protein, (e.g., pECFP-N1) using Lipofectamine 2000 or an equivalent method according to the manufacturer's instructions.
2. After 48 h in culture, measure fluorescence in the CFP and YFP channels for at least 10 cells as described in Subheading 3.3. Make sure stable baselines for the two channels have been reached before recording the measurement (*see* **Note 3**).
3. Calculate the FRET ratios for individual cells as described in Subheading 3.4.
4. Determine the correction factor b by calculating the mean of the acquired FRET ratios.

3.3 Live-Cell Imaging

1. Before placing the cell chamber in the microscope put a small amount of immersion oil on the objective.
2. Select a field of view containing isolated, healthy-looking cells with satisfactory expression of the FRET sensor.
3. Start the imaging software and adjust the LED intensity, exposure time and the interval between the pictures taken (e.g., 5 s) (*see* **Note 4**).
4. Start recording of the cell(s).
5. Before adding a stimulus, wait until the baseline is stable.
6. Add 300 μL of a stimulus prepared in ECS to your sample.
7. Save the recorded images for later analysis, remove the cell chamber from the microscope, clean the used materials, and prepare a new sample.

3.4 Data Evaluation

1. Open the recorded images with "Micromanager" and run a plugin to separate the CFP and YFP channels.
2. Select one or more regions of interests (ROI) and use the plugin "FRET online" (available in the online supplement of [7]) to calculate the FRET (acceptor/donor) ratio for every recorded frame.
3. Correct for the spectral bleedthrough effect by subtracting the correction factor b (*see* Subheading 3.2) from the FRET ratio, resulting in the corrected FRET ratio:

$$\mathrm{FRET}_{\mathrm{corr}} = \mathrm{FRET} - b.$$

4. Normalize the corrected FRET ratios to the average value of the baseline (Fig. 2).

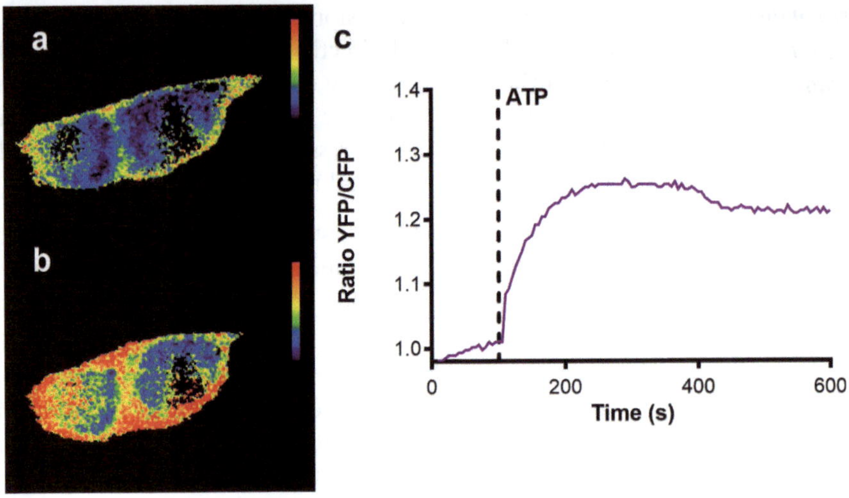

Fig. 2 Visualization of membrane-proximal extracellular ATP levels. Live-cell imaging of the ATP concentration at the cell surface of mouse 3T3 cells stably transfected with the PS3.GPI plasmid. Pseudocolor images (**a**, **b**) and FRET trace (**c**) of a 3T3 cell before (**a**) and 500 s after (**b**) addition of 500 μM ATP

3.5 Generation of Pseudocolor Images

1. Select a region of interest (ROI) in the Venus image using "Edit → Selection → Specify". Store this position with "Analyze → Tools → ROI manager" and apply it to the CFP image. Cut the ROI for CFP and YFP with "Image → Crop".
2. To reduce the background, select "Process → Subtract background" with the rolling ball algorithm within a range of 50–200 (not smaller than the biggest object in the picture).
3. To make sure that both the CFP and the YFP image are aligned use "Plugins → Registration → MultiStackReg". In the "MultiStackReg" menu, select CFP as reference stack and align YFP to it with "Align to First Stack". Additionally use the "Rigid Body" method.
4. For preparation of the threshold convert the image to 32 bit with "ImageType → 32bit".
5. Improve the image quality with "Process → Smooth".
6. For thresholding use "Image → Adjust → Threshold" with "Default", "B&W" and "Dark Background" just for the YFP image. Then apply "Set Background Pixels to NaN" and confirm it.
7. Determine the fluorescence intensities in the donor and acceptor channels for all frames.
8. Calculate the FRET ratio using "Plugins → Ratio Plus", choosing YFP as "Image1" and CFP as "Image2".

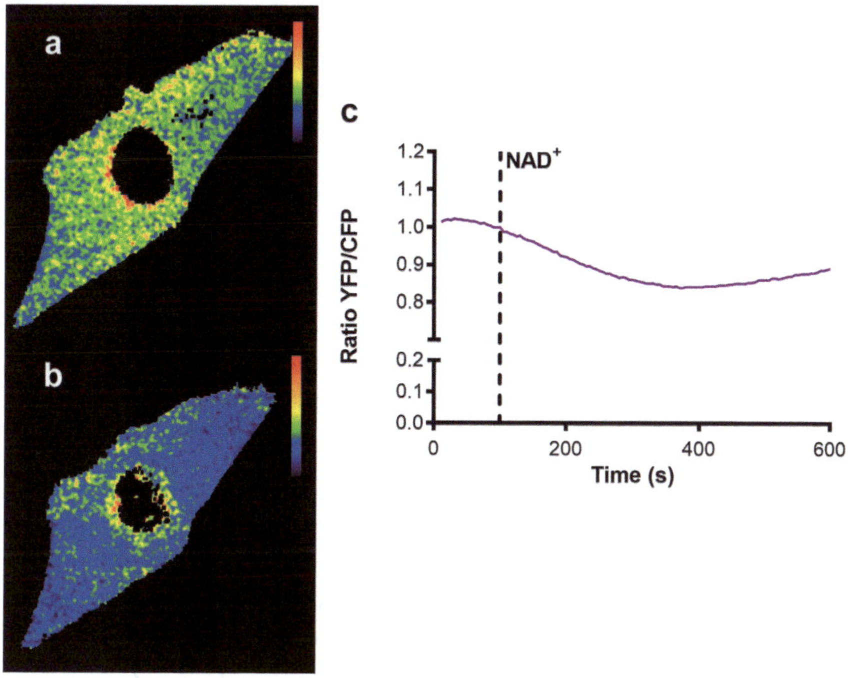

Fig. 3 Visualization of cytosolic ATP levels. 3T3 cells stably transfected with mP2X7 were supertransfected with Bs.cyt and mouse ADP-ribosyltransferase-2b (ARTC2) to permit activation of P2X7 via NAD-dependent ADP-ribosylation [8]. Pseudocolor images (**a**, **b**) and the FRET trace (**c**) of a cell before (**a**) and 500 s after (**b**) stimulation with 20 μM NAD show depletion of cytosolic ATP following gating of P2X7

9. To assign pseudocolors to the generated ratios use "Plugins → Nucmed → Lookup tables" and select a color scheme (e.g., "Blue_Green_Red").
10. Save the image as, for example, "TIFF" and convert it to RGB with "Image → Type → RGB color" (Fig. 3).

3.6 Measurement of Cytosolic ATP Levels by Flow Cytometry

Live cell imaging as described above is the best method to visualize ATP concentrations in or at the surface of individual cells that are reacting to a stimulus. However, flow cytometry can also be used as a sensitive method to quantify the reaction of a large population of similar cells to a given stimulus. The violet laser present in most FACS units (e.g., BD Canto II) excites CFP, and the channel commonly used for measuring the fluorescent dyes such as BV421 or Pacific Blue (i.e., the 450/40 filter) picks up the CFP signal, while the YFP signal is recorded on the channel carrying the 525/50 filter commonly used to measure dyes such as BV510 or Pacific Orange.

1. Prepare a suspension of the cells to be analyzed, stably transfected with the cytoplasmic pBs.cyt plasmid, at a density of approximately 10^7 cells/mL (*see* **Note 5**).
2. Keep the cells on ice until ready for use.

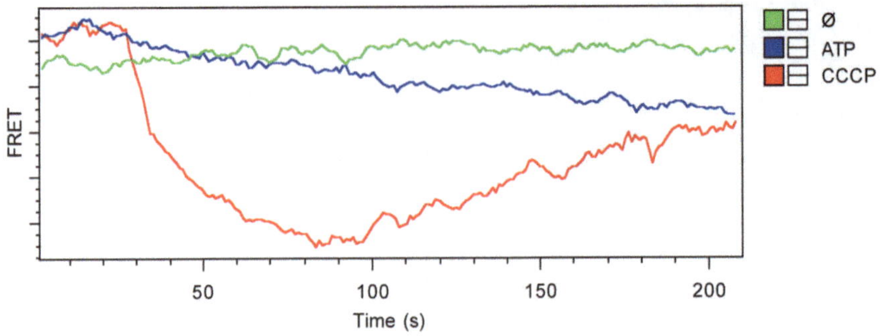

Fig. 4 Visualization of cytosolic ATP levels by flow cytometry. 3T3 cells stably transfected with mouse P2X7 and Bs.cyt were recorded on a FACS Canto2 flow cytometer as described in the text. After 30 s 500 μM ATP or 10 μM CCCP were added and cells were monitored for a further 180 s. The FRET ratio was calculated using FlowJo (version 9.9.4)

3. In a water bath, keep a battery of FACS tubes containing PBS/0.1% BSA at 37 °C.
4. Add 50 μL of cell suspension to a FACS tube, and allow the cells to adjust to the temperature change for 1 min.
5. Begin FACS measurement (*see* **Notes 6** and 7). Record a baseline for 30 s, then add a stimulus and continue the measurement for a further 2–3 min (Fig. 4).

4 Notes

1. Detailed instructions on how to connect all imaging system components, set up the software, and use ImageJ are given in Sprenger et al. [7].
2. Dead and nonadherent cells can provide strong background signals or float by the measured cells and thereby make an experiment unusable. If necessary, wash several times.
3. Measurements to determine the spectral bleedthrough should be carried out at microscope conditions similar to the ones used for genuine FRET measurements, that is, LED intensities and exposure times should mirror "real" experiments as closely as possible.
4. It is a good idea to explore different settings for LED intensity, exposure time of individual frames, and the interval between frames in preliminary experiments. These parameters may have an immense influence on the quality of the measurements. In general, increasing the intensity of the LED and the exposure time increases the signal-to-noise ratio and provides clearer pictures but also increases the propensity for photobleaching. Photobleaching is recognized by a rundown of the measured

FRET ratio in unstimulated (inert) cells. If this occurs, the light intensity and (or) the exposure time should be reduced.

5. Flow cytometry cannot be used to measure extracellular ATP levels, since the cells are flushed through the measuring device in a sheath fluid that will probably wash away any extracellular ATP present.
6. The FITC/EGFP channel can be used to set a gate on transfected cells, excluding untransfected cells from the measurement.
7. Spectral bleedthrough can be determined analogous to live cell microscopy by analysing cells expressing only CFP. Performing compensation between the CFP and YFP channels will correct for bleedthrough.

Acknowledgments

We thank G. Dubberke for technical assistance. This study was supported by the Deutsche Forschungsgemeinschaft (SFB1328 and grant Ha2569/5).

References

1. Trautmann A (2009) Extracellular ATP in the immune system: more than just a "danger signal". Sci Signal 2(56):pe6. https://doi.org/10.1126/scisignal.256pe6
2. Cekic C, Linden J (2016) Purinergic regulation of the immune system. Nat Rev Immunol 16 (3):177–192. https://doi.org/10.1038/nri.2016.4
3. Hou BH, Takanaga H, Grossmann G, Chen LQ, Qu XQ, Jones AM, Lalonde S, Schweissgut O, Wiechert W, Frommer WB (2011) Optical sensors for monitoring dynamic changes of intracellular metabolite levels in mammalian cells. Nat Protoc 6(11):1818–1833. https://doi.org/10.1038/nprot.2011.392
4. Kraft AE, Nikolaev VO (2017) FRET microscopy for real-time visualization of second messengers in living cells. Methods Mol Biol 1563:85–90. https://doi.org/10.1007/978-1-4939-6810-7_6
5. Sprenger JU, Bork NI, Herting J, Fischer TH, Nikolaev VO (2016) Interactions of calcium fluctuations during cardiomyocyte contraction with real-time cAMP dynamics detected by FRET. PLoS One 11(12):e0167974. https://doi.org/10.1371/journal.pone.0167974
6. Imamura H, Nhat KP, Togawa H, Saito K, Iino R, Kato-Yamada Y, Nagai T, Noji H (2009) Visualization of ATP levels inside single living cells with fluorescence resonance energy transfer-based genetically encoded indicators. Proc Natl Acad Sci U S A 106 (37):15651–15656. https://doi.org/10.1073/pnas.0904764106
7. Sprenger JU, Perera RK, Gotz KR, Nikolaev VO (2012) FRET microscopy for real-time monitoring of signaling events in live cells using unimolecular biosensors. J Vis Exp (66):e4081. https://doi.org/10.3791/4081
8. Seman M, Adriouch S, Scheuplein F, Krebs C, Freese D, Glowacki G, Deterre P, Haag F, Koch-Nolte F (2003) NAD-induced T cell death: ADP-ribosylation of cell surface proteins by ART2 activates the cytolytic P2X7 purinoceptor. Immunity 19(4):571–582

Chapter 17

ATP Measurement in Cerebrospinal Fluid Using a Microplate Reader

Laura de Diego-García, Álvaro Sebastián-Serrano, Carolina Bianchi, Caterina Di Lauro, and Miguel Díaz-Hernández

Abstract

Imbalance in extracellular ATP levels in brain tissue has been suggested as a triggering factor for several neurological disorders. Here, we describe the most sensitive and reliable technique for monitoring the ATP levels in mice cerebrospinal samples collected by cisterna magna puncture technique and quantified using a microplate reader.

Key words ATP, Luciferin, Luciferase, Cerebrospinal fluid, In vivo, Neurological disorders, Microplate reader

1 Introduction

Under physiological conditions, extracellular nucleotides levels are finely regulated in the central nervous system (CNS), mainly by the actions of ectonucleotidases, which hydrolyze them to their corresponding nucleosides [1, 2]. However, under pathological conditions a variety of events can compromise cell integrity, favoring the ATP release contained in the cytosolic compartment to the closest brain parenchyma [3], resulting in an extracellular nucleotide concentration increase. In accordance with this idea, several works have reported that both the pharmacological inhibition and depletion of the less ATP-sensitive P2 receptor, P2X7R, delay the progression and abrogated the behavioral alterations associated with different neuronal disorders including Huntington's disease [4], Parkinson's disease [5], Alzheimer's disease [6, 7], epilepsy [8], and hypophosphatasia-associated seizures [9].

Over the last years, different techniques have been developed to measure in vivo extracellular ATP (or ATP released) in interstitial

Laura de Diego-García and Álvaro Sebastián-Serrano contributed equally to this work.

Pablo Pelegrín (ed.), *Purinergic Signaling: Methods and Protocols*, Methods in Molecular Biology, vol. 2041, https://doi.org/10.1007/978-1-4939-9717-6_17, © Springer Science+Business Media, LLC, part of Springer Nature 2020

fluid or in close proximity to the outer surface of the plasma membrane [10–12]. Although these techniques have been shown useful to determine the extracellular ATP concentration in real time employing different cellular models, neither of these methods have been still reported to measure successfully ATP concentration in the brain parenchyma in vivo.

As alternative to this technique limitation, there are ex vivo approaches that can allow to estimate the extracellular ATP levels in CNS, as the determination of ATP concentration in the cerebrospinal fluid (CSF). Using this approach, we can perform longitudinal studies along the disease progression [13], which enables the extracellular ATP levels to be related with disease evolution or with efficacy of treatment used. The most common techniques to collect CSF samples are the cisterna magna puncture in mice [13] and the lumbar puncture in human [14]. Once the CSF samples are obtained, ATP concentration may be determined using different procedures, including high-performance liquid chromatography (HPLC) [15], microelectrodes functionalized with glycerol kinase and glycerol, or bioluminescent techniques using the luciferin–luciferase reagent [16–18]. Here, we describe how to measure extracellular ATP levels in CSF samples collected by cisterna magna puncture technique using a bioluminescence procedure. This method is based on the reaction catalyzed by the luciferase enzyme as follows:

$$\text{ATP} + \text{D-Luciferin} + O_2 \rightarrow \text{Oxyluciferin} + \text{AMP} + \text{PPi} + CO_2 + \text{light (560 nm)}$$

Luciferase is an enzyme that chemically generates light as a by-product of the oxidation of the small-molecule substrate D-luciferin. The reaction catalyzed by luciferase in the presence of magnesium ions implicates the conversion of D-luciferin into oxyluciferin, producing a flash of yellow–green light proportional to the amount of ATP present, with a peak emission at 560 nm [19].

2 Materials

All solutions are prepared using ultrapure water (obtained from deionized water to attain a sensitivity of 17 MΩ-cm at room temperature) and analytical grade reagents. Once prepared, solutions are stored at room temperature unless otherwise indicated. Steel materials, forceps and scissors, are packaged in plastic bags and autoclaved at 121 °C for 20 min and then dried in an oven at 37 °C.

2.1 Cerebrospinal Fluid Collection

1. Isoflurane 1000 mg, liquid for vapor inhalation. Store at room temperature.
2. Stereotaxic equipment adapted to mice and coupled to an anesthesia system.
3. Homeothermic monitoring system.
4. Surgical light with 3.5 diopter magnifier.
5. Hair clipper adapted to mice.
6. Sterilized forceps, scissors, scalpel.
7. 70% ethanol: add to 270 mL water 730 mL of 96% ethanol. Store at room temperature.
8. Borosilicate glass capillaries with an internal diameter of 1.15 mm and external diameter of 1.55 mm.
9. Micropipette puller.
10. 0.5 mL tubes.
11. 100 μM ARL 67156 (ecto-ATPase inhibitor) in water. Store at −20 °C.
12. Dry ice mixed with ethanol: mix 100 g of dry ice with 100 mL of ethanol. The mix is made in a thermal polystyrene box.
13. Autoclip applier.

2.2 Solutions for ATP Measurement

1. Phosphate Buffered Saline (PBS) buffer: Add 500 mL water to a 1 L of graduated cylinder or a beaker glass. Weigh 8.0 g NaCl, 2.7 g KCl, 1.44 g Na_2HPO_4, and 0.24 g KH_2PO_4 and transfer to the cylinder. Add water to a volume of 900 mL. Mix and adjust the pH to 7.4 with HCl. Make up to 1 L with water. Store at room temperature.
2. 1 μM ATP and 10 nM ATP in PBS. Store at −20 °C.
3. ATP standard curve dilutions: 10 μL of PBS for 0 ATP pmol; 1 μL of 10 nM ATP in PBS plus 9 μL of PBS for 0.01 ATP pmol; 10 μL of 10 nM ATP in PBS for 0.1 ATP pmol; 1 μL of 1 μM ATP in PBS plus 9 μL of PBS for 1 ATP pmol; and 10 μL of 1 μM ATP in PBS for 10 ATP pmol (Fig. 1).
4. White polystyrene 96-well plate with round bottom.
5. Plate reader with luminescent measurement technology.
6. rLuciferase/Luciferin (rL/L) reagent: reconstitute lyophilized rL/L reagent in reconstitution buffer that contains purified luciferase, D-luciferin, Tris–acetate buffer (pH 7.75), ethylene-diaminetetraacetic acid (EDTA), magnesium acetate, bovine serum albumin (BSA) and dithiothreitol (DTT), 0.02% sodium azide (*see* **Note 1**). Store at −20 °C.

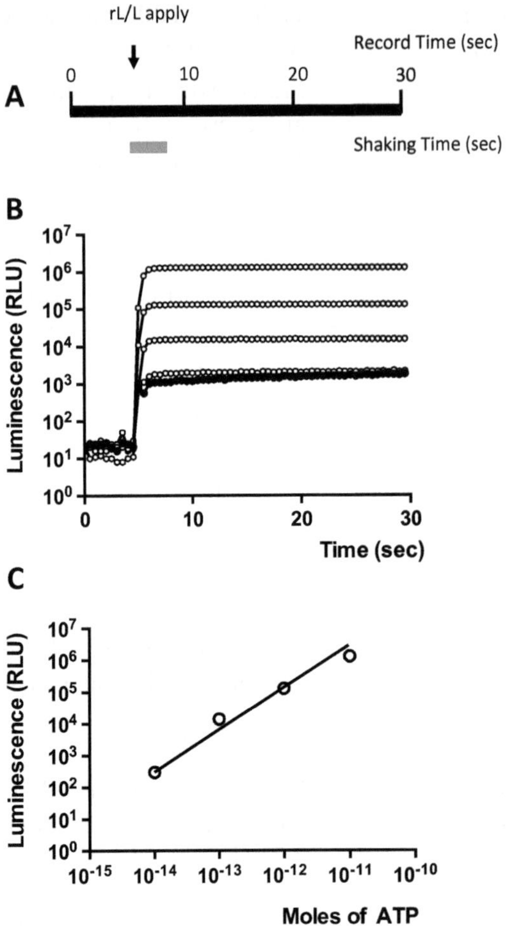

Fig. 1 ATP Standard curve obtained with rL/L reagent and the microplate reader. (**a**) Schematic representation of the protocol used in microplate reader to record the luminescence signal emmitted by luciferase. (**b**) Timecourse of lumisnescence signal recorded by microplate reader using ATP standars containing 0, 0.01, 0.1, 1, 10 ATP pmoles disolved in PBS buffer. (**c**) Graph shows the linear regression line obtained using the 20-s of luminescence signal integration time after the injection of rL/L reagent

3 Methods

Carry out all steps at room temperature, unless otherwise specified.

3.1 Pulling the Glass Capillary Tubes

1. Put the capillary tubes in the capillary puller, with the heat index set at 570, the pull index set at 10, and the velocity index set at 150.
2. Cut the tip of the capillary glass tubes with scissors so that the tip has an inner diameter of about 0.5 mm.

3.2 Puncture of Cisterna Magna to Collect CSF Samples

1. Anesthetize mice by isoflurane inhalation; fill the isoflurane vaporizer with isoflurane 1000 mg/g liquid for vapor inhalation.
2. Set the system to initiate flow to the induction chamber. Turn on the supply O_2 gas, setting the flowmeter between 800 and 1200 mL/min. Turn on the vaporizer at 4% and the passive scavenging system.
3. Place the mouse into the anesthesia chamber and monitor until it is recumbent.
4. When the mouse is anesthetized (*see* **Note 2**), place it in prone position on the stereotaxic frame with direct contact on a heating pad.
5. Insert rectal temperature probe into the rectum and readjust the heating pad in response to changes of body temperature.
6. Switch the system flow to nosecone chamber.
7. Secure the mouse head with the head adaptors, putting its nose into the nosecone chamber adapted to mice (*see* **Note 3**).
8. Restart gas flow with flowmeter at 400–800 mL/min and vaporizer at 2–3%. If the animal starts to respond, gently increase the vaporizer up to 3–4% in nosecone chamber until it reaches a full anesthetized state. During the entire procedure the mouse breathing should be constant; if it is not so, readjust vaporizer up to observe a rhythmic breathing. The mouse is laid down so that the head forms a nearly 135° angle with the body (*see* **Note 4**).
9. Shave the skin of the neck using a hair clipper.
10. Swab the surgical site with 70% ethanol three times.
11. Perform a sagittal incision of the skin in the occiput area.
12. Under a dissection magnifier, separate the subcutaneous tissue and muscles (*M. biventer* cervicis and *M.* rectus capitis dorsalis major) using curved dissection forceps. Cisterna magna appears as a whitish and clear reverse triangle where the medulla oblongata, the major blood vessel (arteria *dorsalis spinalis*), and the CSF space are located.
13. Dry the dura mater of cisterna magna with sterile cotton swab before to be penetrated by the capillary tube. The incision site should be lateral to the arteria *dorsalis spinalis* to avoid the blood vessel rupture and the subsequence blood contamination of the CSF samples collected. Once inside of cisterna magna, the CSF will flow by capillarity into the tube in a couple of seconds.
14. Remove carefully the capillary tube and connect with a 5 mL syringe through the polyethylene tubing with a 1.2 mm inner diameter.

15. Inject the CSF samples in 0.5 mL tube containing 1 μL of 100 μM ARL 67156 (an inhibitor of ectonucleotidases to prevent ATP hydrolysis).
16. Freeze tubes immediately by putting them in a thermal polystyrene box containing dry ice mixed with ethanol 70%.
17. Transfer tubes to a −80 °C freezer to be stored until measurement of ATP concentration.
18. After collecting the CSF samples (*see* **Note 5**), the muscles are realigned and the skin is sutured using surgical autoclips. Mice are kept in a heated incubator at 35 °C until recovery.

3.3 Measure ATP Using a Microplate Reader

1. Turn on microplate reader and select the luminescence mode. The emission light guide is placed on the position that allows to read light emitted at 560 nm by the upper part of the white 96-well plate.
2. Prime microinjection pump with 2 mL of water.
3. Unload dispensing water at a flow rate of 450 μL/s.
4. The same procedure (**steps 2** and **3**) is done with 2 mL of PBS buffer.
5. Prime microinjection pump with 2 mL of freshly reconstituted rL/L reagent. In the case of frozen reconstituted rL/L reagent, it should be previously unfrozen at room temperature.
6. Load the white 96-well plate with the ATP dilutions to be used for the calibration curve using 0, 0.01, 0.1, 1, and 10 pmol of ATP.
7. Then add to the wells of the ATP standard curve 1 μL of 100 μM ARL 67156 and PBS buffer is added to fix the final volume to 11 μL.
8. The rest of the wells of the plate contain 1 μL of CFS sample plus 10 μL of PBS (*see* **Note 6**).
9. Set the software that controls the microplate reader with the following parameters: number of measurements during the kinetic record 60; measurement intervals time 0.5 s; interval time 0.5 s; end time of kinetic window 30 s; Emission filter selected, lens with a gain assigned of 2800 (*see* **Note** 7).
10. Set the thermostated plate supporter to 37 °C.
11. Program pump to inject 100 μL of rL/L reagent at speed of 100 μL/s, 5 s after beginning to record the 560 nm light signal from the wells. After injection, set orbital shaker with a 7 mm of shaking width for 2.3 s for the 96-well plate.
12. Record the 560 nm light signal for an additional 24 s. The emitted light produced by the reaction of ATP with added luciferase and rL/L reagent is proportional to the concentration of ATP (Fig. 1).

4 Notes

1. Once reconstituted, the vial containing the rL/L reagent has to be covered with aluminum foil to protect it from light. Reconstituted rL/L Reagent can be kept at room temperature for 8 h, for long-term storage the reconstituted rL/L reagent can be stored in a single-use aliquot at −20 °C. The activity of reconstituted rL/L reagent decrease 50% after 2 weeks at −20 °C.
2. Prior to beginning any procedure, palpebral reflex and pedal withdrawal reflex are checked. Palpebral reflex and the response to pinching foots are checked. Palpebral reflex may be checked by touching the mouse's eye with an object. If the eyelids do not move on contact with this object, it is considered that the mouse is fully anesthetized. The response to pinching could be checked pressing a foot with a tweezers. If the mouse does not retract its leg, it is considered that the mouse is properly anesthetized.
3. We will have approximately maximum 2–3 min to do all these procedures before animal starts to awaken.
4. Since usually the mice stereotaxic equipment come from rat stereotaxic devices adapted to mice, to put the mouse body into the indicated position we only need to low the mice adapted platform on the U-form base of rat stereotaxic equipment or replace it by another lower one.
5. The whole described process usually takes around 15 min per mouse. The average volume of CSF collected is around 2–7 μL per mouse.
6. The signal from the well containing 0 pmoles is subtracted from the rest of the wells because it corresponds to the background signal [18].
7. Gain should be fixed taking into account the light signal obtained from wells containing 0 and 10 pmoles ATP, allowing to properly scale the signals come from the different standards solutions. It is important that the maximal signal obtained with the 10 pmoles standard does not exceed the maximal capacity of microplate reader sensor.

Acknowledgments

This work was supported by funding from Spanish Ministry of Science and Education BFU2012-31195 to M.D.-H. European Union project H2020-MSCA-ITN-2017 number 766124 to M. D.-H. and from Universidad Complutense of Madrid (UCM)-Santander Central Hispano Bank PR41/17-21014 to M.D.-H.

A.S.-S. was supported by BFU2012-31195 grant, and L.d.D.-G. held a UCM predoctoral fellowship supervised by M.D.-H. C.d.L. and C.B. were hired for an H2020-MSCA-ITN action supervised by M.D.-H.

References

1. Robson SC, Sevigny J, Zimmermann H (2006) The E-NTPDase family of ectonucleotidases: structure function relationships and pathophysiological significance. Purinergic Signal 2 (2):409–430. https://doi.org/10.1007/s11302-006-9003-5
2. Zimmermann H, Zebisch M, Strater N (2012) Cellular function and molecular structure of ecto-nucleotidases. Purinergic Signal 8 (3):437–502. https://doi.org/10.1007/s11302-012-9309-4
3. Burnstock G (2008) Purinergic signalling and disorders of the central nervous system. Nat Rev Drug Discov 7(7):575–590. https://doi.org/10.1038/nrd2605
4. Diaz-Hernandez M, Diez-Zaera M, Sanchez-Nogueiro J, Gomez-Villafuertes R, Canals JM, Alberch J, Miras-Portugal MT, Lucas JJ (2009) Altered P2X7-receptor level and function in mouse models of Huntington's disease and therapeutic efficacy of antagonist administration. FASEB J 23(6):1893–1906. https://doi.org/10.1096/fj.08-122275
5. Wang XH, Xie X, Luo XG, Shang H, He ZY (2017) Inhibiting purinergic P2X7 receptors with the antagonist brilliant blue G is neuroprotective in an intranigral lipopolysaccharide animal model of Parkinson's disease. Mol Med Rep 15(2):768–776. https://doi.org/10.3892/mmr.2016.6070
6. Diaz-Hernandez JI, Gomez-Villafuertes R, Leon-Otegui M, Hontecillas-Prieto L, Del Puerto A, Trejo JL, Lucas JJ, Garrido JJ, Gualix J, Miras-Portugal MT, Diaz-Hernandez M (2012) In vivo P2X7 inhibition reduces amyloid plaques in Alzheimer's disease through GSK3beta and secretases. Neurobiol Aging 33(8):1816–1828. https://doi.org/10.1016/j.neurobiolaging.2011.09.040
7. Martin E, Amar M, Dalle C, Youssef I, Boucher C, Le Duigou C, Bruckner M, Prigent A, Sazdovitch V, Halle A, Kanellopoulos JM, Fontaine B, Delatour B, Delarasse C (2018) New role of P2X7 receptor in an Alzheimer's disease mouse model. Mol Psychiatry. https://doi.org/10.1038/s41380-018-0108-3
8. Engel T, Gomez-Villafuertes R, Tanaka K, Mesuret G, Sanz-Rodriguez A, Garcia-Huerta-P, Miras-Portugal MT, Henshall DC, Diaz-Hernandez M (2012) Seizure suppression and neuroprotection by targeting the purinergic P2X7 receptor during status epilepticus in mice. FASEB J 26(4):1616–1628. https://doi.org/10.1096/fj.11-196089
9. Sebastian-Serrano A, Engel T, de Diego-Garcia L, Olivos-Ore LA, Arribas-Blazquez M, Martinez-Frailes C, Perez-Diaz C, Millan JL, Artalejo AR, Miras-Portugal MT, Henshall DC, Diaz-Hernandez M (2016) Neurodevelopmental alterations and seizures developed by mouse model of infantile hypophosphatasia are associated with purinergic signalling deregulation. Hum Mol Genet 25(19):4143–4156. https://doi.org/10.1093/hmg/ddw248
10. Llaudet E, Hatz S, Droniou M, Dale N (2005) Microelectrode biosensor for real-time measurement of ATP in biological tissue. Anal Chem 77(10):3267–3273. https://doi.org/10.1021/ac048106q
11. Corriden R, Insel PA, Junger WG (2007) A novel method using fluorescence microscopy for real-time assessment of ATP release from individual cells. Am J Physiol Cell Physiol 293 (4):C1420–C1425. https://doi.org/10.1152/ajpcell.00271.2007
12. Hayashi S, Hazama A, Dutta AK, Sabirov RZ, Okada Y (2004) Detecting ATP release by a biosensor method. Sci STKE 2004(258):pl14. https://doi.org/10.1126/stke.2582004pl14
13. Liu L, Duff K (2008) A technique for serial collection of cerebrospinal fluid from the cisterna magna in mouse. J Vis Exp (21). https://doi.org/10.3791/960
14. Vanderstichele H, Demeyer L, Janelidze S, Coart E, Stoops E, Mauroo K, Herbst V, Francois C, Hansson O (2017) Recommendations for cerebrospinal fluid collection for the analysis by ELISA of neurogranin trunc P75, alpha-synuclein, and total tau in combination with Abeta(1-42)/Abeta(1-40). Alzheimers Res Ther 9(1):40. https://doi.org/10.1186/s13195-017-0265-7
15. Czarnecka J, Cieslak M, Michal K (2005) Application of solid phase extraction and high-performance liquid chromatography to qualitative and quantitative analysis of nucleotides and nucleosides in human cerebrospinal fluid. J Chromatogr B Analyt Technol Biomed

Life Sci 822(1–2):85–90. https://doi.org/10.1016/j.jchromb.2005.05.026

16. Xu P, Xu Y, Hu B, Wang J, Pan R, Murugan M, Wu LJ, Tang Y (2015) Extracellular ATP enhances radiation-induced brain injury through microglial activation and paracrine signaling via P2X7 receptor. Brain Behav Immun 50:87–100. https://doi.org/10.1016/j.bbi.2015.06.020
17. Zierhut M, Dyckhoff S, Masouris I, Klein M, Hammerschmidt S, Pfister HW, Ayata K, Idzko M, Koedel U (2017) Role of purinergic signaling in experimental pneumococcal meningitis. Sci Rep 7:44625. https://doi.org/10.1038/srep44625
18. Sebastian-Serrano A, de Diego-Garcia L, Henshall DC, Engel T, Diaz-Hernandez M (2018) Haploinsufficient TNAP mice display decreased extracellular ATP levels and expression of Pannexin-1 channels. Front Pharmacol 9:170. https://doi.org/10.3389/fphar.2018.00170
19. Morciano G, Sarti AC, Marchi S, Missiroli S, Falzoni S, Raffaghello L, Pistoia V, Giorgi C, Di Virgilio F, Pinton P (2017) Use of luciferase probes to measure ATP in living cells and animals. Nat Protoc 12(8):1542–1562. https://doi.org/10.1038/nprot.2017.052

Chapter 18

P2X Electrophysiology and Surface Trafficking in *Xenopus* Oocytes

Eléonore Bertin, Audrey Martínez, and Eric Boué-Grabot

Abstract

Xenopus oocytes serve as a standard heterologous expression system for the study of various ligand-gated ion channels including ATP P2X receptors. Here we describe the whole-cell two-electrode voltage clamp and biotinylation/Western blotting techniques to investigate the functional properties and surface trafficking from P2X-expressing oocytes.

Key words P2X, ATP-gated channels, Double-electrode patch clamp, Electrophysiology, Receptor function, Surface trafficking, *Xenopus*, Oocytes

1 Introduction

Xenopus oocyte expression system and electrophysiology is a powerful system to study ion channels and ligand-gated ion channels. Following the pioneer observation of Gurdon et al. [1] showing that injection of foreign RNA could be translate into proteins, other works demonstrate that receptors and channels could be expressed in the oocytes and characterized [2]. Since the characterization of the functional properties of the cloned acetylcholine receptor expressed into oocytes in 1985 [3], this system was widely and continue to be used for heterologous expression and electrophysiological characterization of channels and receptors including P2X receptors (for reviews [4–6]). Expression cloning using *Xenopus* oocytes has also been proven to be an excellent tool for functional identification of proteins. In 1994, the first member of P2X receptor family was identified and cloned by the functional screening of a cDNA library made from the total population of mRNA isolated from the rat vas deferens [7]. *Xenopus* oocytes present several advantages compared to other recombinant system: (1) Hundreds to thousands of oocytes can be isolated surgically from a given frog without sacrificing the animal. (2) Oocytes are

Pablo Pelegrín (ed.), *Purinergic Signaling: Methods and Protocols*, Methods in Molecular Biology, vol. 2041,
https://doi.org/10.1007/978-1-4939-9717-6_18,

large cells (1 mm in diameter) easy to handle, to inject with DNA, RNA as well as with non-membrane-permeable drugs, and to maintain at room temperature (19 °C) after isolation or injection. (3) Oocytes have few endogenous ionic conductance and no endogenous ligand-gated channels (in contrast to most of the cells which express a plethora of channels and receptors), allowing for the study of a particular receptor channel exogenously expressed in virtual isolation by electrophysiological recordings and/or biochemistry experiments. Functional characterization of ligand-gated ion channels and in particular P2X receptors into oocytes were mainly performed using two-electrodes voltage-clamp recordings in which the membrane potential is clamped at a desired value. Clamped oocytes under constant perfusion can be tested by applications of agonists, antagonists, or drugs. Functional and pharmacological properties of several homomeric and heteromeric P2X receptors from different species were characterized into oocytes by electrophysiology or biochemical methods [7–18] and more recently using FRET-based methods [19]. Combination of mutational approaches and electrophysiological recordings or biochemistry experiments from expressing oocytes produced a powerful system to identified specific protein domains or residues involved in P2X subunit assembly, receptor function, and trafficking [20–39]. The great ability of oocytes to express numerous foreign proteins was also extensively used to identify and characterize cross talk between distinct receptors occurring natively in neurons. Cross-inhibition between several P2X receptors and several members of the "Cys-loop" receptors family including nicotinic, 5-HT3, and $GABA_A$ receptors were characterized into oocytes. Electrophysiological, molecular, and biochemical approaches from coexpressing oocytes revealed that a molecular coupling underlies their activity-dependent cross-inhibition. In addition, the interaction between P2X and $GABA_A$ or 5-HT3 receptors may also regulate receptor trafficking and targeting [40–47]. Recently, coexpression of AMPA and P2X receptors into oocytes was useful to demonstrate that P2X activation regulates AMPA receptors surface trafficking by triggering their internalization. This mechanism identified in oocytes was shown to induce synaptic plasticity in the hippocampal neurons [48–50]. In addition to RNA or DNA, injection of membranes preparations from cells or tissues into oocytes fused within a few hours to the oocyte membrane. Pioneered by the observation that Torpedo electroplaque injection into oocytes caused the appearance of functional Torpedo acetylcholine receptors and chloride channels [51], this approach was developed further to transplant already assembled receptors in membranes from animal or postmortem human tissues to the plasma membrane of oocytes and study their functional properties by electrophysiology [52–54]. Here we describe the two-electrode voltage-clamp method to explore the function from P2X expressing oocytes.

While RNA injection is used in most of the laboratories, we exclusively use in routine nuclear injection of DNA plasmids coding the receptors of interest. Nuclear nanoinjection requires some practice to reach efficiently the nucleus but this method is faster by suppressing time-consuming steps such as in vitro RNA synthesis and quality check and avoid RNAses-free conditions for injection. Biotinylation/Western blotting experiments from oocytes is also a powerful tool to explore surface protein trafficking. Biotin has the capability to bind to primary amines on proteins. Using a non-membrane-permeable sulfo-NHS-SS-biotin, solely surface proteins will be biotinylated and can be subsequently isolated from total protein extracts by affinity purification method. We here also describe this method that allows for measuring the surface/total P2X expression in various conditions.

2 Materials

All solutions for oocytes preparation and maintenance, for electrophysiological recordings and biochemical experiments are prepared with ultrapure deionized water (18 MΩ-cm at 25 °C). Concentrated stock solutions (10×, 20×, or 40×, *see* below) are filtered and stored at 4 °C.

2.1 Anesthetic and Oocytes Media

1. Anesthetic: 1.5 mg/mL ethyl 3-aminobenzoate methanesulfonate (also called tricaine). Prepare 400 mL of solution in cold tap water.
2. Dissection media OR2 (1×): 82.5 mM NaCl, 2 mM KCl, 1 mM $MgCl_2 \cdot 6H_2O$, 5 mM HEPES. The solution is prepared from filtrated 20× OR2 solution. pH is adjusted to 7.5.
3. 1.5 mg/mL collagenase 1A in OR2 media.
4. Barth's solution (1×): 83 mM NaCl, 5 mM NaOH, 1 mM KCl, 0.82 mM $MgSO_4 \cdot 7H_2O$, 2.4 mM $NaHCO_3$, 10 mM HEPES. The complete Barth's solution is prepared from filtrated 20× Barth's stock solution with addition of 0.55 mg/mL pyruvic acid, 10 mg/mL gentamycin, and 1.8 mM $CaCl_2$. The pH is adjusted to 7.4 and then 0.1 mg/mL BSA is added.

2.2 Oocytes Injection

1. Nanoinjector (Drummond, Nanoject II or WPI) mounted on a suitable micromanipulator such as MM33.
2. Pulled micropipettes from the glass adapted to the nanoinjector.
3. Mineral oil.
4. Stereomicroscope (as the Nikon SM2645) with LED lamp, but other similar equipment is also suitable.

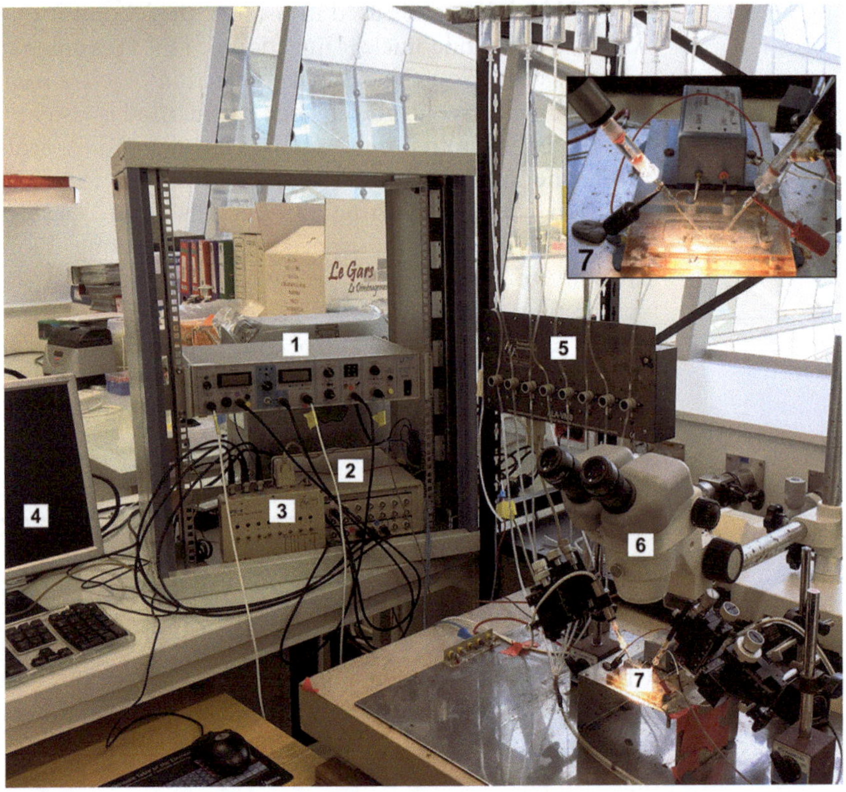

Fig. 1 Overview of a double-electrode voltage clamp station composed of (1) amplifier, (2) digitizer, (3) computer-driven valve controller, (4) personal computer and screen with acquisition software, (5) perfusion system and valves, (6) stereomicroscope, (7) recording chamber (*see* insert). Insert: recording chamber with the two electrodes mounted on holders and silver wire ground electrodes

2.3 Recording Buffers

Ringer solution 1×: 115 mM NaCl, 2.5 mM KCl, 10 mM HEPES, 3 mM NaOH. Prepare a filtrated 40× Ringer stock solution. Dilute the 40× solution to obtain 4 L of Ringer 1× and add 1.8 mM $CaCl_2$. pH adjusted to 7.4 (*see* **Note 1**).

2.4 Recording Set Up (See Fig. 1)

1. Amplifier: OC-725C Oocyte Clamp (Warner Instruments) a main unit connected to the voltage recording probe with electrode holder, current electrode with holder and current sensing bath probe with silver wire electrodes.
2. Digitizer: Analogic data recorded by the amplifier are digitized at 1 kHz using an interface (Instrutech or Digidata) linked to a computer using Axograph X software (Axograph).
3. Computer-driven 4 or 8 valve system (ALA Scientific Instruments, BPS-4 or 8) controlling gravity Ringer perfusion and drug application.
4. Pulled borosilicate glass capillaries adapted to the voltage and current holders (Harvard apparatus, GCISOTF-10 1.5 mm

OD × 1.17 mm ID) mounted on two adapted micromanipulators (MM33).

5. Recording chambers dedicated to oocytes recordings are available commercially. Connect the entry of the chamber to the gravity perfusion system and exit to a vacuum suction system to allow for continuous perfusion of the oocyte during recordings.
6. Stereomicroscope and LED lamp.

2.5 Pharmacology

Adenosine 5′-triphosphate disodium hydrate (ATP) and all other drugs are prepared at their desired final concentrations in the perfusion solution (50 mL). For ATP, prepare stock solution at 100 mM in deionized water and keep aliquots at −20 °C for single use. Useful P2X agonist and antagonists were previously summarized [55].

2.6 Biotinylation

1. 1 mg/mL solution of sulfo-NHS-SS-biotin in cold Ringer 1×. Maintain on ice until use.
2. 100 mM solution of glycine in cold Ringer 1×. Maintain on ice until use.

2.7 Purification

1. High Capacity NeutrAvidin™ Agarose or similar.
2. Spin Column Snap-Cap.

2.8 Protein Isolation Reagents

1. Homogenization Buffer (HB) with 2% Triton: 10 mM HEPES, 300 mM sucrose, pH 7.5, 2% Triton X-100. Add one pellet of Complete™ Protease Inhibitor Cocktail for 40 mL of HB. Vortex well until perfect homogenization.
2. HB 1% Triton: same protocol with 1% Triton.

2.9 Blotting Buffers and Reagents

1. Loading buffer 4× (LB4×): Laemmli buffer 4× with 12% β-mercaptoethanol.
2. Loading buffer 2× (LB2×): dilute one volume of LB4× in one volume of HB 1% Triton.
3. Prestained Protein Ladder.

2.10 Protein Electrophoresis and Western Blot

1. Precast gels (manufactured gels such as Bio-Rad's Ready Gel®, Mini-PROTEAN®, and Criterion™ Precast Gels) or hand-cast gels can be used.
2. Hand-cast gels 12%.
3. Running Buffer 10×: for 1 L mix 100 mL SDS 10%, 30.3 g Trizma base, 144 g glycine.
4. Membrane PVDF. Different techniques such as semidry or tank transfers can be used.

5. Transfer buffer 10× for tank system: for 1 L mix 30.3 g Trizma base and 144 g glycine.
6. Transfer buffer 5× for semidry system: kit transfer.
7. Blocking solution: PBS with 0.1% Tween 20 and 2% of BSA.
8. PBST: PBS with 0.1% Tween 20.
9. Antibodies primary solution: 5 mL for one membrane. Dilute primary antibody in PBST with 1% of BSA.
10. Antibodies secondary HRP solution: 5 mL for one membrane. Dilute antibody in PBST with 5% of milk.
11. Immunological detection: chemiluminescent detection or fluorescence. For HRP use a with ECL Kit Clarity™ Western ECL Substrate, 200 mL. Alternative ECL substrates are also suitable.

2.11 Gel and Blot Imaging Equipment

1. ChemiDoc™ XRS+ System (Bio-Rad) or similar.
2. Trans-Blot® Turbo™ System (Bio-Rad) or similar.
3. Power Supplies for Electrophoretic Transfers.

2.12 Antibodies

Several primary antibodies directed against the different P2X subunits are available commercially from several companies such as Alomone Labs, Abcam, or Antibodies-online. C-terminal tagged-P2X subunits with HA, Flag, His, or Myc tags are commonly used to detect P2X subunits expression by Western blots using specific anti-tag antibodies.

3 Methods

3.1 Oocytes Isolation

Oocytes are surgically removed from female *Xenopus laevis* without sacrificing the animal (Fig. 2a). A frog can be used several times with a minimal period of 4 weeks between two surgeries.

1. Place the animal in a closed box or a 2 L beaker containing 400 mL of cold anesthetic solution in tap water during 15–20 min depending on the size of the frog.
2. Verify the depth of anesthesia by pinching the animal's paw with forceps before operating.
3. Place the animal on the back in a tray containing ice to slow down its metabolism.
4. Make small abdominal incisures of the skin (10 mm) and the underlying fascia with a scalpel. Ovaries are usually directly visible after incision.
5. Gently take out bundles of oocytes with forceps from ovaries and place them in a petri dish containing OR2 solution.

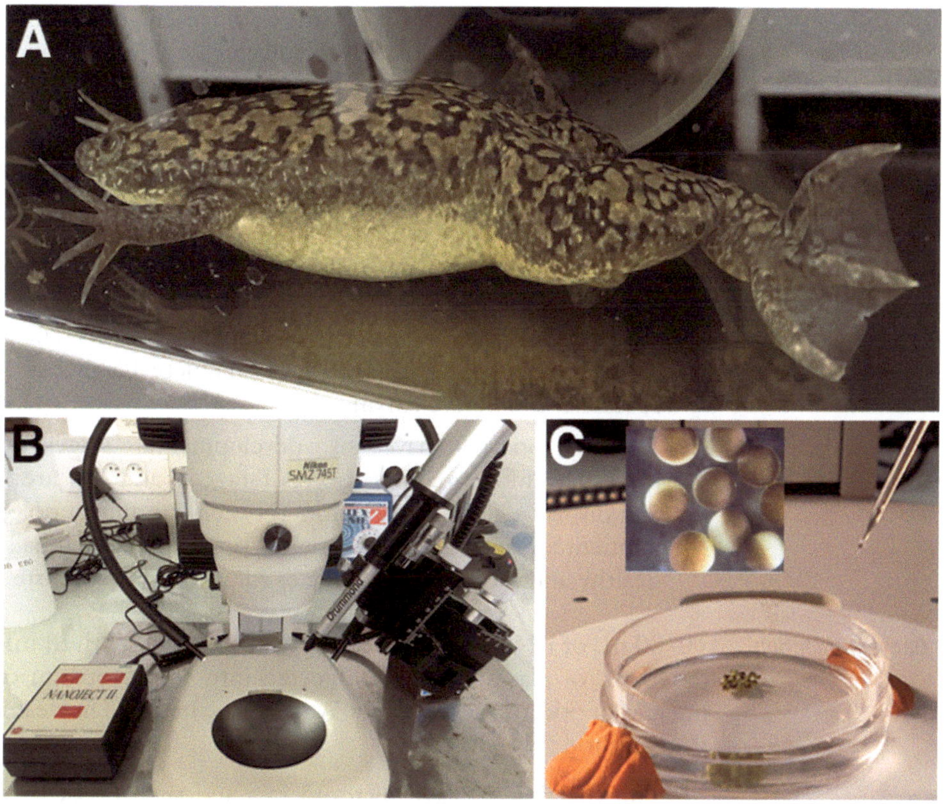

Fig. 2 (**a**) Picture of a female *Xenopus laevis* taken in our amphibian facility. (**b**) Overview of a microinjection station. (**c**) Selected mature oocytes selected for nuclear injection with the glass micropipette containing DNA solution

6. Pushed back the remaining ovary into the abdomen and stitch the muscular layer first to close the excision and then, the skin.
7. Wash briefly the animal in tap water and then transfer to a box or beaker containing only tap water. Very important: Adjust the level of tap water to be sure to not drown the frogs since frogs are air breathers. Let the frog back out of the water to permit skin breath.
8. Once the animal has recovered (1 or 2 h), transfer it to an isolated tank for few days before it is returned to the colony in order to avoid transmission of an eventual infection after operation to other frogs.

3.2 Oocytes Isolation and Sorting

Oocytes extracted from the ovary consist of many cells held together by connective tissue and are each surrounded by follicular cells. For injection and recordings, it is necessary to isolate each oocyte and to remove the follicular layer.

1. Cut bundles of oocytes into small pieces of around 50 cells before being placed into enzymatic treatment.

2. Oocytes are placed into a 50 mL sterile plastic tube containing 15 mL of Calcium-free OR2 collagenase 1A solution under moderate agitation at room temperature during 2 h. After 1 h, replace the enzymatic solution with a fresh collagenase 1A solution to maintain optimal enzymatic activity and to ensure complete dissociation and partial or total removal of follicular layer (*see* **Note 2**).
3. Wash oocytes three times with OR2 to remove both collagenase and lysed cells.
4. Wash two times oocytes with complete Barth's and place oocytes in a petri dish containing complete Barth's at 19 °C at least 1–2 h before sorting.
5. *Xenopus* oocytes have six developmental stages (I–VI) [56]. The more mature oocytes (stage V and VI) are larger and display a clear demarcation between the animal pole, very pigmented, and the light vegetal pole. Select manually the healthy and more mature oocytes for injection (*see* **Note 3**).

3.3 Oocytes Injection

10–80 nL of RNA (mRNA from tissues or cDNA-derived mRNA) or 10 nL of plasmids can be injected into the cytoplasm or the nucleus of oocytes, respectively. We describe here the nuclear injection method (Fig. 2b, c).

1. Pull the tips of glass micropipettes and then break them off with forceps under the stereomicroscope. Ideally the tip size should be between 10 and 30 μm.
2. Backfill the micropipettes with mineral oil using a nonmetallic needle and syringe (*see* **Note 4**).
3. Push the micropipette onto the wire plunger of the nanoinjector and once positioned, tighten the collect securely.
4. Place the tip of the pipette into your DNA sample (2–4 μL of DNA on the top of a drop of mineral oil) and then, fill the pipette with your sample by pressing the FILL button (the wire plunger will retract) (*see* **Note 5**).
5. Place a small number of oocytes in a small petri dish with complete Barth's. Adapted quantity of each DNA plasmid is injected into the nucleus using a single injection of 10 nL (*see* **Note 6**). Touch the top of the animal pole of one oocyte (dark side) with the tips of the micropipette. Important align the vertical axis between the two poles of the oocyte and the pipette axis to reach the nucleus. Go down slowly the micropipette using the micromanipulator and enter into the oocyte. As soon as the micropipette is inside, inject the selected volume using the INJECT button. Because oocyte have a tendency to move away, it is essential to fix the oocytes using a scratched petri dish, for example.

6. Store injected oocytes at 19 °C in small petri dishes containing Barth's medium. Change the medium daily and discard unhealthy oocytes.
7. Recordings will be performed 24–48 h after injection to allow for protein expression.

3.4 Double-Electrode Voltage-Clamp Recordings

1. Backfill two pulled microelectrodes with filtered 3 M KCl using a nonmetallic needle and a syringe. Avoid air bubbles lodged near the tip to ensure electrical coupling. The resistance of the pipette should be between 0.3 and 2 MΩ to ensure fast clamp. A resistance test is available on the amplifier.
2. Install the electrodes into the holders to ensure contact between the chlorided silver wire and the KCl into the pipette for the electrical coupling. The current electrode should be shielded from the voltage electrode and that shield should be grounded to the circuit ground. This is accomplished by wrapping both pipettes with aluminum foil (*see* **Note 7**).
3. Place one oocyte into the chamber containing Ringer solution and insert the voltage and the current electrodes into the oocytes from each side of the oocyte to maintain the oocytes into the chamber when the perfusion will be shut on. Successful impalement of each electrode is easily seen by monitoring the resting potential on the amplifier. For healthy oocytes this value is between −20 and −50 mV.
4. Shut on the perfusion system and control the volume into the chamber with a suction pipette (*see* **Note 8**).
5. Then clamp the oocyte at the desired voltage. Holding potential for ligand-induced current is classically set at −60 mV, but oocytes can be easily clamped between −100 and +40 mV to perform current–voltage relationship (Fig. 3) (*see* **Note 9**).
6. The double electrode voltage-clamp recordings are very stable. You can measure repetitive applications of agonist of diverse drugs on one oocyte expressing ligand-gated ion channels up to several hours (Fig. 3) (*see* **Note 10**).

3.5 Biotinylation of Oocytes

Solutions and oocytes should be kept at 4 °C during all the steps of the procedure.

1. Place 10–15 expressing oocytes per condition in 500 μL of an ice-cold Ringer solution containing 1 mg/mL sulfo-NHS-SS-biotin and incubate at 4 °C during 4 h (or overnight) under gentle agitation (*see* **Note 11**).
2. Wash oocytes thrice with 500 μL of ice-cold Ringer.
3. Wash once with 100 mM glycine to quench free Sulfo-NHS-SS-Biotin.
4. Remove all the liquid and store oocytes at −80 °C (*see* **Note 12**).

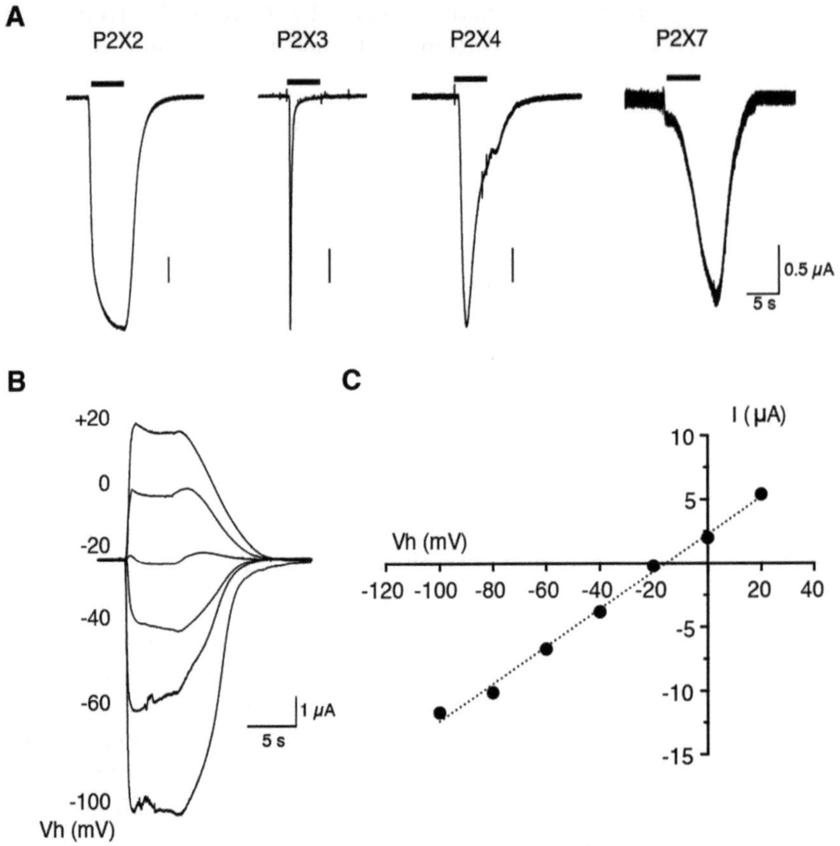

Fig. 3 (**a**) ATP-induced currents recorded by double-electrode voltage clamp of oocytes expressing P2X2, P2X3, P2X4, or P2X7 receptors. Holding potential −60 mV. 100 μM of ATP was applied at a concentration of 100 μM for P2X2, P2X3, and P2X4 and 500 μM for P2X7. (**b**) Superimposed currents evoked on a P2X2-expressing oocytes by successive application of 100 μM of ATP at the indicated holding potential. The Intensity current–voltage relationship of P2X subunits displays a reversion potential around −15 mV in *Xenopus* oocytes recorded in ringer solution

3.6 Protein Extraction

1. Add 20 μL per oocyte of HB 2% Triton to each batch of oocytes (*see* **Note 13**).
2. Lyse cells of each sample on ice by sonication using three pulses of 3 s at 80% of the maximal power with intervals between pulses of 5 s.
3. Incubate at 4 °C during 2 h under gentle agitation to solubilize membrane proteins.
4. Centrifuge at 10,000 × *g* during 15 min at 4 °C.
5. Collect the supernatant containing total proteins and keep a fraction for total protein detection.
6. Add LB4× to the protein extract to be at 1× final in a fume cupboard and mix by pipetting (*see* **Note 14**).
7. Boil samples at 100 °C during 5 min and place them immediately one ice (*see* **Note 15**).

3.7 Surface Proteins Purification

1. Use 20 μL of NeutrAvidin agarose resin (at 50% in the storage solution) per batch of oocytes. Equilibrate the NeutrAvidin agarose resin by washing twice with 500 μL of HB 1% Triton. Add one volume of HB 1% Triton to reconstitute the resin at 50%.
2. Pipet 20 μL of resin (at 50%) in a column for each sample.
3. Apply protein extracts to the column (*see* **Note 16**).
4. Add then the same volume of HB as protein extract volume.
5. Complete with HB 1% Triton to reach a total volume of 500 μL.
6. Leave overnight the columns on a vertical wheel at 4 °C to allow for the binding of biotinylated proteins.
7. Centrifuge columns at 2000 × *g* during 1 min.
8. Wash three times the columns with 500 μL of HB 1% Triton by centrifugation at 2000 × *g* during 1 min.
9. Place the columns in clean 1.5 mL tubes to collect the eluate.
10. Add 20 μL of warm P2X (100 °C) on each column in a fume cupboard.
11. Incubate for 10 min at room temperature.
12. Centrifuge columns at 2000 × *g* during 1 min to elute the bound biotinylated proteins (*see* **Note 17**). Proceed to Western blot analysis or store proteins at −20 °C.

3.8 Western Blot

1. Prepare gels for total and surface proteins including lanes for markers of size.
2. Gels are placed in the electrophoresis cell. Running buffer is added and samples are loaded (20 μL).
3. Run the gel 60 min at 120 V.
4. Transfer separated proteins onto a PVDF membrane either using a tank system (35 min at 120 V) or a semidry (7 min).
5. Wash membrane with PBS–Tween briefly to remove trace of methanol.
6. Saturate PVDF membrane for 1 or 2 h in the blocking buffer at room temperature.
7. Incubate membrane with primary antibody overnight at 4 °C under agitation.
8. Wash five times, 5 min each, with PBS–Tween 0.1%.
9. Incubate with secondary antibody for 1 h at room temperature under agitation.
10. Wash three times, 10 min each, with PBS–Tween 0.1%.
11. Mix the clarity substrate kit components in a 1:1 ratio (5 mL for one membrane).

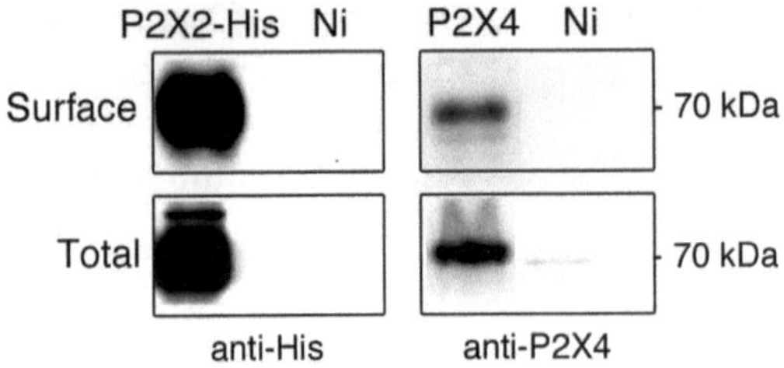

Fig. 4 Detection by Western blotting of total and surface proteins from oocytes expressing P2X2 or P2X4 receptors. Representative immunoblots of total and surface biotinylated proteins from oocytes expressing rat P2X2-HIS or mouse P2X4 wild-type using anti-His (1:1500, Genscript) or anti-P2X4 (1:1000, Alomone) primary antibodies respectively and horseradish peroxidase (HRP)-conjugated secondary antibodies (1:5000, Jackson ImmunoResearch). 10–25 pg and 150–300 pg of pcDNA3 plasmids coding for P2X2-HIS and P2X4 are used for nuclear injection. Biotinylation was performed 48–72 h after injection to allow for receptor expression. Total protein bands corresponded to one-third of oocyte (7 μL of total proteins) and surface or internalized fractions corresponded to seven oocytes (140 μL of total proteins prior purification). No bands are detected in noninjected oocytes (Ni). Note that P2X2 is highly expressed into oocytes and highly targeted to the surface, while P2X4 is mainly detected in intracellular compartments due to its constitutive internalization

12. Incubate the membrane in the substrate solution for 5 min.
13. Place the blotting membrane back on the sample stage of the ChemiDoc MP imager.
14. Start Image Lab software and capture the chemiluminescent signals on the blot (Fig. 4).

4 Notes

1. The continuous gravity perfusion of oocytes during electrophysiological recordings at a flow rate of 10–12 mL/min requires preparing between 4 and 8 L of Ringer for 2 days of recordings and two tanks of 10 L.
2. Prolonged exposure to collagenase could be deleterious for oocytes, therefore adjust the concentration and the time of incubation depending on the purity and enzymatic activity of collagenase in testing on different batches to find ideal conditions.
3. Collagenase treatment sometimes does not fully remove the follicular layer. This can be done directly by means of forceps one oocyte per one oocyte or after treatment with a defolliculation solution causing an osmotic change: 100 mM KCl, 2.5 mM $MgCl_2$, 4 mM K_2HPO_4, 6 mM KH_2PO_4, 1 mM Na_2EGTA, pH 6.5.

4. When backfilling the entire capillaries with mineral oil, take care to avoid bubbles. If there is any, punch the capillaries with the fingertips to make them leave.
5. When filling the capillary with DNA, check the decrease of the drop of DNA under the stereomicroscope and verify there no air bubbles into the pipette. The interface between oil and DNA solution is visible. Before injection of oocytes verify that each push on INJECT button delivers a drop of the same size by placing the pipette tips into a drop of oil for example.
6. The quantity of DNA for nuclear injection to obtain protein expression has to be determined for each plasmid. If several plasmids need to be coexpressed in oocytes prepare first a mixture of the desired plasmids to perform a single nuclear injection of 10 nL. For P2X expression, we routinely inject 150–300 pg of one plasmid for most of the native P2X subunits. P2X2 has a high capability of expression in oocytes, and 5–20 pg is sufficient to record ATP-evoked currents with amplitudes of 5–10 μA. Different DNA quantities need to be tested in order to find the best current amplitude (between 1 and 10 μA). For coexpression of several receptors, it is important to limit the total quantity of DNAs injected (up to 1 ng/oocyte) otherwise the survival of oocyte and/or receptor expression may be altered.
7. Silver wire should be regularly chlorided to reduce noise and prevent exposure of the cell to toxic silver ions. The silver chloride coat can be renewed easily between experiments by incubation into bleach (1 h) and then rinsed with deionized water.
8. Decrease the level of perfusion into the recording chamber as much as possible (be sure to keep the oocyte immerged to avoid disruption of the membrane by flow changes) to increase the speed of solution exchange between ringer and drugs to record fast activation and deactivation of agonist-evoked currents.
9. At a holding potential of −60 mV, healthy oocytes do not display leak current exceeding −100 nA.
10. For P2X receptors, applications of ATP during 2–5 s are routinely used to record ATP-evoked currents (Fig. 3). ATP concentration to evoke the maximal response will be adjusted depending on P2X subtypes. In Fig. 3, 100 μM of ATP is applied during 5 s for P2X2, P2X3, and P2X4, while 500 μM of ATP is used for P2X7.
11. The number of oocytes per condition can be adapted (2–30 oocytes), but we generally use batches of 15 oocytes to limit the variability of the results. Expressing oocytes can be kept

postrecordings and used for biochemistry. Noninjected oocytes should be also considered as negative control.

12. Oocytes can be stored at −80 °C (no alteration of the biotinylation was observed after 1–2 months) or used directly for protein extraction.
13. The oocyte cytoplasm contains a large quantity of yolk platelets challenging the preparation of total protein extract for Western blot analysis of heterologous expressed proteins. For this reason, we preferentially used HB buffer containing Triton (up to 2%) rather than for example RIPA buffer. Oocyte proteins extracted with RIPA solubilized the yolk content altering the protein migration during the electrophoresis.
14. For P2X receptors, total proteins corresponding to 1/3 (7 μL) or 1/2 (10 μL) oocyte loaded onto the gel is sufficient to obtain a strong signal by Western blots using specific anti-P2X subunits.
15. Samples can be directly used for Western blotting or conserved at −20 °C.
16. Detection of surface P2X receptor fraction by Western blotting requires loading onto a gel the purified biotinylated proteins corresponding to 4–10 oocytes according to the surface trafficking on the considered P2X receptors. For constitutively internalized P2X4, a minimum of seven oocytes is generally used (*see* Fig. 4).
17. Samples can be directly used for Western blotting or conserved at −20 °C.

References

1. Gurdon JB, Woodland HR, Lingrel JB (1974) The translation of mammalian globin mRNA injected into fertilized eggs of Xenopus laevis. I. Message stability in development. Dev Biol 39:125–133
2. Gundersen CB, Miledi R, Parker I (1983) Serotonin receptors induced by exogenous messenger RNA in Xenopus oocytes. Proc R Soc Lond B Biol Sci 219:103–109
3. Sakmann B, Methfessel C, Mishina M, Takahashi T, Takai T, Kurasaki M, Fukuda K, Numa S (1985) Role of acetylcholine receptor subunits in gating of the channel. Nature 318:538–543
4. Burnstock G (2012) Discovery of purinergic signalling, the initial resistance and current explosion of interest. Br J Pharmacol 167:238–255
5. Roberts JA, Vial C, Digby HR, Agboh KC, Wen H, Atterbury-Thomas A, Evans RJ (2006) Molecular properties of P2X receptors. Pflugers Arch 452:486–500
6. Sigel E (1990) Use of Xenopus oocytes for the functional expression of plasma membrane proteins. J Membr Biol 117:201–221
7. Valera S, Hussy N, Evans RJ, Adami N, North RA, Surprenant A, Buell G (1994) A new class of ligand-gated ion channel defined by P2x receptor for extracellular ATP. Nature 371:516–519
8. Le KT, Boue-Grabot E, Archambault V, Seguela P (1999) Functional and biochemical evidence for heteromeric ATP-gated channels composed of P2X1 and P2X5 subunits. J Biol Chem 274:15415–15419
9. Brown SG, Townsend-Nicholson A, Jacobson KA, Burnstock G, King BF (2002) Heteromultimeric P2X(1/2) receptors show a novel sensitivity to extracellular pH. J Pharmacol Exp Ther 300:673–680

10. Liu M, King BF, Dunn PM, Rong W, Townsend-Nicholson A, Burnstock G (2001) Coexpression of P2X(3) and P2X(2) receptor subunits in varying amounts generates heterogeneous populations of P2X receptors that evoke a spectrum of agonist responses comparable to that seen in sensory neurons. J Pharmacol Exp Ther 296:1043–1050
11. Xiong K, Li C, Weight FF (2000) Inhibition by ethanol of rat P2X(4) receptors expressed in Xenopus oocytes. Br J Pharmacol 130:1394–1398
12. Boue-Grabot E, Akimenko MA, Seguela P (2000) Unique functional properties of a sensory neuronal P2X ATP-gated channel from zebrafish. J Neurochem 75:1600–1607
13. Bo X, Schoepfer R, Burnstock G (2000) Molecular cloning and characterization of a novel ATP P2X receptor subtype from embryonic chick skeletal muscle. J Biol Chem 275:14401–14407
14. Wang CZ, Namba N, Gonoi T, Inagaki N, Seino S (1996) Cloning and pharmacological characterization of a fourth P2X receptor subtype widely expressed in brain and peripheral tissues including various endocrine tissues. Biochem Biophys Res Commun 220:196–202
15. Seguela P, Haghighi A, Soghomonian JJ, Cooper E (1996) A novel neuronal P2x ATP receptor ion channel with widespread distribution in the brain. J Neurosci 16:448–455
16. Evans RJ, Lewis C, Buell G, Valera S, North RA, Surprenant A (1995) Pharmacological characterization of heterologously expressed ATP-gated cation channels (P2x purinoceptors). Mol Pharmacol 48:178–183
17. Lynch KJ, Touma E, Niforatos W, Kage KL, Burgard EC, van Biesen T, Kowaluk EA, Jarvis MF (1999) Molecular and functional characterization of human P2X(2) receptors. Mol Pharmacol 56:1171–1181
18. Le KT, Paquet M, Nouel D, Babinski K, Seguela P (1997) Primary structure and expression of a naturally truncated human P2X ATP receptor subunit from brain and immune system. FEBS Lett 418:195–199
19. Schneider M, Prudic K, Pippel A, Klapperstück M, Braam U, Müller CE, Schmalzing G, Markwardt F (2017) Interaction of purinergic P2X4 and P2X7 receptor subunits. Front Pharmacol 8:860
20. Wen H, Evans RJ (2009) Regions of the amino terminus of the P2X receptor required for modification by phorbol ester and mGluR1alpha receptors. J Neurochem 108:331–340
21. Codocedo JF, Rodriguez FE, Huidobro-Toro JP (2009) Neurosteroids differentially modulate P2X ATP-gated channels through non-genomic interactions. J Neurochem 110:734–744
22. Low SE, Kuwada JY, Hume RI (2008) Amino acid variations resulting in functional and nonfunctional zebrafish P2X(1) and P2X (5.1) receptors. Purinergic Signal 4:383–392
23. Locovei S, Scemes E, Qiu F, Spray DC, Dahl G (2007) Pannexin1 is part of the pore forming unit of the P2X(7) receptor death complex. FEBS Lett 581:483–488
24. Roberts JA, Evans RJ (2005) Mutagenesis studies of conserved proline residues of human P2X receptors for ATP indicate that proline 272 contributes to channel function. J Neurochem 92:1256–1264
25. Davies DL, Kochegarov AA, Kuo ST, Kulkarni AA, Woodward JJ, King BF, Alkana RL (2005) Ethanol differentially affects ATP-gated P2X (3) and P2X(4) receptor subtypes expressed in Xenopus oocytes. Neuropharmacology 49:243–253
26. Kanjhan R, Raybould NP, Jagger DJ, Greenwood D, Housley GD (2003) Allosteric modulation of native cochlear P2X receptors: insights from comparison with recombinant P2X2 receptors. Audiol Neurootol 8:115–128
27. Paukert M, Hidayat S, Grunder S (2002) The P2X(7) receptor from Xenopus laevis: formation of a large pore in Xenopus oocytes. FEBS Lett 513:253–258
28. Nakazawa K, Ojima H, Ohno Y (2002) A highly conserved tryptophane residue indispensable for cloned rat neuronal P2X receptor activation. Neurosci Lett 324:141–144
29. Ennion SJ, Evans RJ (2002) Conserved cysteine residues in the extracellular loop of the human P2X(1) receptor form disulfide bonds and are involved in receptor trafficking to the cell surface. Mol Pharmacol 61:303–311
30. Ennion SJ, Evans RJ (2002) P2X(1) receptor subunit contribution to gating revealed by a dominant negative PKC mutant. Biochem Biophys Res Commun 291:611–616
31. Eickhorst AN, Berson A, Cockayne D, Lester HA, Khakh BS (2002) Control of P2X(2) channel permeability by the cytosolic domain. J Gen Physiol 120:119–131
32. Clyne JD, LaPointe LD, Hume RI (2002) The role of histidine residues in modulation of the rat P2X(2) purinoceptor by zinc and pH. J Physiol 539:347–359

33. Ennion SJ, Ritson J, Evans RJ (2001) Conserved negatively charged residues are not required for ATP action at P2X(1) receptors. Biochem Biophys Res Commun 289:700–704
34. Ennion S, Hagan S, Evans RJ (2000) The role of positively charged amino acids in ATP recognition by human P2X(1) receptors. J Biol Chem 275:29361–29367
35. Dutton JL, Poronnik P, Li GH, Holding CA, Worthington RA, Vandenberg RJ, Cook DI, Barden JA, Bennett MR (2000) P2X(1) receptor membrane redistribution and down-regulation visualized by using receptor-coupled green fluorescent protein chimeras. Neuropharmacology 39:2054–2066
36. Boue-Grabot E, Archambault V, Seguela P (2000) A protein kinase C site highly conserved in P2X subunits controls the desensitization kinetics of P2X(2) ATP-gated channels. J Biol Chem 275:10190–10195
37. Newbolt A, Stoop R, Virginio C, Surprenant A, North RA, Buell G, Rassendren F (1998) Membrane topology of an ATP-gated ion channel (P2X receptor). J Biol Chem 273:15177–15182
38. Werner P, Seward EP, Buell GN, North RA (1996) Domains of P2X receptors involved in desensitization. Proc Natl Acad Sci U S A 93:15485–15490
39. Becker D, Woltersdorf R, Boldt W, Schmitz S, Braam U, Schmalzing G, Markwardt F (2008) The P2X7 carboxyl tail is a regulatory module of P2X7 receptor channel activity. J Biol Chem 283:25725–25734
40. Emerit MB, Baranowski C, Diaz J, Martinez A, Areias J, Alterio J, Masson J, Boue-Grabot E, Darmon M (2016) A new mechanism of receptor targeting by interaction between two classes of ligand-gated ion channels. J Neurosci 36:1456–1470
41. Jo YH, Donier E, Martinez A, Garret M, Toulme E, Boue-Grabot E (2011) Cross-talk between P2X4 and gamma-aminobutyric acid, type A receptors determines synaptic efficacy at a central synapse. J Biol Chem 286:19993–20004
42. Toulme E, Blais D, Leger C, Landry M, Garret M, Seguela P, Boue-Grabot E (2007) An intracellular motif of P2X(3) receptors is required for functional cross-talk with GABA (A) receptors in nociceptive DRG neurons. J Neurochem 102:1357–1368
43. Khakh BS, Fisher JA, Nashmi R, Bowser DN, Lester HA (2005) An angstrom scale interaction between plasma membrane ATP-gated P2X2 and alpha4beta2 nicotinic channels measured with fluorescence resonance energy transfer and total internal reflection fluorescence microscopy. J Neurosci 25:6911–6920
44. Boue-Grabot E, Toulme E, Emerit MB, Garret M (2004) Subunit-specific coupling between gamma-aminobutyric acid type A and P2X2 receptor channels. J Biol Chem 279:52517–52525
45. Khakh BS, Zhou X, Sydes J, Galligan JJ, Lester HA (2000) State-dependent cross-inhibition between transmitter-gated cation channels. Nature 406:405–410
46. Boue-Grabot E, Barajas-Lopez C, Chakfe Y, Blais D, Belanger D, Emerit MB, Seguela P (2003) Intracellular cross talk and physical interaction between two classes of neurotransmitter-gated channels. J Neurosci 23:1246–1253
47. Boue-Grabot E, Emerit MB, Toulme E, Seguela P, Garret M (2004) Cross-talk and co-trafficking between rho1/GABA receptors and ATP-gated channels. J Biol Chem 279:6967–6975
48. Pougnet JT, Compans B, Martinez A, Choquet D, Hosy E, Boue-Grabot E (2016) P2X-mediated AMPA receptor internalization and synaptic depression is controlled by two CaMKII phosphorylation sites on GluA1 in hippocampal neurons. Sci Rep 6:31836
49. Pougnet JT, Toulme E, Martinez A, Choquet D, Hosy E, Boue-Grabot E (2014) ATP P2X receptors downregulate AMPA receptor trafficking and postsynaptic efficacy in hippocampal neurons. Neuron 83:417–430
50. Boue-Grabot E, Pankratov Y (2017) Modulation of central synapses by astrocyte-released ATP and postsynaptic P2X receptors. Neural Plast 2017:9454275
51. Marsal J, Tigyi G, Miledi R (1995) Incorporation of acetylcholine receptors and Cl− channels in Xenopus oocytes injected with Torpedo electroplaque membranes. Proc Natl Acad Sci U S A 92:5224–5228
52. Miledi R, Eusebi F, Martinez-Torres A, Palma E, Trettel F (2002) Expression of functional neurotransmitter receptors in Xenopus oocytes after injection of human brain membranes. Proc Natl Acad Sci U S A 99:13238–13242
53. Bernareggi A, Reyes-Ruiz JM, Lorenzon P, Ruzzier F, Miledi R (2011) Microtransplantation of acetylcholine receptors from normal or denervated rat skeletal muscles to frog oocytes. J Physiol 589:1133–1142
54. Belujon P, Baufreton J, Grandoso L, Boue-Grabot E, Batten TF, Ugedo L, Garret M,

Taupignon AI (2009) Inhibitory transmission in locus coeruleus neurons expressing GABAA receptor epsilon subunit has a number of unique properties. J Neurophysiol 102:2312–2325

55. North RA, Jarvis MF (2013) P2X receptors as drug targets. Mol Pharmacol 83:759–769
56. Dumont JN (1972) Oogenesis in Xenopus laevis (Daudin). I. Stages of oocyte development in laboratory maintained animals. J Morphol 136:153–179

Chapter 19

Heterologous Expression and Patch-Clamp Recording of P2X Receptors in HEK293 Cells

Lin-Hua Jiang and Sébastien Roger

Abstract

P2X receptors (P2XRs) are ligand-gated ion channels gated by extracellular adenosine 5′-triphosphate (ATP) and play a critical role in mediating ATP-induced purinergic signaling in physiological and pathological processes. Heterologous expression of P2XR in human embryonic kidney 293 (HEK293) cells and measurement of P2XR-mediated currents using patch-clamp recording technique have been widely used to study the biophysical and pharmacological properties of these receptors. Combination of electrophysiology with site-directed mutagenesis and structural information has shed light on the molecular basis for receptor activation and mechanisms of actions by receptor antagonists and modulators. It is anticipated that such methodologies will continue helping us to provide more mechanistic understanding of P2XRs and to test novel receptor antagonists and allosteric modulators for therapeutical purposes. In this chapter, we describe protocols of transiently or stably expressing the P2XR in HEK293 cells and measuring P2XR-mediated currents by using whole-cell recording.

Key words P2X receptors, HEK293 cells, Heterologous expression, Patch-clamp recording

1 Introduction

P2X receptors (P2XRs) for extracellular ATP comprise the ionotropic P2 purinergic receptor family. There are seven different subunits, P2X1–P2X7, which can assemble in homotrimeric/heterotrimeric ligand-gated ion channels gated by ATP as being the physiological agonist [1]. P2XRs show wide expression in mammalian cells, both excitable and nonexcitable cells, and have an important role in mediating ATP-induced purinergic signaling in a diversity of physiological processes ranging from neuromodulation [2], immune response [3] to regulation of stem cell functions [4, 5]. Compelling evidence from both preclinical and clinical studies supports critical engagement of P2XRs in the pathogenesis of numerous diseases, including chronic pain [6], neurodegenerative diseases [7], mood disorders [8], inflammatory diseases [9], metabolic disorders [10], and cancers [11, 12].

Pablo Pelegrín (ed.), *Purinergic Signaling: Methods and Protocols*, Methods in Molecular Biology, vol. 2041, https://doi.org/10.1007/978-1-4939-9717-6_19,

Since its invention [13], the patch-clamp recording technique has been the golden experimental tool in the study of ion channels. Studies using this technique of the P2XRs recombinant in heterologous expression cell systems have revealed distinctive biophysical and pharmacological properties of these receptors, which has been useful in elucidating their physiological and pathological roles [1]. Human embryonic kidney (HEK) 293 cells express no endogenous P2XR and, in addition, they are readily transfected with plasmids with good transfection efficiency, and robust in protein expression, membrane trafficking and posttranslational modifications. The membrane of HEK293 cells is also amenable to formation of seal with patch-clamp electrodes. Therefore, HEK293 cells represent a widely used heterologous mammalian cell expression system to express the mammalian P2XRs. Electrophysiological studies of the P2XRs expressed in HEK293 cells, in combination with site-directed mutagenesis and more recently with structural information, have shed light on the ion-permeating pore [14], ATP binding [15, 16] and conformational changes accompanying channel gating [17–22] as well as residues coordinating the actions of antagonists and modulators [23–26]. As have been nicely summarized in recent reviews [27–30], electrophysiological studies using HEK293 cells and other heterologous expression cell systems, such as *Xenopus* oocytes, have provided substantial insights into the molecular basis that determine ATP-induced activation of the P2XRs and actions of P2XR antagonists and allosteric modulators. In addition, measurements of agonist-induced currents in HEK293 cells expressing P2XR carrying disease-associated mutations have been helpful in informing the mutational effects on receptor functions and thereby furthering our understanding of the disease mechanisms [31–33]. It is expected that electrophysiology using HEK293 cells as a mammalian cell expression system will continue to be a very useful tool to develop a better understanding of the P2XRs, particularly to test novel receptor antagonists and allosteric modulators for therapeutical purposes [8].

In this chapter, we describe protocols of using HEK293 cells to transiently and stably express P2XRs and using whole-cell patch-clamp recording to measure P2XR-mediated currents.

2 Materials

2.1 Cells and Reagents

1. HEK293 cells (American Type Cell Collection).
2. Culture medium: Dulbecco's modified Eagles medium (DMEM)/F-12, with penicillin and streptomycin and 10% fetal bovine serum (FBS).
3. Dulbecco's Ca^{2+}-/Mg^{2+}-free phosphate buffer saline (D-PBS).
4. 0.05% trypsin-EDTA solution.

5. 400 μg/mL G418 in culture medium.
6. Opti-MEM I serum free medium (Invitrogen) or similar.
7. Lipofectamine 2000 transfection reagent (Invitrogen) or similar.
8. Plasmids encoding wild-type (WT) and mutant P2XRs, generated in-house, most often using pcDNA3.1 vector (Invitrogen) or similar.

2.2 Equipment

1. Benchtop pH meter.
2. Advanced Instruments Osmol, single sample micro-osmometer.
3. Sterile cell culture plastics: T25 vented flasks, 6- and 96-well plates; 35-mm petri dishes; 5-, 10- and 25-mL plastic pipettes; 15-mL conical centrifuge tubes; 1.5-mL microcentrifuge tubes.
4. 2-, 10-, 20-, 200-, and 1000-μL pipettes and tips.
5. Benchtop swing-out centrifuge.
6. Hemocytometer.
7. 13-mm glass coverslips.
8. CO_2 tissue/cell incubator.
9. 50- and 10-mL syringes.
10. 0.22-μm-pore diameter filters.
11. Borosilicate glass capillaries with 1.5-mm outer diameter and 1.12-mm inner diameter.
12. AgCl-coated Ag pellet reference electrode.
13. PP-830 glass micropipette puller (*see* **Note 1**).
14. PC computer with 24-in. thin film transistor monitor.
15. Axopatch 200B patch-clamp amplifier, and 1332A Digidata (Molecular Devices) (*see* **Note 2**).
16. Data acquisition and analysis software: pClamp (Clampex and Clampfit; Molecular Devices), and Origin (OriginLab).
17. Inverted microscope (we use Axiovert-200 from Zeiss) (*see* **Note 3**).
18. MP-85 manual micromanipulator (*see* **Note 4**).
19. RSC-160 rapid solution changer (*see* **Note 5**).

2.3 Recording Solutions

Prepare recording solutions with deionized water, adjust to pH 7.3, and measure the osmolarity. Store extracellular solutions at 4°C and warm to room temperature before use. Filter intracellular recording solution using a 50-mL syringe with a 0.22-μm-pore diameter filter attached and aliquot in 1-mL volume and kept at −20°C. Frost one aliquot of intracellular solution and thawed to room temperature before use.

1. Standard extracellular recording solution: 147 mM NaCl, 2 mM KCl, 1 mM $MgCl_2$, 2 mM $CaCl_2$, 13 mM D-glucose, 10 mM HEPES; pH 7.3 with NaOH; ~300 mOsm (*see* **Note 6**).
2. Standard intracellular solution: 145 mM NaCl, 10 mM EGTA and 10 mM HEPES; pH 7.3 with NaOH; ~290 mOsm (*see* **Note 7**).
3. 100 mM ATP and its analogs (e.g., BzATP) in water as stock solutions. Adjust ATP stock solution to pH 7.3 with NaOH if used in mM to elicit P2X7R activation. Aliquot ATP and BzATP stock solutions in small volumes, and store at −20°C. Prepare working solutions with desired concentrations by diluting the stock solution in extracellular recording solutions.

3 Methods

Carry out the following procedures for cell culture, transfection and cell plating in a tissue culture fume hood at room temperature. Maintain HEK293 cells in culture medium (*see* **Note 8**) in a tissue culture incubator at 37°C and 5% CO_2 under humidified conditions. Passage HEK293 cells or P2XR-expressing stable cells every 3–4 days or when cell confluency reaches ~80%. Warm all culture media to room temperature (20–22°C) before use.

3.1 Cell Passage

1. Remove media and rinse cells with 1 mL D-PBS. Add 1–2 mL trypsin-EDTA, and incubate cells at 37°C for 1–2 min (*see* **Note 9**).
2. Once cells are detached, add 1–2 mL fresh culture medium, and transfer the cell suspension in a 15-mL conical centrifuge tube.
3. Collect cells by centrifugation using a benchtop centrifuge at ~150 × *g* for 5 min.
4. Discard the supernatant, resuspend the cell pellet gently and thoroughly in 2 mL fresh culture medium (*see* **Note 10**). Transfer 10–20% of cells (0.2–0.4 mL) to a new T25 flask, depending on the confluency of starting cell cultures, frequency of using cells, and cell proliferation.

3.2 Transient Transfection

1. For each transfection, seed ~10^6 cells for one well in a 6-well plate or a 35-mm petri dish and incubate cells overnight or until reaching 70–80% confluency.
2. For each transfection, dilute 1 μg plasmid encoding P2XR (and 0.1 μg plasmid encoding enhanced green florescence protein (eGFP) for transient expression of P2XR; *see* **Note 11**) in 100 μL Opti-MEM medium in one 1.5-mL microcentrifuge tube, and 3 μL Lipofectamine 2000 into 100 μL Opti-MEM

medium in a second tube. Incubate them for 5 min at room temperature.

3. Combine the contents into one single tube and incubate further 20 min.
4. Add 800 μL fresh culture medium into the tube and mix the content.
5. Remove media from cell-containing well or petri dish and replace with the transfection medium (*see* **Note 12**).
6. Return cells to the CO_2 incubator and incubate for 24–48 h before use for patch-clamp recording or 48 h for generating stable cell lines (*see* **Note 13**).

3.3 Generation of Stable Cell Lines Expressing P2XR

1. Following the steps described above in Subheading 3.2, replace the transfection medium with 1 mL fresh culture media containing G418 antibiotic used as the selecting agent when using expression vectors such as pcDNA3.1 bearing the G418 resistance gene.
2. Culture cells for 1–2 weeks, with replacing G418 every 2–3 days.
3. Treat cells with 1 mL trypsin–ETDA, plate individual islands of cells in separate wells in 96-well plates, and incubate the cells until confluent in the presence of G418 (*see* **Note 14**).
4. Detach cells in each well with 100 μL trypsin-ETDA, transfer cells to T25 flasks, and grow cells in 5 mL fresh culture medium until confluent.
5. Validate stable P2XR-expressing cells by plating cells on coverslips and measuring agonist-induced currents using patch-clamp recording described below (*see* **Note 15**).

3.4 Cell Plating for Electrophysiological Recordings

1. To prepare cells for recording, remove transfection medium (for transient expression) or culture media, wash with D-PBS, and detach cells using trypsin-EDTA as described above.
2. Place 13-mm glass coverslips (up to 4) per well in a 6-well plate or per 35-mm petri dish, and add 1 mL fresh culture medium in one well or petri dish.
3. Count cell number using a hemocytometer. Transfer 40,000 cells per well or petri dish and incubate cells for 12–24 h before use (*see* **Note 16**).

3.5 Whole-Cell Recordings

1. Switch on the patch-clamp rig, and open Clampex for recording and RSC software that controls solution changes.
2. Connect the recording chamber to solution reservoir provided by a 50-mL syringe, add extracellular recording solution into the syringe, and adjust solution flow at a rate of 1–2 mL/min.

3. Connect the recording chamber to RSC solution tubes provided by 10-mL syringes. Add agonists, antagonists, or other agents into syringes and register them in the RSC control system.
4. Check AgCl-pellet reference electrode in contact with extracellular recording solution and connection to the ground via the headstage of the amplifier.
5. Prepare recording pipettes from glass capillaries using PP-830 puller. Back-fill the pipette with intracellular recording solution, and mount it onto the holder of the headstage that is connected to the amplifier via AgCl-coated silver wire (*see* **Note 17**).
6. Lower the pipette into the extracellular solution in the recording chamber, and apply a 5 mV test pulse to determine the resistance of the pipette, which should be in the range of 3–5 MΩ.
7. Identify single cells under the microscope (*see* **Note 18**).
8. Adjust the basal current level to zero. Maneuver the pipette using a micromanipulator under the microscope to bring its tip into contact with cell surface, which is indicated with slightly increased resistance of the pipette.
9. Apply suction to the cell membrane, through a syringe connected via a tube to the recording pipette, to form a giga-ohm seal (in cell-attached configuration). Compensate the transient capacitive currents (refer to the amplifier's manuscript for details).
10. Apply additional suction to break-through the cell membrane to achieve the whole-cell configuration, which is indicated by appearance of relatively slow capacitive currents (*see* **Note 19**). Compensate the capacitive currents. Set the membrane potential to −60 mV or desired holding potential (*see* **Note 20**).
11. Compensate the capacitive current by ~60–70% (*see* **Note 21**).
12. Start recording agonist-induced P2XR-mediated currents. Apply agonists for 2–10 s through RSC at an interval of 2–4 min, depending on the activation and desensitization properties of the P2XR under investigation, which requires optimization experimentally. The following are some commonly used protocols in the study of the P2XRs:
13. To determine the potency or EC_{50} value of an agonist by constructing the concentration-current amplitude relationship curve, apply agonist at 2–4 min intervals from low to high concentrations (*see* an example in Fig. 1).
14. To determine the potency or IC_{50} value of an antagonist at P2XR by constructing the concentration-inhibition relationship curve, choose the concentration of agonist to elicit

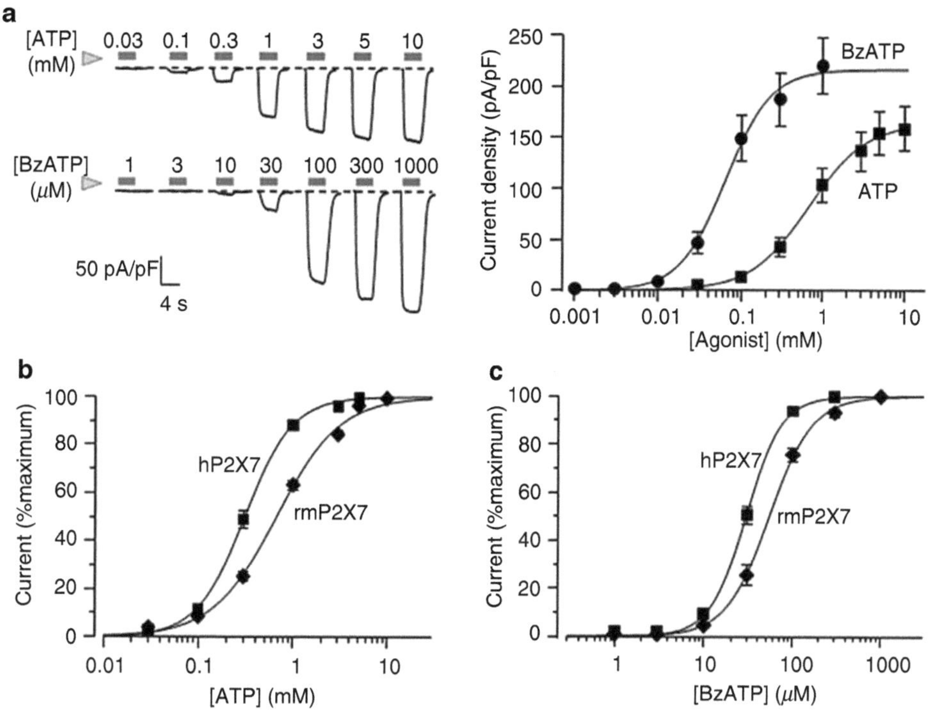

Fig. 1 Comparison of the sensitivity of rhesus macaque (rm) and human P2X7Rs to ATP and BzATP. (**a**) *Left*, representative currents evoked by different concentrations of ATP or BzATP in HEK293 cells expressing rmP2X7R. *Right*, agonist concentration–current amplitude curves. Each data point represents mean from 8 to 11 cells for ATP and 4–6 cells for BzATP. (**b**, **c**) Comparison of agonist concentration–current relationship curves between rmP2X7R and hP2X7R. The data for rmP2X7R are from panel (**a**). Each data point represents mean from 12 cells for ATP (**b**) and 5 cells for BzATP (**c**) for hP2X7R. The smooth curves show the fit of the mean data to the Hill equation. Take from [34]

measureable currents (e.g., EC_{50} or EC_{90-100}), apply agonist at 2–4 min intervals to establish stable current response, and treat the patched cell with antagonist between agonist applications at increasing concentrations (*see* an example in Fig. 2). For the antagonists with fast dissociation kinetics, coapply antagonist with agonist as well as treatment with antagonist between agonist applications (*see* an example in Fig. 3).

15. To examine the effect of exposure to reagent, for example, dithiothreitol (DTT) used in cysteine substitution studies, on P2XR-mediated currents. Apply agonist at a chosen concentration of agonist at 2–4 min to establish a stable current response before exposing the patched cell with DTT between agonist applications (*see* an example in Fig. 4).

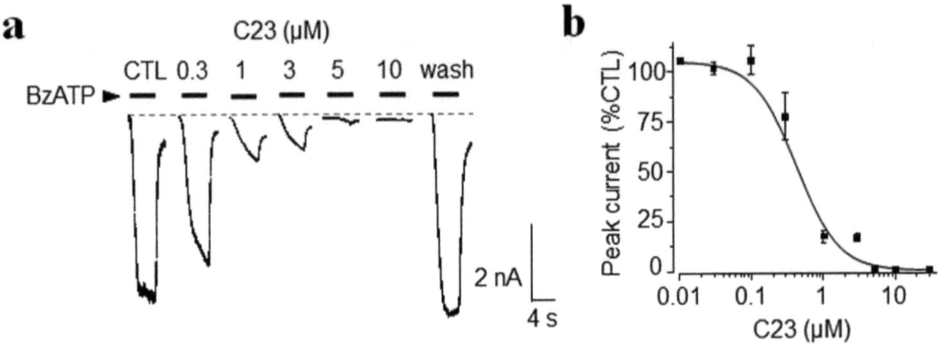

Fig. 2 Inhibition of hP2X7R by a novel antagonist. (**a**) Representative currents from a HEK293 cell expressing hP2X7R before (CTL) and after treatment with compound C23 at different concentrations. Currents were elicited in extracellular low divalent cation solution by 4-s application of 300 μM BzATP at an interval of 2–4 min. Current inhibition was reversed upon washing. (**b**) Antagonist concentration-current inhibition relationship curves. Each data point represents mean from four cells. The smooth line shows the fit of the mean data to the Hill equation. Modified from [35]

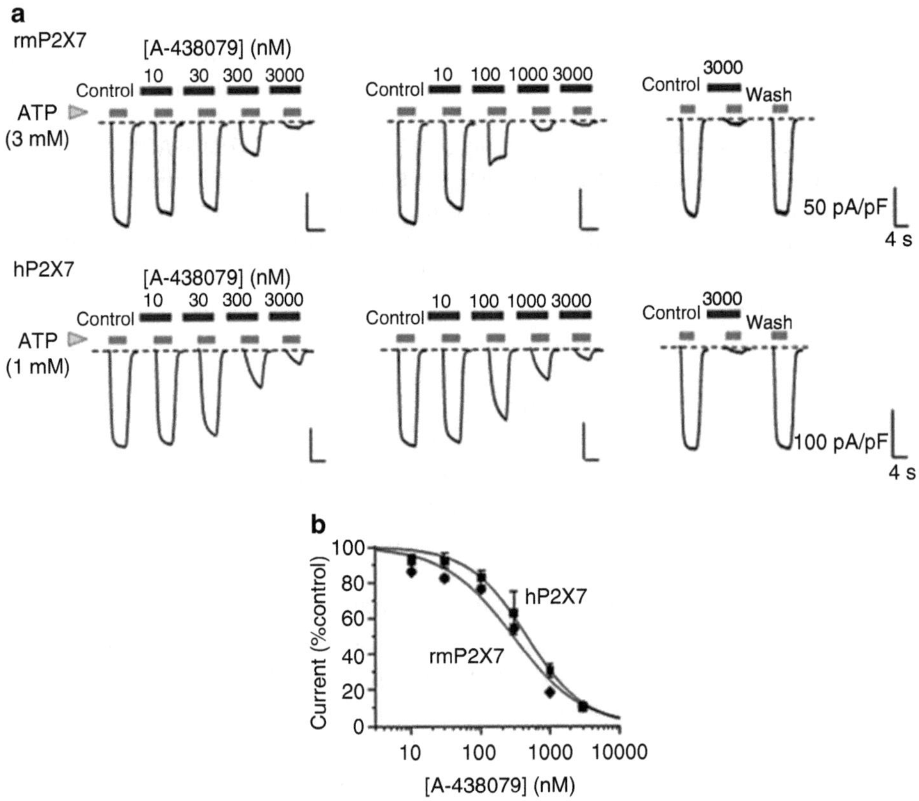

Fig. 3 Inhibition of rhesus macaque and human P2X7Rs by A-438079. (**a**) Representative ATP-induced currents in HEK293 cells expressing rmP2X7R or hP2X7R in the absence (control) or presence of different concentrations of A-438079. Currents were evoked by 4-s application of 3 mM ATP at rmP2X7R and 1 mM at hP2X7R every 4 min. Cells were exposed to A-438079 between and during ATP applications. (**b**) Comparison of antagonist concentration–current inhibition curves between rmP2X7R and hP2X7R. The smooth lines show the fit of the mean data to the Hill equation. Modified from [34]

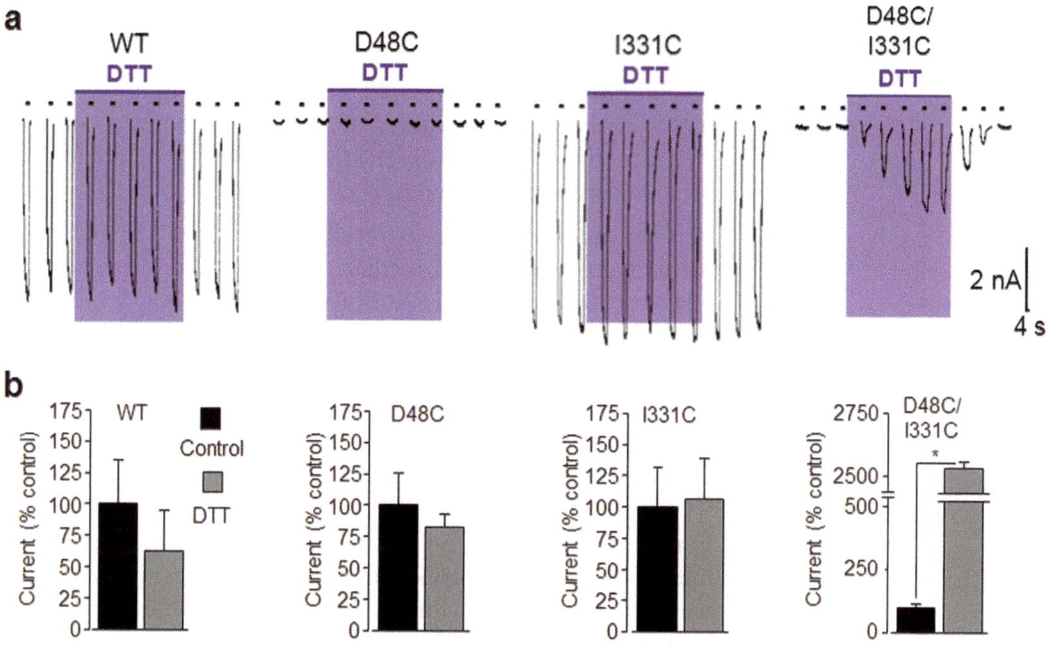

Fig. 4 DTT-induced reversible increase in BzATP-induced currents in HEK293 cells expressing hP2X7R carrying D48C/I331C double cysteine mutations. (**a**) Representative currents induced by 4-s application of 300 μM BzATP every 2 min before, during, and after exposure to 10 mM DTT in cells expressing WT, D48C, I331C, or D48C/I331C mutant hP2X7R. (**b**) Mean BzATP-induced currents at the end of 10-min exposure to DTT (gray) as % of control currents (black). DTT reversibly increased BzATP-induced currents in cells expressing D48C/I331C mutant hP2X7R, but WT, D48C, or I331C mutant hP2X7R. Such results are interpreted to indicate that D48 and I331 residues, located in the outer ends of the first and second transmembrane segments, respectively, are in close vicinity in closed state, and these parts undergo substantial conformational change during P2X7R activation. Modified from [21]

4 Notes

1. PP-830 is discontinued and replaced by PC-100. Other models of micropipette pullers from different vendors can be used.
2. Molecular Devices provides Axopatch 200B amplifier, which offers low-noise recordings, particularly suitable for single-channel recording, and MultiClamp 700B amplifier (https://www.moleculardevices.com/products/axon-patch-clamp-system/amplifiers/axon-instruments-patch-clamp-amplifiers). Harvard Bioscience provides HEKA EPC 10 USA amplifier (https://www.heka.com/products/products_main.html#physiol_epc10single). These are the most common models.
3. Several manufactures provide inverted microscopes that are suitable for building a manual patch-clamp rig. A fluorescence system incorporated into the microscope is required to identify eGFP-positive cells, if cells are cotransfected with plasmids for P2XR and eGFP.

4. There are many choices of manual and automatic micromanipulators.
5. A computer-controlled rapid solution changing system is highly desired for application of agonists for a few seconds to P2XR to avoid receptor desensitization. Several manufacturers provide fast solution change systems.
6. P2X7Rs are inhibited by extracellular Ca^{2+} and Mg^{2+}. Experiments recording P2X7R-mediated currents often use extracellular low divalent cation solution: 147 mM NaCl, 2 mM KCl, 0.3 mM $CaCl_2$, 23 mM D-glucose, 10 mM HEPES, pH 7.3 with NaOH; ~300 mOsm. Please note that it is possible but difficult to form gigaohm seal in low divalent cation solution. It is recommended to establish the gigaohm seal or whole-cell configuration in standard extracellular solution and then change to low divalent cation solution.
7. For not fully understood reasons, fluoride ions in the intracellular solution can improve seal formation and stabilize the cell membrane, favoring longer and more stable recordings. Replace NaCl in the intracellular solution in part or whole with NaF, if required.
8. Supplementing the culture medium with antibiotics is not essential, but it is recommended to add 50 U/mL penicillin and 50 μg/mL streptomycin to prevent contamination, if required.
9. Avoid prolonged incubation with trypsin–EDTA, particularly when plating cells for recording. Gentle tapping of the flask, plate, or petri dish from side or bottom can help to dislodge loosely attached cells.
10. Use a 1000-μL pipette to pipette up and down 10–20 times and, if required, then a 200-μL pipette to pipette up and down 10–20 times to break down cell clumps to single cells.
11. The efficiency of transfection has been improved using commercially available reagents but falls far away from 100%. Therefore, including 0.1 μg plasmid encoding eGFP during transfection is highly recommended, if transfected cells are used for recording and the microscope in the patch-clamp rig has a fluorescence system. To select eGFP-positive cells post-transfection for recording.
12. The protocol in detail using Lipofectamine2000 reagent to transiently transfect the cells is available at: https://www.ecu.edu/cs-dhs/biochemistry/upload/Transfection-Protocol.pdf). Several other transfection reagents or kits are also suitable for transient expression.

13. Optional: if 48 h incubation time is preferred, replace transfection medium with culture medium 24 h posttransfection, and incubate cells further 24 h before use.
14. Repeat this subcloning step, if necessary.
15. Other methods can be used to validate the stable cell lines, such as measurement of agonist-induced calcium responses, or through molecular biology and biochemical techniques (e.g., RT-PCR, western blotting).
16. Cells can be used for patch-clamp recording after shorter incubation times, if cells adhere to coverslip faster or earlier. In addition, plate cells at lower cell density if cells are used 36–48 h posttransfection.
17. Poor chloride coating of the silver wire can result in a drift of the basal or zero current level over the time. Coat the silver wire regularly in bleach for 30–60 min, rinse with water, and dry before use.
18. As mentioned above, if HEK293 cells are transiently cotransfected with eGFP and P2XR, a fluorescence microscope is required to identify eGFP-positive cells.
19. If it is difficult to establish the whole-cell configuration following seal formation, use the "zap" button on the front panel of the amplifier to rupture the cell membrane. However, this practice is not recommended for routine use, as zapping often reduces sealing or leads to loss of sealing.
20. Holding the cell membrane at a negative potential can help stabilize or improve the sealing.
21. Determine the series resistance and cell capacitance. The cell capacitance can be used to derive the current density, which can mitigate the effect of the cell size on the current amplitude (*see* an example in Fig. 1). Check the series resistance during the recording and use data with a series resistance of no more than 10 MΩ.

Acknowledgments

The work from Jiang's laboratory was supported by Biotechnology and Biological Sciences Research Council and Wellcome Trust.

References

1. North RA (2002) Molecular physiology of P2X receptors. Physiol Rev 82:1013–1067
2. Khakh BS, North RA (2012) Neuromodulation by extracellular ATP and P2X receptors in the CNS. Neuron 76:51–69
3. Di Virgilio F, Sarti AC, Grassi F (2018) Modulation of innate and adaptive immunity by P2X ion channels. Curr Opin Immunol 52:51–59

4. Oliveira Á, Illes P, Ulrich H (2016) Purinergic receptors in embryonic and adult neurogenesis. Neuropharmacology 104:272–281
5. Jiang LH, Hao Y, Mousawi F, Peng H, Yang X (2017) Expression of P2 purinergic receptors in mesenchymal stem cells and their roles in extracellular nucleotide regulation of cell functions. J Cell Physiol 232:287–297
6. Bernier LP, Ase AR, Séguéla P (2018) P2X receptor channels in chronic pain pathways. Br J Pharmacol 175:2219–2230
7. Tewari M, Seth P (2015) Emerging role of P2X7 receptors in CNS health and disease. Ageing Res Rev 24:328–342
8. Wei L, Syed Mortadza SA, Yan J, Zhang L, Wang L, Yin Y, Li C, Chalon S, Emond P, Belzung C, Li D, Lu C, Roger S, Jiang LH (2018) ATP-activated P2X7 receptor in the pathophysiology of mood disorders and as an emerging target for the development of novel antidepressant therapeutics. Neurosci Biobehav Rev 87:192–205
9. Burnstock G (2016) P2X ion channel receptors and inflammation. Purinergic Signal 12:59–67
10. Tozzi M, Novak I (2017) Purinergic receptors in adipose tissue as potential targets in metabolic disorders. Front Pharmacol 8:878
11. Roger S, Jelassi B, Couillin I, Pelegrin P, Besson P, Jiang LH (2015) Understanding the roles of the P2X7 receptor in solid tumour progression and therapeutic perspectives. Biochim Biophys Acta 1848:2584–2602
12. Di Virgilio F, Sarti AC, Falzoni S, De Marchi E, Adinolfi E (2018) Extracellular ATP and P2 purinergic signalling in the tumour microenvironment. Nat Rev Cancer 18:601–618
13. Hamill O, Marty A, Neher E, Sakmann B, Sigworth F (1981) Improved patch-clamp techniques for high-resolution current recording from cells and cell-free membrane patches. Pflugers Arch 391:85–100
14. Rassendren F, Buell G, Newbolt A, North RA, Surprenant A (1997) Identification of amino acid residues contributing to the pore of a P2X receptor. EMBO J 16:3446–3454
15. Marquez-Klaka B, Rettinger J, Bhargava Y, Eisele T, Nicke A (2007) Identification of an intersubunit cross-link between substituted cysteine residues located in the putative ATP binding site of the P2X1 receptor. J Neurosci 27:1456–1466
16. Stelmashenko O, Lalo U, Yang Y, Bragg L, North RA, Compan V (2012) Activation of trimeric P2X2 receptors by fewer than three ATP molecules. Mol Pharmacol 82:760–766
17. Stelmashenko O, Compan V, Browne LE, North RA (2014) Ectodomain movements of an ATP-gated ion channel (P2X2 receptor) probed by disulfide locking. J Biol Chem 289:9909–9917
18. Browne LE, Nunes JP, Sim JA, Chudasama V, Bragg L, Caddick S, North RA (2014) Optical control of trimeric P2X receptors and acid-sensing ion channels. Proc Natl Acad Sci U S A 111:521–526
19. Zhao WS, Wang J, Ma XJ, Yang Y, Liu Y, Huang LD, Fan YZ, Cheng XY, Chen HZ, Wang R, Yu Y (2014) Relative motions between left flipper and dorsal fin domains favour P2X4 receptor activation. Nat Commun 5:4189
20. Wang J, Sun LF, Cui WW, Zhao WS, Ma XF, Li B, Liu Y, Yang Y, Hu YM, Huang LD, Cheng XY, Li L, Lu XY, Tian Y, Yu Y (2017) Intersubunit physical couplings fostered by the left flipper domain facilitate channel opening of P2X4 receptors. J Biol Chem 292:7619–7635
21. Caseley EA, Muench SP, Jiang LH (2017) Conformational changes during human P2X7 receptor activation examined by structural modelling and cysteine-based cross-linking studies. Purinergic Signal 13:135–141
22. Jiang R, Taly A, Lemoine D, Martz A, Cunrath O, Grutter T (2012) Tightening of the ATP-binding sites induces the opening of P2X receptor channels. EMBO J 31:2134–2143
23. Sim JA, Broomhead HE, North RA (2008) Ectodomain lysines and suramin block of P2X1 receptors. J Biol Chem 283:29841–29846
24. Karasawa A, Kawate T (2016) Structural basis for subtype-specific inhibition of the P2X7 receptor. Elife 5:e22153
25. Kasuya G, Yamaura T, Ma XB, Nakamura R, Takemoto M, Nagumo H, Tanaka E, Dohmae N, Nakane T, Yu Y, Ishitani R, Matsuzaki O, Hattori M, Nureki O (2017) Structural insights into the competitive inhibition of the ATP-gated P2X receptor channel. Nat Commun 8:876
26. Wang J, Wang Y, Cui WW, Huang Y, Yang Y, Liu Y, Zhao WS, Cheng XY, Sun WS, Cao P, Zhu MX, Wang R, Hattori M, Yu Y (2018) Druggable negative allosteric site of P2X3 receptors. Proc Natl Acad Sci U S A 115:4939–4944
27. Browne LE, Jiang LH, North RA (2010) New structure enlivens interest in P2X receptors. Trends Pharmacol Sci 31:229–237

28. Coddou C, Yan Z, Obsil T, Huidobro-Toro JP, Stojilkovic SS (2011) Activation and regulation of purinergic P2X receptor channels. Pharmacol Rev 63:641–683
29. Pasqualetto G, Brancale A, Young MT (2018) The molecular determinants of small-molecule ligand binding at P2X receptors. Front Pharmacol 9:58
30. Schmid R, Evans RJ (2018) ATP-gated P2X receptor channels: molecular insights into functional roles. Annu Rev Physiol. https://doi.org/10.1146/annurev-physiol-020518-114259
31. Roger S, Mei ZZ, Baldwin JM, Dong L, Bradley H, Baldwin SA, Surprenant A, Jiang LH (2010) Single nucleotide polymorphisms that were identified in affective mood disorders affect ATP-activated P2X7 receptor functions. J Psychiatr Res 44:347–355
32. Aprile-Garcia F, Metzger MW, Paez-Pereda M, Stadler H, Acuña M, Liberman AC, Senin SA, Gerez J, Hoijman E, Refojo D, Mitkovski M, Panhuysen M, Stühmer W, Holsboer F, Deussing JM, Arzt E (2016) Co-expression of wild-type P2X7R with Gln460Arg variant alters receptor function. PLoS One 11:e0151862
33. Metzger MW, Walser SM, Dedic N, Aprile-Garcia F, Jakubcakova V, Adamczyk M, Webb KJ, Uhr M, Refojo D, Schmidt MV, Friess E, Steiger A, Kimura M, Chen A, Holsboer F, Arzt E, Wurst W, Deussing JM (2017) Heterozygosity for the mood disorder-associated variant Gln460Arg alters P2X7 receptor function and sleep quality. J Neurosci 37:11688–11700
34. Bradley HJ, Browne LE, Yang W, Jiang LH (2011) Pharmacological properties of the rhesus macaque monkey P2X7 receptor. Br J Pharmacol 164:743–754
35. Caseley EA, Muench SP, Fishwick CW, Jiang LH (2016) Structure-based identification and characterisation of structurally novel human P2X7 receptor antagonists. Biochem Pharmacol 116:130–139

Chapter 20

Recording P2X Receptors Using Whole-Cell Patch Clamp from Native Monocytes and Macrophages

Leanne Stokes

Abstract

Investigating ion channels in their native cell type is important when striving to understand their regulation and function, but this comes with added complexities due to the plethora of channels and receptors present. Details of recording ATP-gated ion channels in macrophages are presented together with information on how to prepare the primary cells for electrophysiological analysis.

Key words Whole-cell patch clamp, P2X receptor, Macrophage, ATP, EPC10, P2X7, P2X4, Monocyte

1 Introduction

Ligand-gated ion channels such as P2X receptors are known to control multiple cellular responses [1]. Studying the detailed kinetic and pharmacological properties of individual ion channels is best performed using a heterologous expression system as outlined in an earlier chapter because overexpression in HEK293 cells typically allows for isolation of the ion channel of interest. Investigating ion channels expressed at endogenous (and typically low) levels in primary isolated cells brings added complexity but also important information about the interplay between a multitude of channels and receptors. Depending on the information required in the study, it is often possible to separate P2X channels based on their pharmacology or by the use of cells isolated from transgenic mice [2, 3]. The investigation of polymorphic variants of P2X ion channels is also possible from cells isolated from human volunteers [4]. Indeed, if studying isolated cells from human subjects for measuring P2X4 or P2X7 channel activity it is pertinent to acknowledge that many single nucleotide polymorphisms have been identified in these genes that can dramatically affect their response to ligand [5].

Pablo Pelegrín (ed.), *Purinergic Signaling: Methods and Protocols*, Methods in Molecular Biology, vol. 2041,
https://doi.org/10.1007/978-1-4939-9717-6_20,

Whole-cell patch clamping allows for access to the entire plasma membrane complement of P2X channels and can be used to assess channel trafficking to and from the membrane. This technique is also useful if one would like to compare ion channel pharmacology between a heterologously expressed P2X ion channel and a cell type endogenously expressing the P2X ion channel of interest. Here I describe how to measure P2X ion channel activity in primary peritoneal macrophages from rodents, primary human monocytes, and monocyte-derived macrophages.

2 Materials

1. Native cells of interest (e.g., peritoneal macrophages, peripheral blood monocytes, monocyte-derived macrophages) (*see* **Note 1**).
2. Media for culturing cells: RPMI1640 medium containing 10% fetal bovine serum, L-glutamine, penicillin and streptomycin.
3. Recombinant human M-CSF endotoxin free.
4. Glass coverslips 13 mm (*see* **Note 2**).
5. 35 mm petri dishes (sterilized for cell culture).
6. HEKA EPC10 amplifier with Patchmaster software.
7. MP-283 micromanipulator.
8. Isolation Table (TMC) with a Faraday cage.
9. Inverted microscope (e.g., Nikon Ti-U 2000).
10. Narishige PC-10 pipette puller.
11. World Precision Instruments borosilicate glass capillaries catalog number TW150F-4 or Harvard Clark glass catalog number GC150TF-10.
12. Fast solution exchanger (e.g., BioLogic Instruments RSC-200).
13. Perfusion chamber suitable for 13 mm coverslips (e.g., QE-1 from Warner Instruments).
14. Intracellular pipette solution: 145 mM NaCl, 10 mM HEPES, 10 mM EGTA pH 7.3 (*see* **Note 3**).
15. Extracellular bath solution: 145 mM NaCl, 2 mM KCl, 2 mM $CaCl_2$, 1 mM $MgCl_2$, 13 mM D-glucose, 10 mM HEPES pH 7.35 (*see* **Note 4**).
16. Sterile phosphate buffered solution (PBS): 137 mM NaCl, 2.7 mM KCl, 10 mM Na_2HPO_4, 1.8 mM KH_2PO_4, pH 7.4.
17. MACS buffer: PBS containing 0.5% bovine serum albumin and 2 mM EDTA.

18. CD14 microbeads (e.g., Miltenyi Biotec catalog number 130-050-201).
19. Miltenyi Biotec MS or LS columns or other compatible columns.

3 Methods

3.1 Preparation of Peritoneal Macrophages

The preparation of native macrophages from the peritoneal cavity of rodents is quick and easy to perform and similar protocols have been published [6–8].

1. Following sacrifice the abdomen should first be swabbed down with a 70% ethanol solution to clean the area.
2. Using forceps lift the skin upward away from the body; scissors can be used to cut the skin, allowing for access to the peritoneal cavity. It is easier to proceed if an area of skin plus fur is excised. The transparent peritoneal membrane should be visible underneath the skin.
3. Prepare a 5 ml syringe with sterile cold PBS solution.
4. Use forceps to lift the peritoneal membrane upward and scissors to cut an incision. Hold the edge of the peritoneal membrane in forceps in one hand and use the other hand to add the 5 ml of PBS into the peritoneal cavity. You may need to tilt the animal so that the solution does not spill out (*see* **Note 5**).
5. Use a rocking movement of the animal to move the liquid around in the cavity to allow the cells to be flushed from their location into the PBS solution. You can also perform gentle massage of the sides of the abdomen.
6. Following a couple of minutes of flushing, the solution needs to be removed from the cavity using the 5 ml syringe to suction the solution up. This is helped by tilting the animal to collect the fluid. Recovery of at least 4–5 ml of solution is possible.
7. Transfer the peritoneal PBS solution to a 15 ml centrifuge tube and keep on ice until ready to use.
8. Centrifuge the cells at 300 × *g* for 5 min and resuspend the pellet in 1 ml of fresh RPMI complete media.
9. Plate 10–50 μl of cells per coverslip (four coverslips per 35 mm petri dish), add additional media to a total volume of 100 μl per coverslip and allow the cells to attach for 2 h at 37 °C in a humidified CO_2 incubator.
10. Following cell attachment 2 ml of media can be added to each petri dish. The cells are cultured overnight to allow for recovery and good attachment to the coverslip.

3.2 Preparation of Human Monocyte-Derived Macrophages

Preparation of human monocytes from heparinized peripheral venous blood can be performed using a standard procedure. Monocyte-derived macrophages can be generated by culture for 6–7 days in media containing M-CSF.

1. Collect peripheral venous blood in a heparin Vacutainer or obtain heparinized whole blood in a blood bag.
2. Dilute the blood 1:1 with sterile PBS.
3. Add 15 ml of Ficoll Hypaque into a sterile 50 ml centrifuge tube.
4. Layer 30–35 ml of diluted blood carefully on top of the Ficoll solution. Tilt the tube and allow one or two drops to run down the side of the tube and form a thin layer on the Ficoll. Slowly release more of the diluted blood allowing it to float atop the Ficoll. Gradually straighten the tube as you fill it.
5. Centrifuge the tube for 20–30 min at 400 × *g* with no brake. The layering will stay intact if the centrifuge decelerates slowly.
6. Using a sterile transfer pipette or serological pipette, remove the mononuclear layer into a fresh 50 ml centrifuge tube. Add 30–40 ml sterile PBS to wash the cells. Centrifuge at 300 × *g* for 5 min.
7. Resuspend cells in PBS.
8. Human monocytes can be labelled with CD14 microbeads for positive selection with a magnetic column. Peripheral blood mononuclear cells (PBMCs) from the Ficoll separation should be resuspended in cold MACS buffer at 80 μl per 10^7 cells.
9. Following the manufacturer's instructions, 20 μl of CD14 microbeads can be used per 10^7 cells to label monocytes. Cells are incubated for 15 min at 4 °C.
10. Wash off unbound microbeads by adding 2 ml MACS buffer and centrifuge at 300 × *g* for 10 min.
11. Resuspend cells in 500 μl of MACS buffer.
12. Prepare the appropriate column (MS or LS columns) by rinsing with cold MACS buffer. Add the labelled cells to the column and wash three times collecting the flow-through into a sterile 15 ml tube.
13. Remove the column from the magnetic separator and place in a collection tube. Add buffer to the column and push buffer through the column by using the supplied plunger. This releases all magnetically labelled cells from the column into the collection tube.

14. Centrifuge the cells and resuspend in appropriate media. For differentiation, cells are resuspended in complete RPMI media supplemented with 25 ng/ml M-CSF. Culture for 6 days on 13 mm glass coverslips in 35 mm petri dishes using a standard humidified 37 °C 5% CO_2 incubator.

3.3 Recording P2X Channels

1. Standard electrophysiology equipment is required to record P2X channels in native cells (*see* **Note 6**).
2. Insert a coverslip of cells into the recording chamber and allow washing of the cells with extracellular bath solution to occur for a few minutes to remove all traces of cell media.
3. Prepare the glass microelectrode using the Narishige pipette puller (or equivalent) and backfill with syringe-filtered intracellular pipette solution. Attach the microelectrode to the HEKA amplifier pipette holder.
4. Using the microscope with high magnification choose a suitable cell for patching based on morphological features or by staining with specific fluorescent antibodies to known cell surface markers (e.g., CD14 or CD16 on human monocytes) (*see* **Note 7**).
5. Navigate and attach the microelectrode to the cell membrane and make a gΩ seal by applying gentle suction. Seal formation is assessed by watching the R-series value (pipette resistance) as suction is applied to a 1 ml syringe (or 3 ml syringe).
6. Once a high resistance seal is acquired apply the correct voltage clamp to the cell (typically −60 mV).
7. Select the "whole cell" recording mode and apply C-fast compensation. The seal is broken by applying short bursts of suction using the syringe. A successful rupture will result in appearance of capacitative transients indicative of access to the whole cell membrane. C-slow compensation should then be performed to calculate cell size (typically 10–20 pF).
8. Check series resistance is not too large (>30 mΩ).
9. Once the seal has been broken at least 1–2 min should be left to allow for dialysis of cell cytoplasmic contents with the intracellular pipette solution. This can be checked by performing a voltage ramp protocol (−130 mV to +30 mV) to assess whether outward K^+ currents remain.
10. Currents in response to specific agonists can then be recorded by utilizing a computer controlled fast flow trigger system (e.g., RSC-200 system).
11. Treatment with specific antagonists can be performed by bath applying an extracellular solution containing antagonist and/or by applying antagonist (±agonist) solution through the fast flow tubing (*see* **Note 8**).

4 Notes

1. Donor variability is an important factor to consider when studying human subjects.
2. Glass 13 mm coverslips should be sterilized with 70% ethanol and air-dried under sterile conditions. Once coverslips have been prepared in sterile petri dishes (4 per dish), place under UV lamp in a tissue culture hood for sterilization (30 min).
3. Alternative intracellular pipette solutions may be used. This is a standard NaCl-based pipette solution typically used to record P2X currents.
4. A standard bath solution containing calcium and magnesium is used. Again, this may be varied according to experimental design. For recording P2X7 currents agonists can be prepared in a low divalent bath solution (0.2 mM $CaCl_2$ and zero $MgCl_2$) to maximize ATP responses.
5. Other published protocols for isolating peritoneal macrophages inject the PBS into the peritoneal cavity using a syringe needle; however, from experience it is possible to inject the PBS into organs. Furthermore, some protocols use the needle to pull out the peritoneal fluid but the needle tends to get clogged rather easily.
6. Standard electrophysiology equipment is required. The microscope is placed on an antivibration table and surrounded with a Faraday cage. Positioning of the micromanipulator for navigating the microelectrode should be to the right hand side and the fast flow tubing control box and micromanipulator on the left-hand side.
7. The microscope will need a fluorescent xenon lamp attachment and relevant filter blocks to observe fluorescence of stained cells.
8. Use of selective antagonists is important when attempting to distinguish different components of an ATP-induced response (*see* Figs. 1, 2, and 3). AZ11645373 is useful for human P2X7 [9], or AZ10606120 [10] for rodent P2X7. Ivermectin is a useful tool to demonstrate the presence of the P2X4 channel [3], and recently discovered antagonists will also be useful in future studies for P2X4 [11].

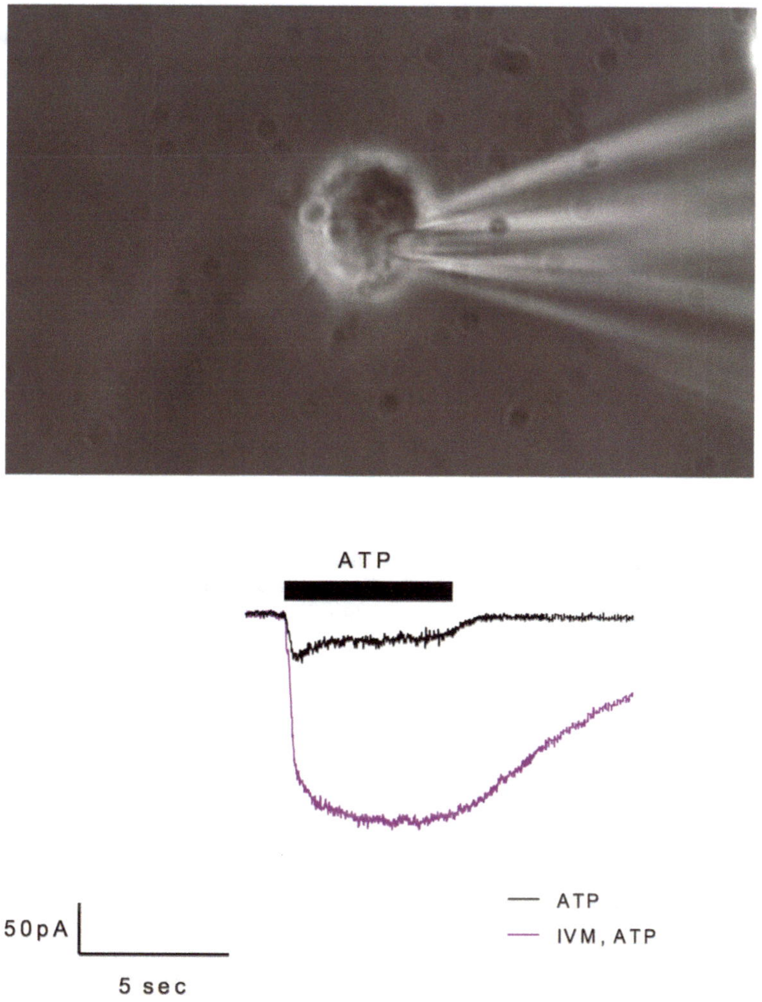

Fig. 1 Responses recorded from mouse peritoneal macrophages. Image of the patched cells and inward current response to a 5-s pulse of 100 μM ATP (black) followed by application of ivermectin (IVM; 3 μM) for 1 min and reapplication of 100 μM ATP (magenta). Responses are similar to those recorded in NR8383 and J774 macrophage cell lines [3] and are typical of a P2X4 response

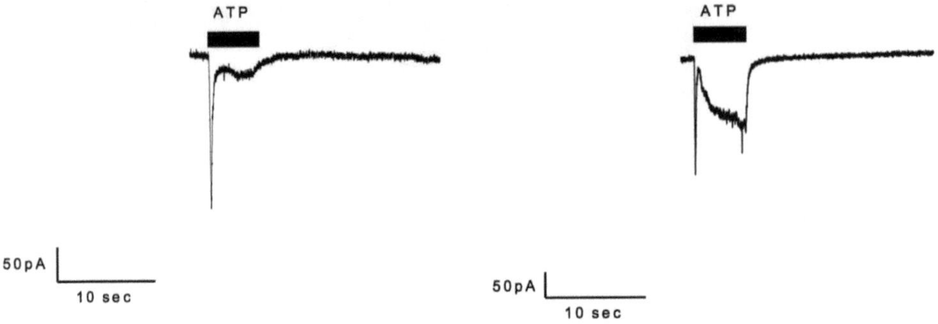

Fig. 2 Responses recorded from human monocyte-derived macrophages. Inward currents were recorded in response to a 5-s pulse of 1 mM ATP in low divalent bath solution. Traces represent two cells from the same donor

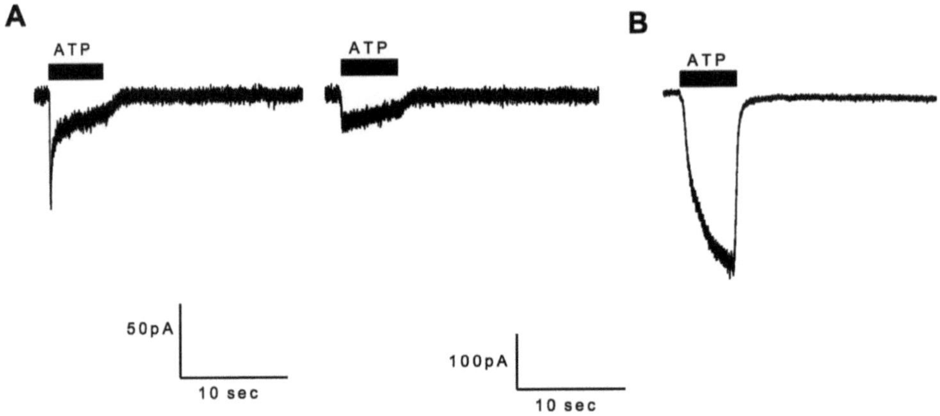

Fig. 3 Responses recorded from human monocyte-derived macrophages. Inward currents were recorded in response to a 5-s pulse of (**a**) 100 μM ATP in low divalent bath solution or (**b**) 3 mM ATP in low divalent bath solution. This demonstrates the initial spike response on first encounter with ATP, the disappearance of this fast response in the second application of ATP, and the presence of a P2X7 response to a high concentration of ATP. The onset of the P2X7 response is much slower than other P2X currents. Traces shown are from the same cell

References

1. North RA (2002) Molecular physiology of P2X receptors. Physiol Rev 82:1013–1067
2. Brone B et al (2007) P2X currents in peritoneal macrophages of wild type and P2X4−/− mice. Immunol Lett 113:83–89
3. Stokes L, Surprenant A (2009) Dynamic regulation of the P2X4 receptor in alveolar macrophages by phagocytosis and classical activation. Eur J Immunol 39:986–995
4. Stokes L et al (2011) A loss-of-function polymorphism in the human P2X4 receptor is associated with increased pulse pressure. Hypertension 58:1086–1092
5. Bartlett R, Stokes L, Sluyter R (2014) The P2X7 receptor channel: recent developments and the use of P2X7 antagonists in models of disease. Pharmacol Rev 66:638–675
6. Layoun A, Samba M, Santos MM (2015) Isolation of murine peritoneal macrophages to carry out gene expression analysis upon toll-like receptors stimulation. J Vis Exp 2015:52749

7. Ray A, Dittel BN (2010) Isolation of mouse peritoneal cavity cells. J Vis Exp 2010:1488
8. Zhang X, Goncalves R, Mosser DM (2008) The isolation and characterization of murine macrophages. Curr Protoc Immunol. Chapter 14:Unit 14.1
9. Stokes L et al (2006) Characterization of a selective and potent antagonist of human P2X (7) receptors, AZ11645373. Br J Pharmacol 149:880–887
10. Allsopp RC et al (2017) Unique residues in the ATP gated human P2X7 receptor define a novel allosteric binding pocket for the selective antagonist AZ10606120. Sci Rep 7:725
11. Stokes L et al (2017) P2X4 receptor function in the nervous system and current breakthroughs in pharmacology. Front Pharmacol 8:291

Chapter 21

Automated Planar Patch-Clamp Recording of P2X Receptors

Carol J. Milligan and Lin-Hua Jiang

Abstract

P2X receptors are a structurally and functionally distinctive family of ligand-gated ion channels that play important roles in mediating extracellular adenosine 5′-triphosphate (ATP) signaling in diverse physiological and pathophysiological processes. For several decades, the "manual" patch-clamp technique was regarded as the gold standard assay for investigating ion channel properties. More recently, breakthroughs in the development of automated patch-clamp technologies are enabling the study of ion channels, with much greater throughput capacities. These automated platforms, of which there are many, generate consistent, reliable, high-fidelity data. This chapter demonstrates the versatility of one of these technologies for ligand-gated ion channels, with a particular emphasis on protocols that address some of the issues of receptor desensitization that are commonly associated with P2X receptor-mediated currents.

Key words Automated electrophysiology, Planar patch-clamp, Planar chip, Microfluidics, Stacked solution application, Ligand-gated ion channels, P2X receptors, Voltage-clamp

1 Introduction

P2X receptors (P2XRs) belong to the ligand-gated ion channel superfamily with distinctive structural and functional properties [1]. In the mammalian system, there are seven different receptor subunits ($P2X_1R$-$P2X_7R$), each of which contains intracellular N- and C-termini and two transmembrane segments linked by a large extracellular domain. P2XR subunits assemble into homotrimeric or heterotrimeric complexes to form functional channels. The second transmembrane domain from each of the three subunits, in the complex, forms the ion-permeating pathway [2, 3]. Extracellular ATP binding specifically activates P2XRs leading to variable permeation of small inorganic cations such as Ca^{2+}, Na^+, and K^+, with the exception of the human $P2X_5R$, which also conducts Cl^- ions [4]. There are three phases associated with gating of P2XRs: the ATP-evoked activation or rising phase that induces a rapid current, the desensitization or decay phase that occurs in the presence of ATP and develops slowly, and a relatively quick deactivation phase elicited by removal of the agonist. The kinetics of these three phases

Pablo Pelegrín (ed.), *Purinergic Signaling: Methods and Protocols*, Methods in Molecular Biology, vol. 2041,
https://doi.org/10.1007/978-1-4939-9717-6_21,

vary considerably among the different P2XR subtypes [5]. In addition, recurrent application of ATP, results in the attenuation of current responses [6]. P2XRs exhibit widespread expression patterns in neuronal and nonneuronal tissue, where they play important roles in mediating a diverse range of physiological functions [2, 7]. While there is evidence for localization in intracellular organelles such as lysosome, P2XRs are predominantly expressed within the plasma membranes or on the cell surface [8, 9]. A large body of evidence supports the involvement of P2XRs in a number of human pathologies and diseases [2, 7, 10], identifying them as attractive therapeutic targets for precision medicine [11].

The manual patch-clamp technique was developed as the benchmark to measure ionic currents flowing through open channel pores [12, 13], thereby enabling the investigation of the biophysical properties of ion channels, testing of therapeutic compounds, and examination of mechanisms of action. The manual patch-clamp recording technique has certain limitations, in that extensive technical training is required, and data acquisition is generally low, necessitating a large time investment. Automated technologies have been extensively used for pharmaceutical drug discovery for some time now and as research groups combine resources, automated patch-clamp systems are becoming a more commonplace feature in academic laboratories. Numerous first and second generation platforms have been manufactured over the last two decades, including QPatch HTX, QPatch II, Qube 384 (Sophion A/S, Copenhagen); PatchXpress® 7000A, IonWorks® Quattro, IonWorks Barracuda™ (Molecular Devices, LLC); NPC-16 Patchliner®, SyncroPatch® 96 SyncroPatch® 384PE (Nanion Technologies GmbH, Munich); CytoPatch™ (Cytocentrics AG, Rostock); Dyna flow® HT (Cellectricon AB, Mölndal); and IonFlux HT (Fluxion Bioscience Inc. USA). Data output from these platforms, is generally highly comparable to the high-fidelity data acquired using the manual patch-clamping technique [14]. Here we describe protocols using the first-generation platform Patchliner, as an example, to measure agonist-induced P2XR-mediated whole-cell currents from stably transfected or transiently transfected cells. For most automated patch-clamp systems on the market it is challenging to use transiently transfected cells, because of the blind approach of capturing cells. Interestingly, robust currents from HEK293 cells transiently expressing $P2X_4Rs$ were recorded using the Patchliner [15]. Patchliner has built-in robotic, microfluidic liquid handling capabilities [16], allowing for the complete application and washout of agonist in a millisecond time frame [17], making it an ideal platform for measuring rapidly activating, desensitizing, ligand-gated currents. It should, however, be noted that many of the features and notes given in this chapter can be applied to other automated systems, similarly.

2 Materials

2.1 Reagents and Cells

1. Human embryonic kidney (HEK) 293 host cells, HEK293 cells stably expressing the human $P2X_7$ receptor ($hP2X_7R$) and human astrocytoma cells (1321N1) stably expressing $P2X_2R$ and $P2X_3R$ subunits ($P2X_{2/3}R$).
2. HEK293 cell culture media: Dulbecco's Modified Eagle's Medium (DMEM), 2 mM L-glutamate, 10% fetal bovine serum (FBS).
3. DMEM/F-12 complete media: DMEM/F-12 liquid, 2 mM L-glutamate, 10% FBS.
4. 1321N1 cell culture media: DMEM, 10% FBS, 50 U/mL penicillin, and 50 μg/mL streptomycin.
5. D-PBS: Dulbecco's phosphate buffer saline Ca^{2+} and Mg^{2+} free.
6. Accutase® cell detachment solution (Invitrogen) or similar.
7. Opti-MEM I reduced serum media (Invitrogen) or similar.
8. Transfection reagent, Lipofectamine 2000 (Invitrogen) or similar.
9. DNA plasmids: empty plasmid vector for mammalian expression of GFP, full-length $P2X_4$-WT-GFP (WT), mutant $P2X_4$-G135S (G135S) constructs. DNA constructs were generated in-house.
10. 100 mM adenosine triphosphate (ATP) and its analogues (e.g., BzATP) in water as stock solutions. Adjust ATP stock solution to pH 7.3 with NaOH. Aliquot ATP and BzATP stock solutions in small volumes, and store at −20 °C. Prepare working solutions with desired concentrations by diluting the stock solution in extracellular recording solutions.

2.2 Equipment

1. Vented flasks for cell culture (T25). Conical centrifuge tubes (15 mL). Microcentrifuge tubes 1.5 mL.
2. Pipettes and tips 10-, 20-, 200-, and 1000 μL and Easypet pipetting aid and Maxitip tips 5 mL.
3. High-speed benchtop centrifuge.
4. Countess™ automated cell counter (ThermoFisher Scientific) or a hemocytometer and trypan blue.
5. NPC®-16 Patchliner Probe Selector/Quattro/Octo (*see* **Note 1**), PatchControl^{HT} software, single-hole (Fig. 1a), 4-hole ensemble (Fig. 1b) or 8-hole ensemble NPC®-16 chips (*see* **Note 2**). NPC®-16 electrode set (*see* **Note 3**) (Nanion Technologies GmbH).

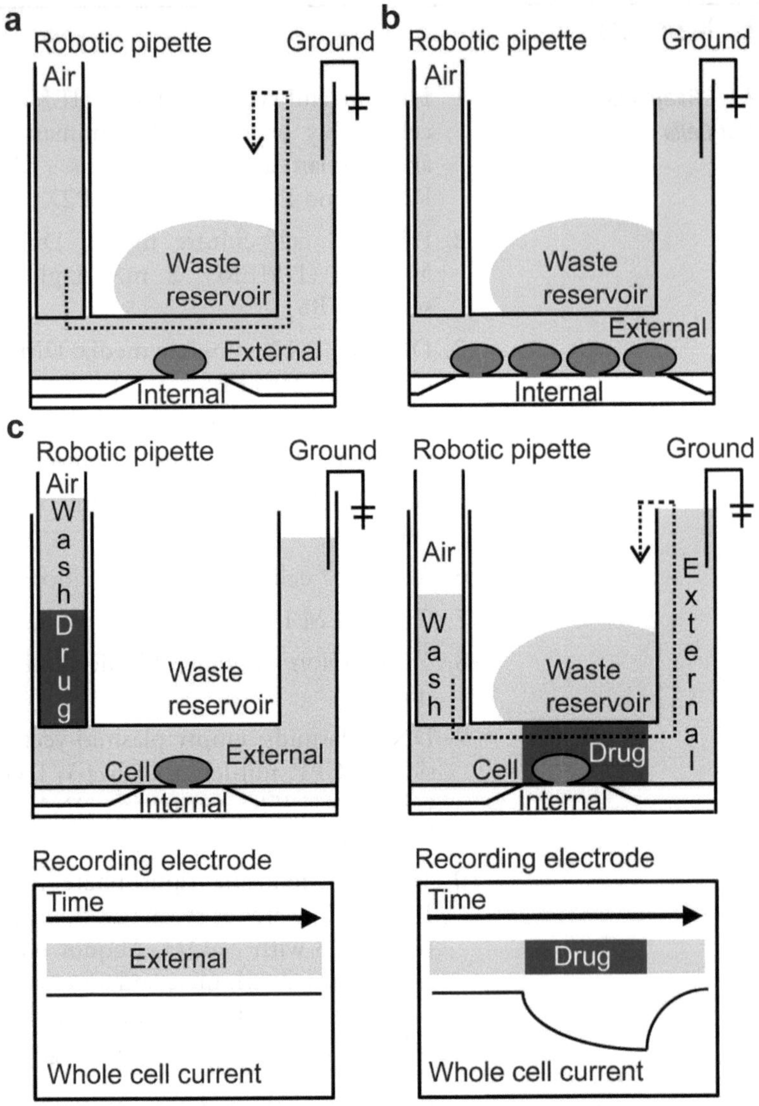

Fig. 1 Schematic of the cross section of chip chamber microdomains, illustrating single and double stacked solution application. (**a**) Example of a chip chamber with a cell sealed on the single chip aperture. The robotic pipetting arm dispenses solutions, cell suspensions, and compounds into the chip. With every addition to the external microchannel, the solution is completely replaced, overflowing into the waste reservoir. The waste reservoir is emptied throughout the experiment. (**b**) Cross section of a 4-hole ensemble chip, illustrating the arrangement of four cells simultaneously sealed onto four individual apertures in a single chip chamber. (**c**) Illustration of double stacked solution application. The robotic pipette aspirates external solution (Wash) followed by ligand of interest (Drug) before precisely timed application into the external microchannel. Output from the recording electrode showing the brevity of drug contact with the cell during a continuous recording (Modified from [17])

6. Patch-clamp amplifiers (PATCHMASTER, HEKA Instruments) (*see* **Note 4**).
7. Computer with 24-in. thin film transistor monitor (*see* **Note 5**).
8. Multichannel stimulation/acquisition software with programmable experiment control and automation (HEKA Instruments) and analysis packages (such as Microsoft Excel, MatLab R2018a), Igor Pro 6.37 (WaveMetrics Inc.), Adobe Illustrator CS5.1 (Adobe Systems) and GraphPad Prism 7 (Molecular Devices).
9. 50 mL syringe with 0.22-μm-pore diameter filter.
10. Bench top pH meter.
11. Advanced Instruments Osmol, single sample micro-osmometer.

2.3 Planar Patch-Clamp Solutions

Prepare solutions with deionized water, filter (*see* **Note 6**) and measure the osmolarity (*see* **Note 7**) and the pH (*see* **Note 8**). Solutions can be stored at 4 °C for up to 5 days. Solutions should be warmed to room temperature (20–22 °C) before use. The built-in robotic pipette manages all liquid handling requirements. Solutions are aspirated from the preprogrammed positions on the work-station and dispensed into the appropriate chip chamber. The pipette is capable of aspirating and dispensing a single solution (Fig. 1a, b), double stacked (Fig. 1c) or triple stacked solutions [17], the later allowing the application of a channel modulator prior to agonist.

1. Standard extracellular solution for recording with HEK293 cells stably expressing $hP2X_7Rs$: 147 mM NaCl, 2 mM KCl, 1 mM $MgCl_2$, 2 mM $CaCl_2$, 13 mM D-glucose, 10 mM HEPES (pH 7.3 with NaOH; ~298 mOsm).
2. Standard extracellular solution for recording with HEK293 cells transiently expressing $P2X_4Rs$: 145 mM NaCl, 2 mM KCl, 1 mM $MgCl_2$, 2 mM $CaCl_2$, 13 mM D-glucose, 10 mM HEPES (pH 7.4 with NaOH; ~298 mOsm).
3. Standard extracellular solution for recording with 1321N1 cells stably expressing $P2X_{2/3}Rs$: 140 mM NaCl, 4 mM KCl, 1 mM $MgCl_2$, 2 mM $CaCl_2$, 5 mM D-glucose, 10 mM HEPES (pH 7.4 with NaOH; ~298 mOsm).
4. Standard extracellular solution for enhancing seals: 80 mM NaCl, 3 mM KCl, 10 mM $MgCl_2$, 35 mM $CaCl_2$, 10 mM HEPES (pH 7.4 with HCl) (*see* **Note 9**).
5. Standard intracellular solution for whole-cell recordings with HEK293 cells stably expressing $hP2X_7Rs$: 85 mM NaCl, 60 mM NaF (*see* **Note 10**), 10 mM EGTA, 10 mM HEPES (pH 7.2 with NaOH; ~285 mOsm).

6. Standard intracellular solution for whole-cell recordings with HEK293 cells transiently expressing $P2X_4Rs$: 85 mM NaCl, 60 mM NaF, 10 mM EGTA, 10 mM HEPES (pH 7.2 with NaOH; ~285 mOsm).
7. Standard intracellular solution for whole-cell recordings with 1321N1 cells stably expressing $P2X_{2/3}Rs$: 110 mM KF, 10 mM KCl, 10 mM EGTA, 10 mM HEPES (pH 7.2 with KOH; ~285 mOsm).

3 Methods

3.1 Cell Culture and Transient Transfection

The following procedures should be conducted in a sterile tissue culture fume hood.

1. Culture HEK293 host cells in standard T25 tissue culture flasks in HEK293 cell culture media in a humid 37 °C, 5% CO_2 tissue culture incubator, until 60% confluent.
2. Culture HEK293 cells, stably expressing $hP2X_7R$, in standard T25 tissue culture flasks in DMEM/F-12 complete media and incubate in a humid 37 °C, 5% CO_2 tissue culture incubator, until subconfluent (70–80%).
3. Culture 1321N1 cells, stably expressing $P2X_{2/3}R$, in standard T25 tissue culture flasks in 1321N1 cell culture media at 37 °C in a humidified atmosphere composed of 95% air and 5% CO_2, until subconfluent (70–80%).
4. Cells should be passaged every 2–3 days using Accutase® cell detachment solution once they become subconfluent (70–80%) (*see* **Note 11**).
5. For each transfection, dilute 3 μg of selected cDNA:pAcGFP-N1 empty vector (Mock); $P2X_4$-WT-AcGFP wild-type (WT) or $P2X_4$-G135S (G135S) with Opti-MEM I reduced serum media (total volume 200 μL) in a 1.5 mL tube and in a second tube, add Lipofectamine 2000 Reagent (30 μL) with Opti-MEM I reduced serum media (170 μL), mix contents of each tube thoroughly by gentle pipetting.
6. Incubate both tubes at room temperature for 5 min.
7. Combine contents of the two tubes, mix thoroughly by gentle pipetting and incubate at room temperature for 20 min.
8. Add this transfection media to HEK293 host cells cultured to 60% confluency in a standard T25 flask and return to the incubator.
9. After 24 h replace the transfection media with normal culture media and return to the incubator for a further 24 h (*see* **Note 12**).

3.2 Harvesting Cells for Planar Patch-Clamp

A critical feature determining the success rate of planar patch-clamp recordings is that the healthy cells are maintained at the optimum subconfluency. The process of capturing cells is completely random, so unlike conventional patch-clamp recording, it is not possible to visually select the healthiest looking cell. In addition, the quality of the seal formed between the cell membrane and the planar chip, which ultimately influences the quality of the recording, is reliant on the health of the cell.

1. Discard the media from the culture flask and, using 5 mL D-PBS, gently wash the cells twice.
2. Discard the D-PBS and dissociate with 0.5 mL prewarmed Accutase cell detachment solution and gently tilt the flask from side to side to cover all the cells. Incubate for 3 min in a humid 37 °C, 5% CO_2 tissue culture incubator (*see* **Note 13**).
3. To neutralize the Accutase, add 5 mL of culture media and pipette up and down gently in order to lift and separate the cells.
4. Transfer the cell suspension to a 15 mL conical centrifuge tube and centrifuge at 180 × *g* for 2 min at room temperature and then discard the supernatant by decanting (*see* **Note 14**).
5. Resuspend the cell pellet by gently pipetting in a mixture of extracellular recording solution and culture media (50:50 ratio, *see* **Note 15**) at a density of 1×10^6–5×10^7 mL (*see* **Note 16**). Cells can be counted using a Countess automated cell counter, although a standard hemocytometer is quite adequate.
6. Transfer the cells, in suspension, to the cell hotel (*see* **Note 17**).

3.3 Automated Planar Patch-Clamp Using Patchliner

The Patchliner software (PatchControl^{HT}) is coupled to the HEKA amplifier software (Patchmaster), via the optical PCI card in the computer, and when PatchControl^{HT} is opened, Patchmaster opens automatically. This allows the experimental protocols (PatchControl^{HT} *Trees*), programmed in PatchControl^{HT}, to communicate with the amplifiers. Preprogrammed *Trees* can be loaded and modified for optimization with different cell types/characteristics. Selection of appropriate chip cartridges, with the resistance suitable for cells of a particular size, is important (*see* **Note 18**). The motorized stage on the workstation of the Patchliner ("chip-wagon") has the capacity to hold three chip cartridges and each cartridge allows data acquisition from sixteen cells (eight at any given time), hence allowing for 48 recordings without operator intervention. Chip cartridges are embedded with microfluidic chambers and when a PatchControl^{HT} *Tree* is activated, the pipetting robot dispenses appropriate recording solutions into the microfluidic chambers of the chip and the cartridge is moved into the measuring head, which contains the pneumatic and electric contacts and moves up and down to address the chip cartridges. The recording head houses eight headstages, allowing acquisition of data from eight cells simultaneously. Once the intracellular and extracellular solutions have been dispensed, a slight positive pressure is applied to each chip chamber, independently, and the offsets are corrected. Cells from the cell hotel are dispensed into the extracellular chamber and a small suction (−50 mBar) is applied, to attract cells onto each of the eight individual chip apertures, leading to a small increase in the seal resistance. Seal enhancing solution is then added and further suction pulses applied, together with application of negative voltage to aid in the formation of a gigaohm seals. The seal enhancing solution is then replaced with recording solution before additional short suction pulses are applied to achieve the whole-cell access. In some circumstances, it may be necessary to support this process using the zap function to encourage the patch of membrane to rupture (*see* **Note 19**). The pressure applied during this process is controlled by PatchControl^{HT} parameter settings (e.g., chip resistance, series resistance, and slow capacitance), that the user can adjust according to the cell type/characteristics. It is also possible to adjust quality control parameters (e.g., seal resistance, series resistance), so cells that do not meet the specifications are disabled at this stage. Once the whole-cell configuration has been established, the experimental part of the protocol will commence.

1. Load a preprogrammed PatchControl^{HT} ligand *Tree* (File → load → *Tree*) and select the edit mode (☒ Edit) to make modifications according to cell/channel/receptor type/characteristics, and experimental paradigm (*see* **Note 20**).
2. Select chips with the desired resistance and format for cells/receptors and place three chips onto the chip-wagon.
3. Prepare compound solutions directly before each experiment (*see* **Note 21**).
4. Place recording solutions and compounds in position, according to those defined in the joblist (*see* **Note 22**).
5. Place the cells into the cell hotel, where they will be aspirated every 30 s throughout the experiment to prevent clumping and sedimentation (*see* **Note 23**).
6. Select and activate the initialization folder to initialize the robot and wash the pipette. This also generates a new data file within Patchmaster and sets all amplifier and robot parameters to default starting values. This folder only needs to be activated once at the beginning of each day of experiments.
7. The robot will start when the *Tree* is activated. At the end of each run, it will loop back to the start and continue this process until all chips on the chip-wagon have been used.

Illustrative automated planar patch-clamp recordings from HEK293 cells stably expressing the WT hP2X$_7$R (Fig. 2). Cells

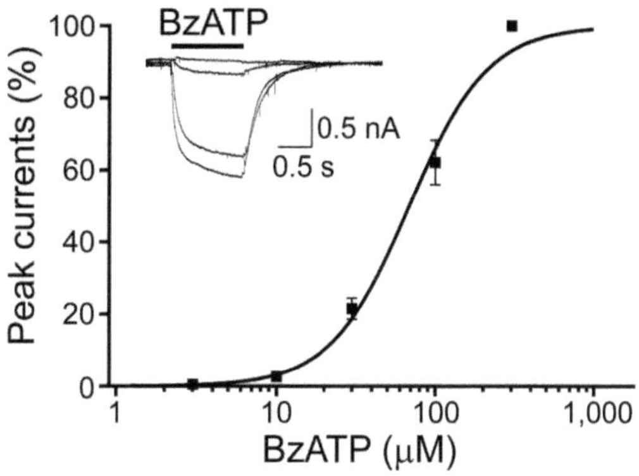

Fig. 2 BzATP concentration–response curve obtained from WT hP2X$_7$Rs stably expressed in HEK293 cells. Data are fit to the Hill equation with the following EC$_{50}$ value: 69.7 ± 7.5 μM and Hill coefficient: 2.3 ± 0.5 ($n = 8$ for each data point). Inset shows representative whole-cell currents evoked by BzATP (3–100 μM). Cells were voltage-clamped at a holding potential of −60 mV (Reproduced from [16])

were captured on single-hole chips and voltage-clamped at a holding potential of −60 mV while a continuous recording performed data acquisition. BzATP, applied at a rate of 86 μL/s (speed 15), activated inward currents in a concentration-dependent manner. No desensitization was observed in the continued presence of BzATP. These recordings did not utilize a stacked solution application, ligand was applied independently of the external solution.

3.4 Stacked Solution Application

P2XRs exhibit receptor desensitization [5], which is a common characteristic feature of ligand-gated ion channels. The kinetics and level of desensitization of ligand-gated ion channels are determined by ligand concentration and exposure time, or both. For rapidly desensitizing ion channels, it is important that compound application is rapid and short-lived, so that the entire ion channel population is exposed to maximum concentration before entering the desensitized state. Therefore, rapid solution exchange combined with brief drug exposure times can minimize or correct for the deleterious effect caused by receptor desensitization. This is achieved by using a stacked solution application, whereby two or three zones of solution are aspirated into the pipette before they are dispensed into the chamber, where they have brief and rapid contact with the cell/s (Fig. 1, *see* **Note 24**).

1. Load a preprogrammed ligand *Tree* for double stacked solution application. Volumes and speeds of applications can be adjusted (*see* **Note 25**). In addition, if required, temperature can also be controlled (*see* **Note 26**).
2. Select medium resistance chips (2–4 MΩ). For HEK293 stably expressing P2X$_4$Rs, single-hole chips were utilized and for

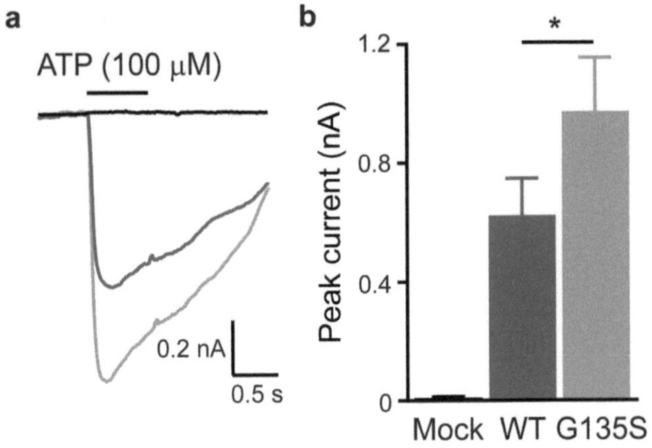

Fig. 3 $P2X_4R$ functional assay using HEK293 cells transiently transfected with AcGFP, WT $P2X_4$-AcGFP, or $P2X_4$-G135S-AcGFP. (**a**) Representative whole-cell current traces from WT $P2X_4$ (dark gray) and G135S mutant $P2X_4$ (light gray) in response to 100 μM ATP as indicated by the bar above the current traces. No currents were elicited for cells transfected with AcGFP empty vector alone (Mock; black). (**b**) Mean peak current amplitude for Mock (black; $n = 6$), WT $P2X_4$ (dark grey; $n = 6$ cells) and G135S $P2X_4$ (light grey; $n = 10$ cells) in response to ATP (100 μM). A volume of 60 μL ATP was applied at a rate 86 μL/s; (speed 15), followed immediately by wash (120 μL) at a rate of 19 μL/s (speed 24), using a double stacked protocol. Peak currents were compared and statistical significance marked as $^{*}p < 0.01$ (Reproduced from [15] with permission from Wiley)

1321N1 cells stably expressing $P2X_{2/3}$Rs 4- or 8-hole ensemble chips were selected. Load three chips onto the chip-wagon.

3. Within the *Tree*, adjust the holding potential and the required speed of drug/wash delivery. The joblist will contain a continuous recording protocol, for fast ligand activated currents. The duration of the continuous recording can be modified in the Patchmaster pulse generator file.
4. Follow **steps 3–6** in Subheading 3.3.

In the example shown in Fig. 3a, HEK293 cells transiently expressing AcGFP vector, WT $P2X_4$-AcGFP, or $P2X_4$-G135S-AcGFP were voltage-clamped at a holding potential of −50 mV, utilizing single-hole chips. 100 μM ATP applied for 700 ms, rapidly activated robust inward current for WT and mutant $hP2X_4R$, which desensitized slowly in the presence of ATP. No currents were elicited for cells transfected with mock vector control alone, which is reflected in the mean peak current amplitudes (Fig. 3b).

Figure 4 shows the reproducibility of responses to repetitive activation of robust inward currents, from 1321N1 cells stably expressing the $P2X_{2/3}R$, in response to 30 μM ATP applied using

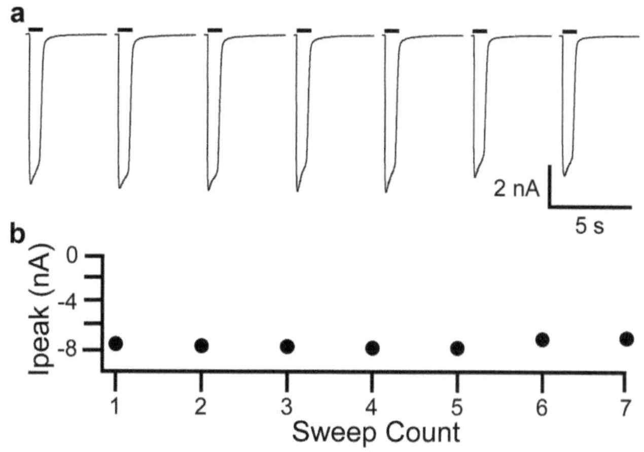

Fig. 4 Repetitive activation of $P2X_{2/3}R$. (**a**) Current traces induced, in a single 1321N1 cell, by repetitive application of ATP (30 μM) using a double stacked solution application protocol. Black bars above the current traces illustrate ATP (30 μM) contact time. (**b**) Timecourse of the experiment showing 7 reproducible inward currents with consistent current amplitudes of approximately 8 nA

a stacked application protocol. The time course demonstrates, very elegantly, that there is little effect on the peak current amplitude in response to brief repetitive application of agonist.

Representative current responses of an individual cell expressing $P2X_{2/3}Rs$ to increasing concentrations of ATP are shown in Fig. 5a. The currents exhibit a slow desensitization phase in the continued, but brief, presence of 30 μM ATP. The concentration response curve revealed an EC_{50} for ATP activation of 7.8 ± 1.0 μM (Fig. 5b). $P2X_{2/3}Rs$ could be repetitively activated by 30 μM ATP and blocked by suramin (Fig. 5c) in a dose-dependent manner. A full concentration response curve to suramin was performed, generating an IC_{50} of 28.0 ± 5.3 μM (Fig. 5d). To achieve short exposure times, solutions were stacked in the robotic pipette. First, wash solution (155 μL) was aspirated, followed by aspiration of the agonist-containing solution (40 μL) and then application to the cell at a speed of 57 μL/s. The cells were pre-incubated with suramin before coapplication with 30 μM ATP.

4 Notes

1. NPC®-16 Patchliner Quattro or Octo are being used here, as an example of a validated automated patch-clamp system, but it should be noted that other systems have comparable capabilities.
2. NPC®-16 chips (single use, disposable) are manufactured with different specifications. In general, chips are manufactured with

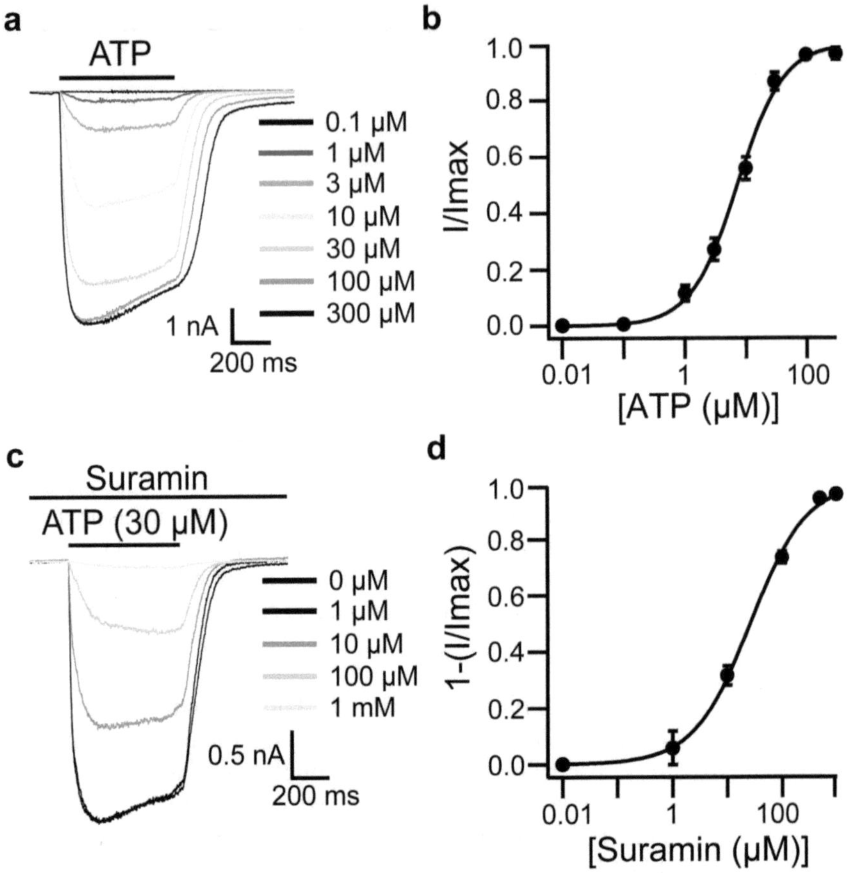

Fig. 5 Effects of ATP and suramin on $P2X_{2/3}Rs$ stably expressed in 1321N1 cells. (**a**) Representative current traces showing activation of $P2X_{2/3}Rs$ by increasing concentrations of ATP (0.1–300 μM). (**b**) Mean dose–response curve for ATP activation. Data are fit to the Hill equation with the following EC_{50} value: 7.8 ± 1.0 μM and Hill coefficient: 1.2 ± 0.065 ($n = 10$ for each data point). (**c**) Representative traces showing the concentration-dependent block of ATP-induced $P2X_{2/3}Rs$ currents by suramin (1 μM–1 mM). Suramin at increasing concentrations was preincubated and then coapplied with 30 μM ATP. (**d**) Concentration–response curve for suramin block. Data are fit to the Hill equation with the following EC_{50} value: 28.0 ± 5.3 and Hill coefficient: 0.82 ± 0.063 ($n = 7$ for each data point)

three different megaohm (MΩ) resistance ranges: low (1–2 MΩ), medium (2–4 MΩ) and high (5–6 MΩ). It is also possible to have bespoke chips manufactured according to the specific requirements of the user and/or cell type. Single-hole chips are ideal for use with voltage-gated channels. For ligand-gated channels, ensemble chips are generally preferred, especially if currents are small in amplitude. Ensemble chips are manufactured with either four or eight holes per chip, which is ideal for enhancing current size, because the currents are summated.

3. Electrodes need to be chloridated in bleach filled chambers for 30–60 min and then rinsed with deionized water and air dried before use.

4. Patchliner Quattro and Octo, use either EPC-10 USB Quadro multi-headstage patch-clamp amplifiers or the EPS 16 Probe Selector headstage multiplexer, combined with an EPC 10 Plus amplifier (HEKA Instruments).
5. Any brand of computer can be purchased, but minimum specifications include Windows 10 (Microsoft) with a 64-bit operating system and an optical PCI card for communication with the HEKA amplifiers.
6. Sterile filter all recording solutions using a 50 mL syringe with a 0.22-μm-pore diameter filter attached. This is particularly important for the internal recording solution.
7. Measure the osmolarity of all recording solutions using a freezing-point osmometer. The internal solutions should measure ~285 mOsm/L and the external solutions should measure ~298 mOsm/L. The osmolarity of external solutions should always be higher than the osmolarity of the internal solution.
8. When pH adjusted stock solutions (1 M), of each salt, are used to make the final recording solutions, then the osmolarity should not require further adjustment.
9. A high Ca^{2+}-containing external solution helps in the formation of a strong seal between the cell membrane and the planar chip. This seal enhancing solution is replaced once a gigaohm seal is achieved and before establishing whole-cell access. Calcium in the extracellular solution can be replaced by barium as the charge carrier, to avoid calcium-dependent inactivation of voltage-gated calcium channels and increase calcium channel currents, since most voltage-gated calcium channels also conduct barium ions [18, 19].
10. Fluoride ions in the intracellular solution improve gigaohm seal formation and stabilize the cell membrane, which in turn results in longer, more stable recordings [20]. The mechanism of this effect is poorly understood.
11. To avoid cells growing in clusters or adhering too tightly to the support substrate, it is vital that they are passaged every 2–3 days. If cells are allowed to grow to a confluency greater than 80%, aggregates commonly form, which in turn lead to lower capture rates, poor seal formation and difficulties breaking into the whole-cell configuration.
12. A Lipofectamine 2000 Reagent protocol, available at (https://www.ecu.edu/cs-dhs/biochemistry/upload/Transfection-Protocol.pdf), was used to transiently transfect the cells. It should be noted that other transfection kits are also suitable for transient expression of ion channel proteins.
13. Accutase® cell detachment solution gently dissociates mammalian cells from support substrates and from each other. It

should be stored at 4 °C and used at room temperature in sterile conditions.

14. To prevent shearing of the cell membranes during centrifugation, it is recommended to set the centrifuge acceleration speed to 6 and deceleration speed to 3. The value 0 is equivalent to the lowest acceleration and the value 9 is equivalent to the highest acceleration. The value 9 is equivalent to the shortest possible brake time and the value 0 to longest possible brake time.
15. When cells are resuspended in a mixture of recording solution and culture media (50:50), their bench life is greatly improved, and they remain viable for up to 4 h at room temperature.
16. A standard density of 1×10^6–5×10^7 cells/mL, for most cell types, works well for use on the Patchliner. When working with primary cells, it is often challenging to harvest such large numbers of cells. It is possible to maintain a good cell capture rate with as few as 1000 cells/mL [21].
17. Cells are housed in the "cell hotel," where they are kept from clumping and sedimenting by gentle automated pipetting, which improves cell viability. The user can set the pipetting volume and speed.
18. The capacitance of the cell, determines the size of the aperture (chip resistance) required. As an example, medium resistance chip are ideal for use with HEK293 cells and CHO cells [16, 22] and high resistance chips are ideal for use with mouse osteoblasts [23]. For example, if cells are very small, they will pass through a low resistance chip hole when suction is applied, and conversely, large cells will not form good seals on high resistance chips.
19. If it is difficult to establish the whole-cell configuration, following seal formation, then try harvesting the cells one day after plating. In addition, adjust the size of the high voltage pulses to 600–800 mV ("zap"), which can be applied to help rupture the patch of membrane, thus establishing the whole-cell configuration.
20. PatchControl^{HT} comes with a range of preprogrammed protocols ("*Trees*"), which have numerous experimental paradigms (e.g., Ligand, IV, Pharm), for specific cell characteristics (e.g., cell capacitance, membrane fragility) and channel types (e.g., P2XR, Na^+ or K^+ channels). Features of a *Tree* can be adjusted, enabling the user to modify amplifier and suction parameters and also allows access to commands that can be inserted into a *Tree* via a generic drag-and-drop function. In addition, once a *Tree* has been optimized for a particular cell type, it will not normally require modification for future use with the same cell type. Application notes are available at http://www.nanion.de.

21. Prepare fresh compound solutions, directly prior to use, to avoid precipitation. Compound solutions, where possible should be stored in glassware, because some compounds will adhere to other substrates. An underestimation of concentration may result from adhesion or precipitation of a compound solution.
22. The joblist, within the *Tree*, defines the position of the compounds, the volume and speed of application and selection of the appropriate pulse generator file in Patchmaster.
23. Regular aspiration of cells retains viability for at least 4 h after they have been prepared in suspension, although some deterioration in success rates has been observed after 3 h.
24. For a double stack with two zones of solution, the wash buffer is aspirated first, directly followed by the ligand resulting in the ligand zone being applied to the cells first, followed immediately by the wash buffer (Fig. 1c). To examine the effects of a channel modulator, a triple stack with three solution zones should be selected [17]. The wash buffer is aspirated first, followed directly by solution containing both ligand and modulator, followed immediately by modulator.
25. The volumes of the different solution zones can be adjusted, as well as the speed of application, enabling exposure times of as little as 100 ms. Typical volumes for a double stack are 200 μL wash buffer zone, 10–60 μL ligand zone, but these can be optimized for different ligands, accordingly. The maximum volume that the pipette can aspirate at one time is 350 μL. The zone containing the ligand should be applied more rapidly (e.g., 171 μL/s; speed 12) than the other zones (e.g., 19 μL/s; speed 24). In addition, it is possible to set the volumes to be dispensed at different rates and these do not need to correspond the volume of the two phases (e.g., "Vol fast" 120 μL (speed 12) and "Vol slow" 140 μL (speed 24).
26. The recording chamber, chip-wagon, and pipette can be heated simultaneously or independently. These modules can be heated to 80 °C; however, it is not recommended to increase the temperature above 55 °C. The temperature of the cells and the solutions can also be cooled via an add-on cooling plate. Cooling the cells and compounds improves viability and stability.

Acknowledgments

Many thanks to Dr. Sonja Stoelzle and Dr. Alison Obergrussberger (Nanion Technologies GmbH) for $P2X_{2/3}R$ data.

References

1. Khakh BS, North RA (2006) P2X receptors as cell-surface ATP sensors in health and disease. Nature 442:527–532
2. North RA (2002) Molecular physiology of P2X receptors. Physiol Rev 82:1013–1067
3. Browne LE, Jiang LH, North RA (2010) New structure enlivens interest in P2X receptors. Trends Pharmacol Sci 31:229–237
4. Bo X, Jiang LH, Wilson HL, Kim M, Burnstock G, Surprenant A, North RA (2003) Pharmacological and biophysical properties of the human $P2X_5$ receptor. Mol Pharmacol 63:1407–1416
5. Coddou C, Yan Z, Obsil T, Huidobro-Toro JP, Stojilkovic SS (2011) Activation and regulation of purinergic P2X. receptor channels. Pharmacol Rev 63:641–683
6. Valera S, Hussy N, Evans RJ, Adami N, North RA, Surprenant A, Buell G (1994) A new class of ligand-gated ion channel defined by P_{2X} receptor for extracellular ATP. Nature 371:516–519
7. Jiang LH (2012) P2X receptor-mediated ATP purinergic signaling in health and disease. Cell Health Cytoskel 4:83–101
8. Robinson LE, Murrell-Lagnado RD (2013) The trafficking and targeting of P2X receptors. Front Cell Neurosci 7:1–6
9. Murrell-Lagnado RD (2018) A role for $P2X_4$ receptors in lysosome function. J Gen Physiol 150:185–187
10. Wei L, Mortadza SAS, Yan J, Zhang L, Wang L, Yin Y, Li C, Chalon S, Emond P, Belzung C, Li D, Lu C, Roger S, Jiang LH (2018) ATP-activated P2X7 receptor in the pathophysiology of mood disorders and as an emerging target for the development of novel antidepressant therapeutics. Neurosci Biobehav Rev 87:192–205
11. North RA, Jarvis MF (2013) P2X receptors as drug targets. Mol Pharmacol 83:759–769
12. Neher E, Sakmann B, Steinbach JH (1978) The extracellular patch clamp: a method for resolving currents through individual open channels in biological membranes. Eur J Phys 375:219–228
13. Hamill O, Marty A, Neher E, Sakmann B, Sigworth F (1981) Improved patch-clamp techniques for high-resolution current recording from cells and cell-free membrane patches. Eur J Phys 391:85–100
14. Bell DC, Dallas ML (2018) Using automated patch clamp electrophysiology platforms in pain-related ion channel research: insights from industry and academia. Br J Pharmacol 175:2312–2321
15. Sadovnick AD, Gu BJ, Traboulsee AL, Bernales CQ, Encarnacion M, Yee IM, Criscuoli MG, Huan X, Ou A, Milligan CJ, Petrou S, Wiley JS, Vilarino-Guell C (2017) Purinergic receptors *P2RX4* and *P2RX7* in familial multiple sclerosis. Hum Mutat 38:736–744
16. Milligan CJ, Li J, Sukumar P, Majeed Y, Dallas ML, English A, Emery P, Porter KE, Smith AM, McFadzean I, Beccano-Kelly D, Bahnasi Y, Cheong A, Naylor J, Zeng F, Liu X, Gamper N, Jiang L-H, Pearson HA, Peers C, Robertson B, Beech DJ (2009) Robotic multiwell planar patch-clamp for native and primary mammalian cells. Nat Protoc 4:244–255
17. Milligan CJ, Möller C (2013) Automated planar patch-clamp. Methods Mol Biol 998:171–187
18. Veselovskii NS, Fedulova SA (1986) Effects of substituting barium for calcium ions during research into inward currents in mammalian neurons. Neurophysiology 18:227–231
19. Ferreira G, Yi J, Rios E, Shirokov R (1997) Ion-dependent inactivation of barium current through L-type calcium channels. J Gen Physiol 109:449–461
20. Kostyuk PG, Krishtal OA, Pidoplichko VI (1975) Effect of internal fluoride and phosphate on membrane currents during intracellular dialysis of nerve cells. Nature 23:691–693
21. Becker N, Stoelzle S, Göpel S, Guinot D, Mumm P, Haarmann C, Malan D, Bohlen H, Kossolov E, Kettenhofen R, George M, Fertig N, Brüggemann A (2013) Minimized cell usage for stem cell-derived and primary cells on an automated patch clamp system. J Pharmacol Toxicol Methods 68:82–87
22. Richards K, Milligan CJ, Richardson RJ, Jancovski N, Grunnet M, Jacobson LH, Undheim EAB, Mobli M, Chow CY, Herzig V, Csoti A, Panyi G, Reid CA, King GF, Petrou S (2018) A selective $Na_V1.1$ activator rescues Dravet Syndrome mice from seizures and premature death. Proc Natl Acad Sci U S A 115:E8077–E8085
23. Petty SJ, Milligan CJ, Todaro M, Richards KL, Kularathna PK, Pagel CN, French CR, Hill-Yardin EL, O'Brien TJ, Wark JD, Mackie EJ, Petrou S (2016) The antiepileptic medications carbamazepine and phenytoin inhibit native sodium currents in murine osteoblasts. Epilepsia 57:1398–1405

Chapter 22

Controlling Engineered P2X Receptors with Light

Benjamin N. Atkinson, Vijay Chudasama, and Liam E. Browne

Abstract

This chapter details methods to express and modify ATP-gated P2X receptor channels so that they can be controlled using light. Following expression in cells, a photoswitchable tool compound can be used to covalently modify mutant P2X receptors, as previously demonstrated for homomeric P2X2 and P2X3 receptors, and heteromeric P2X2/3 receptors. Engineered P2X receptors can be rapidly and reversibly opened and closed by different wavelengths of light. Light-activated P2X receptors can be mutated further to impart ATP-insensitivity if required. This method offers control of specific P2X receptor channels with high spatiotemporal precision to study their roles in physiology and pathophysiology.

Key words P2X, ATP, Ion channel, Synthetic optogenetics, Optochemicals, Photoswitchable, Azobenzene

1 Introduction

P2X receptors are trimeric transmembrane proteins that function as ATP-gated ion channels [1]. They have been implicated in afferent sensation, neuroeffector transmission, central control of respiration, and inflammation [2]. However, many of their roles in physiology and pathophysiology are not precisely understood, partly due to a lack of available tools to modulate their activity specifically. Optogenetic approaches employ light for the manipulation of genetically encoded light-sensitive proteins to study the function of molecules, synapses, cells, and systems [3, 4]. Using light for the remote control of biological processes provides high spatiotemporal precision.

P2X receptors can be engineered to impart optical control (Fig. 1) as previously demonstrated for homomeric P2X2 receptors and P2X3 receptors, and heteromeric P2X2/3 receptors [5]. A single cysteine mutation is introduced in each P2X receptor subunit to provide a new functional handle within the protein channel. A photoswitchable tool compound, 4,4′-bis(maleimido)azobenzene (BMA), is then used to bridge cysteines between two of the

Pablo Pelegrín (ed.), *Purinergic Signaling: Methods and Protocols*, Methods in Molecular Biology, vol. 2041,
https://doi.org/10.1007/978-1-4939-9717-6_22, © Springer Science+Business Media, LLC, part of Springer Nature 2020

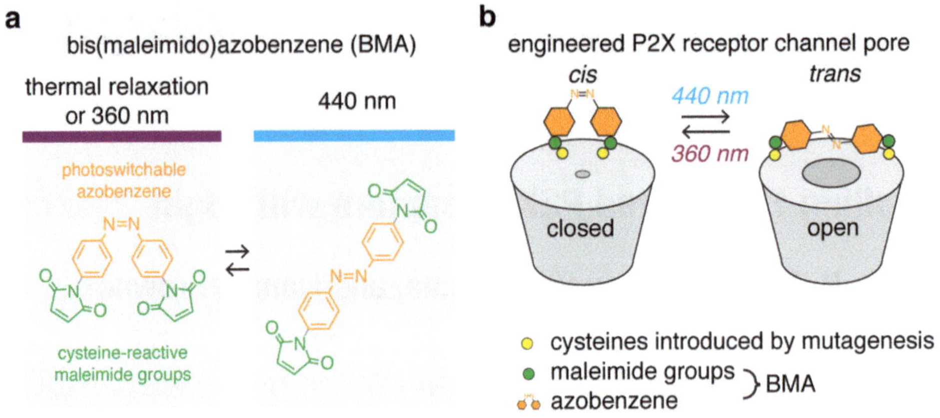

Fig. 1 (**a**) Light activation of engineered P2X receptors, for high spatiotemporal control of their activity in cells. Photoswitching is carried out using bis(maleimido)azobenzene (BMA), a compound that contains two cysteine-reactive maleimide groups bridged by a photoisomerizable azobenzene. BMA is in a *cis* state in the dark or under near-UV light (360 nm) and rapidly and reversibly converts to an extended *trans* state under blue light (440 nm). (**b**) BMA covalently attaches to introduced cysteines at the pore of trimeric P2X receptor channels. The bridged P2X receptor subunits are pushed apart as BMA enters a *trans* state (440 nm light), leading to channel opening. The channel can be closed by returning BMA to a *cis* state (360 nm light)

mutated P2X receptor subunits via maleimide moieties at either end of the compound (Fig. 1a). This covalent attachment does not affect the P2X receptor function or its sensitivity to ATP but crucially allows for the channel to be opened and closed with different wavelengths of light [5]. Light at a wavelength of 440 nm isomerizes the azobenzene at the center of BMA molecule from *cis* to *trans*. The increased distance between the maleimide groups attached to the pore push the channel open (Fig. 1b). The location of the cysteine (e.g., rP2X2[P329C]) is crucial for effective channel opening. As thermal relaxation back to the *cis* state is relatively slow the channel is essential bistable when channel desensitization is limited [5]. Illumination with 360 nm light reverses isomerization of BMA, shortening the length of the BMA modification and thus facilitating the channel to close. Here we describe the methods (Fig. 2) for transient expression of mutant P2X receptors in HEK293T cells and protocols for chemical modification and light switching of engineered P2X receptors during patch clamp recordings.

2 Materials

2.1 Cell Culture and Transient Transfection

1. Cell culture medium: Dulbecco's Modified Eagle's Medium (DMEM) containing 2 mM L-glutamine and 10% heat-inactivated fetal bovine serum (*see* **Note 1**). Aliquot to 50 mL and store at 4 °C.

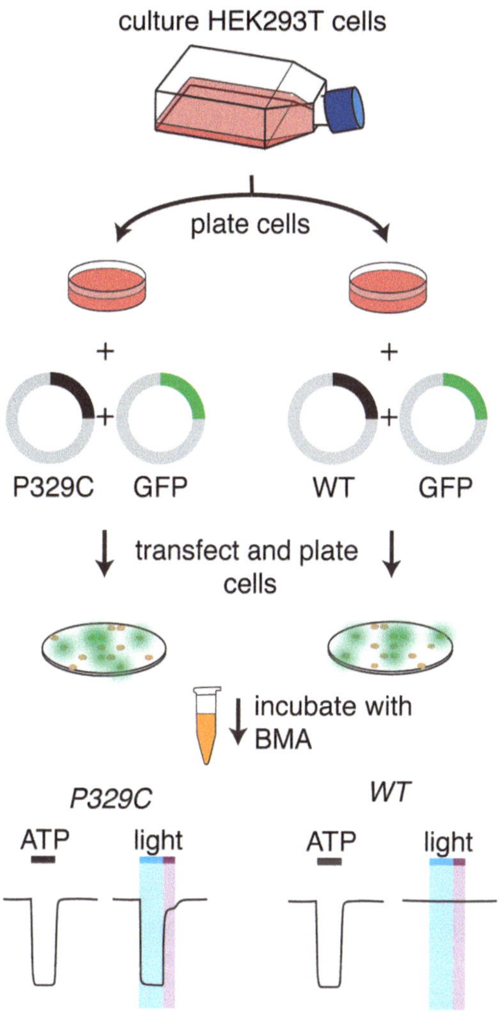

Fig. 2 Simplified workflow for the generation of engineered P2X receptors that can be rapidly and reversibly opened and closed with light. HEK293T cells are cultured, plated for transient transfection with a P2X receptor plasmid DNA (P329C and WT as an example) and fluorescent reporter plasmid DNA (GFP as an example), and plated for experiments. Fluorescent reporter-positive cells are identified, incubated with BMA, and recorded using whole-cell patch clamp electrophysiology. Engineered P2X receptor channels respond to ATP and can be rapidly opened and closed with two different wavelengths of light

2. Reduced serum medium: Opti-MEM, for example. Aliquot to 5 mL and store at 4 °C.
3. Dulbecco's phosphate-buffer solution (DPBS).
4. Trypsin–EDTA: 0.05% solution in 5 mL aliquots and store at −20 °C.
5. Lipofectamine 2000 transfection reagent or similar transfection reagent. Store at 4 °C.

6. Mammalian DNA expression plasmids encoding wild-type and mutant P2X receptors (e.g., 0.1 μg/μL rP2X2[WT] and 0.1 μg/μL rP2X2[P329C]), and a fluorescent reporter (e.g., 0.05 μg/μL GFP). Plasmids for expression of engineered homomeric P2X2 and P2X3 receptors, and heteromeric P2X2/3 receptors are available [5], including some that are further modified for ATP insensitivity. Store at −20 °C.
7. Round borosilicate glass coverslips, 12 mm diameter (*see* **Note 2**).
8. HEK293T cells.
9. 25 cm^2 cell culture flasks.
10. Cell culture incubator with temperature, CO_2 and humidity control.

2.2 Light Switching

1. Extracellular solution: 147 mM NaCl, 2 mM KCl, 2 mM $CaCl_2$, 1 mM $MgCl_2$, 10 mM HEPES, 13 mM D-glucose, pH 7.3. To 800 mL ultrapure water (*see* **Note 3**), add 8.59 g NaCl, 0.149 g KCl, 2.38 g HEPES, and 2.34 g glucose, and dissolve at room temperature with stirring (*see* **Note 4**). To this, add 2 mL of 1 M $CaCl_2$ solution and 1 mL of 1 M $MgCl_2$ solution. Adjust to pH 7.28–7.33 with dropwise addition of 5 M NaOH solution with stirring. Make this solution up to 1000 mL in a volumetric flask using ultrapure water. Store at 4 °C and use within 7 days.
2. Adenosine 5′-triphosphate (ATP) solution: 100 mM ATP, pH 7.3 in extracellular solution. Weigh 1 g adenosine 5-′-triphosphate disodium salt hydrate and add extracellular solution filtered through a 200 μm membrane filter to yield 200 mM ATP, considering the hydrate number of the batch. Mix gently until fully dissolved (*see* **Note 4**), adjusting to pH 7.28–7.33 with dropwise addition of 5 M NaOH solution. Make to 100 mM using extracellular solution. Aliquot to 5 mL and store at −20 °C.
3. 30 μM ATP working solution: 3 μL of the ATP solution and 10 mL extracellular solution.
4. 10 mM BMA stock solution: 10 mM solution of (*E*)-4,4′-bis (maleimido)azobenzene (BMA) in dimethyl sulfoxide (DMSO). BMA can be synthesized according to the published procedure [5] (*see* **Note 5**) or commercial sources are also available (*see* **Note 6**). To make the stock solution, add 300 μL DMSO to solid BMA to yield a 10 mM orange solution with vortexing. Store at room temperature in the dark. Dispose of after 12 h.
5. 1 mM BMA: Dilute 10× the 10 mM BMA stock solution in extracellular solution.

6. 10 μM BMA working solution: 120 μL of 1 mM BMA solution into 11.88 mL of extracellular solution in a Falcon tube.
7. Illumination system. A monochromator or multiple LEDs can be used to provide rapid switching (<3 ms) of narrow bandpass (14 nm full width at half maximum) blue light (440 nm, 7.5 mW/mm^2) and near-ultraviolet light (360 nm, 2 mW/mm^2). This can be applied through an optical fibre or a microscope fluorescence objective above the patched cell. The light source(s) should be computer-controlled to allow illumination to be timed accurately.

2.3 Electrophysiological Recordings

1. Intracellular solution: 147 mM NaCl, 10 mM EGTA, 10 mM HEPES, pH 7.3. To 80 mL ultrapure water (*see* **Note 3**), add 0.859 g NaCl, 0.380 g EGTA, and 0.238 g HEPES, and dissolve at room temperature with stirring (*see* **Note 4**). Adjust to pH 7.28–7.33 with dropwise addition of 5 M NaOH solution with stirring. Make this solution up to 100 mL in a volumetric flask using ultrapure water. Filter through a 200 μm membrane filter, aliquot to 5 mL and store at −20 °C.
2. Patch-pipettes. Use a pipette puller (e.g., Sutter P-97 or P-2000), to pull thin-wall borosilicate capillary glass (outer diameter = 1.50 mm, inner diameter = 1.17 mm, length = 10 cm) to give a 3–5 mm taper and 2–4 MΩ tip resistances when filled with intracellular solution.

3 Methods

All cell culture solutions are warmed to 37 °C in a water bath at least 20 min before use unless otherwise stated and manipulations are carried out under sterile conditions in a laminar flow hood. Light switching of P2X receptors does not require electrophysiological recordings, these can be considered optional; however, this method includes the details to carry out such experiments.

3.1 Cell Subculture

1. HEK293T cells are cultured in 25 cm^2 flasks containing 5 mL cell culture medium in an incubator at 37 °C with 5% CO_2. HEK293T cells that reach 90% confluence should be subcultured and plated for DNA plasmid transfection when required for the experiment.
2. Remove and discard culture medium from the 25 cm^2 flask.
3. Briefly rinse cells by carefully applying 1 mL warm DPBS along the side of the flask, minimizing cell disturbance, and gently move the solution across the cell monolayer before removing and discarding the DPBS (*see* **Note 7**).

4. Add 1 mL warm trypsin–EDTA solution to the cell monolayer, gently move the solution across the cell monolayer and incubate for 2 min at 37 °C.
5. Confirm that the majority of cells have detached from the flask and add 5 mL of warm cell culture medium using a 10 mL Stripette. Gently pipette up and down to release any remaining cells while being careful not to introduce air bubbles (*see* **Note 8**).
6. Transfer the HEK293T cell suspension to a 15 mL Falcon tube and centrifuge at room temperature at 90 × *g* for 5 min.
7. Meanwhile, add 5 mL cell culture medium into two new 25 cm^2 flasks and add 1.8 mL cell culture medium into two 35 mm petri dishes placed inside of a 100 mm petri dish.
8. Take the centrifuged cells and remove the medium so as not to dislodge the pellet at the bottom.
9. Add 5 mL of warm cell culture medium to the cell pellet using a 10 mL Stripette and gently pipette up and down (~20–25 times) to a homogeneous cell suspension, minimizing air bubble formation.
10. Transfer 0.3 mL of the cell suspension into each of the 25 cm^2 flasks containing the previously added 5 mL and ensure coverage by slowly moving the flask in a figure of eight motion (*see* **Note 9**).
11. To prepare cells for next-day transfection, transfer 0.4 mL of the cell suspension into each of the two 35 mm petri dishes (*see* **Note 10**).

3.2 Transient Transfection and Plating of Cells

1. Visually assess the preplated cells in 35 mm petri dishes. The cells should be 60% confluent and healthy (*see* **Note 11**). Return the petri dishes to the incubator to await transfection.
2. Allow the wild-type and mutant P2X receptor plasmids, fluorescent reporter plasmid, and Lipofectamine 2000 to reach room temperature. Briefly vortex and minicentrifuge to draw liquid to the bottom of each tube.
3. Add 0.2 mL warm reduced serum medium into a sterile 1.5 mL tube and then 3 μL of Lipofectamine 2000. Mix gently and incubate at room temperature for 5 min.
4. Meanwhile, add 0.1 mL of reduced serum medium into two additional sterile 1.5 mL tubes. Each tube will contain fluorescent reporter plasmid and one of the two P2X receptor plasmids. Label accordingly.
5. Add 1 μL of the fluorescent reporter plasmid along with either 1 μL wild-type P2X receptor plasmid, or 1 μL mutant P2X receptor plasmid, to one of the two tubes.

6. Remove the media from preplated cells. Briefly rinse in warm DPBS before immediately replacing with 800 μL warm reduced serum medium.
7. After the 5 min incubation is complete, transfer 0.1 mL of the Lipofectamine 2000 mix to both of the tubes containing reduced serum medium and plasmids, and incubate for 20 min at room temperature.
8. Into each 35 mm petri dish containing preplated cells carefully add the contents of one of the Lipofectamine-DNA tubes. Slowly move the dishes in a figure of eight motion.
9. Incubate for 6 h at 37 °C with 5% CO_2, and transfer five prewashed glass coverslips (*see* **Note 2**) into new 35 mm petri dishes (*see* **Note 12**).
10. Add 1.8 mL of warm cell culture medium into each of the 35 mm petri dishes containing coverslips. Ensure the coverslips are fully covered by pushing them to the bottom of the dish with a sterile pipette tip.
11. Incubate for 6 h at 37 °C with 5% CO_2 (*see* **step 8**), then remove the all the reduced serum medium from the transfected cells in petri dishes.
12. Immediately rinse with 175 μL of warm DPBS being careful not to detach cells.
13. Remove the DPBS and carefully add 175 μL of trypsin–EDTA solution, ensuring complete coverage of the cells. Incubate at 37 °C for 2 min.
14. After 2 min, check that the majority of cells have detached and add 1.8 mL of cell culture medium and gently pipette up and down >20 times to ensure an even suspension in the medium (*see* **Note 8**), avoiding the creation of air bubbles.
15. Carefully add the cell suspension dropwise over each coverslip in the petri dishes. For whole-cell patch clamp, we use cells at approximately 10–30% confluence on the day of the recordings so that cells are not attached to one another (*see* **Note 13**).

3.3 Light Switching and Electrophysiology

1. Allow light sources and electrophysiology equipment to warm up.
2. Allow the extracellular solution, intracellular solution, and ATP solution to reach room temperature.
3. Meanwhile, pull glass electrodes required for the day.
4. Transfer 1 mL intracellular solution to a 2 mL syringe containing a 200 μm membrane filter and MicroFil.
5. Make the 30 μM ATP working solution, vortex, and store on ice until required.

6. Prepare the perfusion system, setting up and testing separate flows of extracellular solution and 30 μM ATP working solution.
7. Make the 1 mM BMA stock solution and immediately vortex. Keep at room temperature and protect from light.
8. Immediately before the experiment, make the 10 μM BMA working solution and immediately vortex. No precipitate should be visible by eye. Protect from light.
9. Add the 10 μM BMA working solution to a third channel on the perfusion system (*see* **Note 14**), protect from light and test flows from all three channels (*see* **Note 15**).
10. From a petri dish containing transfected cells, transfer a coverslip to the perfusion bath (*see* **Note 16**).
11. Under the microscope, identify an isolated, healthy, normal-sized fluorescent reporter-positive cell using epifluorescence with blue light excitation and green light emission.
12. Perfuse 10 μM BMA working solution over the cell for 12 min.
13. Half-fill a pulled glass pipette with intracellular solution using the prepared syringe and carry out whole-cell patch clamp recordings, filtering at 5 kHz and sampling at 10 kHz. Recordings may be made before, during and after the application of ATP or light is be applied (*see* **Note 17**).

4 Notes

1. Store fetal bovine serum as 50 mL aliquots at −80 °C and avoid freeze–thaw cycles.
2. Submerge glass coverslips in ethanol overnight and air dry under a class II hood. Store in a sterile 35 mm petri dish.
3. Ultrapure water is obtained by purifying deionized water to resistivity of 18.2 MΩ.cm at 25 °C.
4. Salts to do not dissolve immediately on adding water. The solution becomes homogenous nearing the target pH of 7.3.
5. A slight modification in the synthesis can be made, rather than column chromatography purification of the BMA may be out by recrystallization from 1:1 Ethanol:1,4-Dioxane.
6. *Toronto Research* Chemicals (B497250), *Sigma-Aldrich Partner Product* (ENAH049F1C70), *Enamine* (EN300-190205).
7. This step removes dead cells and dilutes any divalent cations contained in the media that can inhibit the action of trypsin.
8. The media will inactive the trypsin.

9. In our hands, HEK293T cells approximately double in number every 24 h and should be split twice a week at around 1/17. The volume of cell suspension in this step might need adjusting depending on actual cell growth rate and viable cell numbers.
10. The volume of cell suspension should give 60% confluence by the time of transfection. Again, this volume depends on actual cell growth rate and viable cell numbers.
11. For transfection, healthy cells will be well adhered to the petri dish and not completely rounded.
12. Typically, transfection of cells in one 35 mm petri dish yields enough cells for the following day at 10–30% confluence in four 35 mm petri dishes.
13. In our hands, transient transfection using this protocol provides cells for the following 3 days. However, there are fewer isolated cells on day 3 and cell shape can sometimes result in poor space clamping.
14. Alternatively, rather than perfusing BMA over the cells a coverslip can be placed in a 12-well plate along with 1 mL of the 10 μM BMA working solution for 12 min. If this option is employed it is important to minimize subsequent blue light illumination of cells during the identification of a transfected cells. For example, very brief UV illumination is compatible with the excitation spectrum of GFP. In this case, **steps 11** and **12** are no longer required.
15. To minimize photoisomerization, reduce ambient lighting and cover the solution with blackout tape.
16. Minimize the time the petri dish is spent outside of the incubator and the time the coverslip is transferred from the petri dish to the perfusion bath.
17. A typical light-switching protocol might involve illumination of the cells for 2 s with 440 nm light followed immediately with 2 s illumination with 360 nm light before returning to dark. However, at the suggested illumination intensities, a 440 nm light pulse as short as 100 ms gives 80% of the maximum light activated current [5].

References

1. North RA (2002) Molecular physiology of P2X receptors. Physiol Rev 82:1013–1067
2. Khakh BS, North RA (2006) P2X receptors as cell-surface ATP sensors in health and disease. Nature 442:527–532
3. Kramer RH, Mourot A, Adesnik H (2013) Optogenetic pharmacology for control of native neuronal signaling proteins. Nat Neurosci 16:816–823
4. Szobota S, Isacoff EY (2010) Optical control of neuronal activity. Annu Rev Biophys 39:329–348
5. Browne LE, Nunes JP, Sim JA, Chudasama V, Bragg L, Caddick S, North RA (2014) Optical control of trimeric P2X receptors and acid-sensing ion channels. Proc Natl Acad Sci U S A 111:521–526

Chapter 23

Intracellular Calcium Recording After Purinoceptor Activation Using a Video-Microscopy Equipment

Maria Teresa Miras-Portugal, Felipe Ortega, Javier Gualix, Raquel Perez-Sen, Esmerilda G. Delicado, and Rosa Gomez-Villafuertes

Abstract

Calcium is one of the most important intracellular messengers, triggering a wide range of cellular responses. Changes in intracellular free calcium concentration can be measured using calcium sensitive fluorescent dyes, which are either EGTA- or BAPTA-based organic molecules that change their spectral properties in response to Ca^{2+} binding. One of the most common calcium indicators is the ratiometric dye Fura-2. The main advantage of using ratiometric dyes is that the ratio signal is independent of the illumination intensity, dye concentration, photobleaching, and focus changes among others, allowing for the concentration of intracellular calcium to be determined independently of these artifacts. In this protocol, we describe the use of Fura-2 to measure intracellular calcium elevations in single cultured cells after purinoceptor activation using a video-microscopy equipment. This method, usually known as calcium imaging, allows for real-time quantification of intracellular calcium dynamics and can be adapted to measure agonist mediated intracellular calcium responses due to the activation of different purinergic receptors in several cellular models using the appropriate growth conditions.

Key words Calcium imaging, Fura-2, Ratiometric calcium dye, Purinergic receptor, Intracellular free calcium concentration, Video microscopy, Calcium responses

1 Introduction

Extracellular ATP and other nucleotides are sensed by specific plasma membrane receptors called purinergic P2 receptors, which are further classified into two subgroups based on their sequence homology, membrane topology, and pharmacological profile: ionotropic P2X and metabotropic P2Y receptors [1, 2]. Currently, seven P2X subunits (P2X1-P2X7) and eight P2Y ($P2Y_{1, 2, 4, 6, 11, 12, 13, 14}$) receptors have been cloned and characterized in humans. P2X receptors are nonselective ATP-gated ion channels that mediate Na^+/Ca^{2+} influx and K^+ efflux, leading to cell membrane depolarization. P2Y receptors, which can be activated by adenine and uridine nucleotides, are seven-membrane-spanning G protein-

Pablo Pelegrín (ed.), *Purinergic Signaling: Methods and Protocols*, Methods in Molecular Biology, vol. 2041,
https://doi.org/10.1007/978-1-4939-9717-6_23,

coupled receptors. Most of P2Y receptors are able to activate phospholipase C, eliciting Ca^{2+} mobilization from intracellular calcium stores such as the endoplasmic reticulum. Independently of the purinergic receptor involved in a biological process, calcium signaling is critical for all cells. Intracellular free Ca^{2+} is a ubiquitous signaling ion that serves as both a charge carrier and a chemical intermediate linking many physiological stimuli to their intracellular effectors [3, 4]. For this reason, different techniques/methods to measure intracellular Ca^{2+} levels have been developed during the last decades. In this context, intracellular free calcium measurement has been successfully used to characterize the presence of functional purinergic receptors in a wide variety of cellular models [5–12].

Fluorescent indicators are particularly useful to determine the intracellular free Ca^{2+} levels, since they are commercially available with different affinities, brightness or spectral characteristics. Fura-2-acetoxymethyl ester (Fura-2 AM), is a membrane-permeable derivative of the ratiometric calcium indicator Fura-2 [13]. In general, most cells can be loaded by incubation with dilute aqueous solutions of the cell-permeant AM esters. Nonspecific esterases present in most cells hydrolyze the AM esters, liberating the Ca^{2+}-sensitive probe. The low leakage rate of the polyanionic indicator results in a final intracellular concentration that is much higher than the incubation concentration. Ratiometric methods are based on the use of a ratio between two fluorescence intensities linked to the physicochemical properties of the probe. Thus, in contrast to non-ratiometric indicators such as Fluo-3, Fluo-4, and Calcium-Green-3, the excitation maximum of Fura-2 shifts from 363 nm for the Ca^{2+}-free form to 335 nm for the Ca^{2+}-bound state. The wavelength of maximum fluorescence emission is independent of calcium concentration, being about 510 nm in both states. The largest dynamic range for Ca^{2+}-dependent fluorescence signals is obtained by using excitation at 340 and 380 nm and rationing the fluorescence intensities detected at 510 nm. Thus, the F_{340}/F_{380} ratio correlates directly with the amount of intracellular free Ca^{2+}. The use of ratiometric indicators makes measurements and data processing more complicated, since they require a more expensive equipment with the possibility to change the wavelength emission in a very rapid way [14]. However, by using ratiometric probes, factors such as uneven dye distribution, photobleaching, focus changes, intracellular probe concentration, optical path length and light intensity variations are minimized because they should affect both measurements to the same extent. This protocol describes how to quantify intracellular Ca^{2+} levels in cell cultures using Fura-2 as a ratiometric calcium indicator coupled to a video-microscopy equipment. This method, usually known as calcium imaging, allows for real-time quantification of intracellular calcium dynamics and can also be adapted to measure agonist mediated intracellular calcium mobilization in different subcellular areas using the appropriate equipment and cell growth conditions [15].

2 Materials

2.1 Solutions and Reagents

1. Locke's solution: 140 mM NaCl, 4.7 mM KCl, 1.2 mM $MgSO_4$, 1.2 mM KH_2PO_4, 10 mM HEPES, 5.5 mM glucose, 2.5 mM $CaCl_2$, pH 7.4 (*see* **Notes 1–4**).
2. Locke's solution supplemented with 0.1% bovine serum albumin (BSA) (*see* **Note 5**).
3. Tris buffer (1 M): Tris base 1 M, adjusting solution to pH 7.4 using HCl.
4. 1 mM Fura-2-acetoxymethyl ester (Fura-2 AM) stock solution: 50 μL of DMSO to a 50 μg vial of Fura-2 AM, and keep it in a dark place. Fura-2 AM in DMSO is stable at RT for 24 h and is stable at −20 °C for longer storage.
5. Dimethyl sulfoxide (DMSO), high-quality anhydrous.
6. Fura-2 Calcium Imaging Calibration Kit.

2.2 Equipment

1. Inverted fluorescence microscope (*see* **Note 6**).
2. Xenon 75Watt High Pressure XBO 100 Lamp (as the one manufactured by Nikon).
3. VC-8 channel perfusion valve control system (as the one manufactured by Warner Instruments).
4. Excitation filter wheel Lambda 10-2 (as the one manufactured by Sutter Instrument Company).
5. Dichroic mirror 430 nm.
6. Band-pass filter 510 nm.
7. Cooled charge-coupled device (CCD) camera (like the Hamamatsu ORCA-ER C4742-80).
8. Solution inline heater (as the SH-27B from Warner Instruments).
9. Automatic temperature controller (as the TC 324-B from Warner Instruments).
10. MetaFluor 6.2r & PC software (Universal Solutions) or similar analysis software.

2.3 Other Material

1. Culture dishes (35 mm).
2. Glass coverslips (15 mm).
3. Silicone tubes.

3 Methods

3.1 Cell Culture

Calcium imaging allows for the measure of intracellular free calcium levels in real time in a variety of cells that can be cultured using established techniques. Cells must be plated on 35 mm culture dish containing three 15 mm glass coverslips (*see* **Notes** 7 and **8**). Avoid a high-density cell culture. Depending on the cell morphology, a 50–75% subconfluent culture is convenient in order to better analyze single cellular responses.

3.2 Loading of Fura-2 AM Calcium Dye

The exact parameters for Fura-2 AM loading vary widely across cell types. We recommend testing various conditions by preparing several loading solutions containing multiple concentrations of Fura-2 AM ranging from 1 to 10 μM and incubating cells in the loading solution for a variety of times from 15 min to 1 h. Use the minimum concentration of AM ester necessary to obtain an adequate signal. The protocol described in this chapter is optimized for cerebellar neurons, astrocytes and neuroblastoma cell lines.

1. Prepare 1 mM Fura-2 AM stock solution.
2. Prepare 35 mm dish containing 2 mL Locke's solution supplemented with BSA at 37 °C and transfer a coverslip with the cells into the dish in order to wash once the cells.
3. Prepare 35 mm dish containing 2 mL Locke's solution supplemented with BSA at 37 °C containing 10 μL Fura-2 AM stock in order to obtain 5 μM Fura-2 AM solution. Next, transfer the coverslip with the washed cells into the dish.
4. Incubate the cells at 37 °C for 45 min in a dark microplate incubator shaker to allow for Fura-2 AM loading and de-esterification (*see* **Note 9**). Although incubation does not necessarily require either shaking or agitation, a slight balancing is recommended.
5. Prepare 35 mm dish containing 2 mL fresh Locke's solution at 37 °C. Remove the coverslip from the loading solution and place it into the new dish.
6. Remove the coverslip from the 35 mm dish and rapidly mount on the bottom of the superfusion chamber making sure to prevent drying of the cells (*see* **Note 10**). We use a RC-25F open chamber (bath volume 133 μL/mm) placed on a PH-1 heated platform, both manufactured by Warner Instruments (Fig. 1). The input line is connected to an eight-channel perfusion valve control system that works by gravity and the output line is coupled to a vacuum trap. The flow rate is kept constant at 1.5 mL/min.

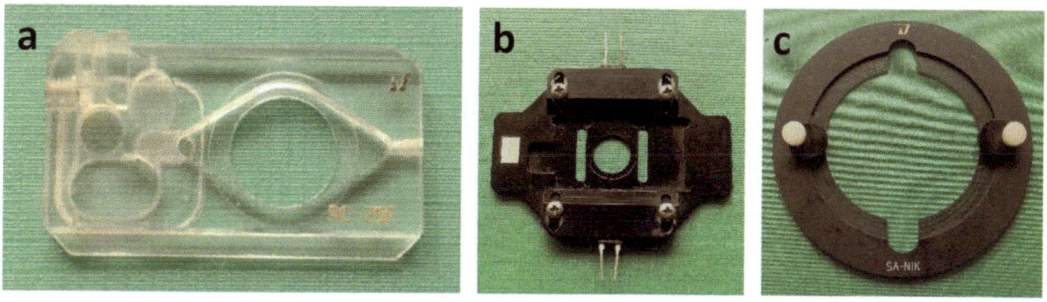

Fig. 1 The superfusion system. (**a**) Open diamond bath imaging chamber. The coverslip fits in the round hollow remaining in the center of the chamber. (**b**) The aluminum platform provides clamping between the chamber and the bottom glass coverslip. The platform has a pair of resistive heaters attached to control chamber temperature. (**c**) The stage adapter allows for the mounting of the platform onto the microscope stage

3.3 Imaging Protocol

1. Load perfusion solutions into the input lines taking care to prevent the formation of air bubbles inside the silicone tubes in order to prevent cells from detaching or damage.
2. Connect the chamber to the perfusion lines and perfuse Locke's solution through the chamber. A solution in-line heater controlled by an automatic temperature controller warms perfusion solutions to 37 °C.
3. When necessary, place a drop of oil on the objective, place the chamber on the microscope stage and focus on the cells using transmitted light.
4. Examine the fluorescence of the cells using illumination at 340 and at 380 nm using the eyepieces. Resting cells should be dark at 340 and bright at 380 nm (*see* **Note 11**).
5. Examine the cells using the camera and set the gain and exposure conditions of the camera to generate an image that is almost saturated when illuminated at 380 nm and well below saturation when illuminated at 340 nm. Keep the exposure time below 300 ms if possible.
6. Set up the time-lapse interval to collect images between 0.1 and 10 s depending on the type of signals that you expect to see. We usually fix the interval in 1 s.
7. Find the fields that you plan to image and start the experiment. Depending on the objective used and the cell size, 40–200 cells per field can be monitored in real time.

3.4 Data Analysis

Once the experiment is complete, you can convert the set of ratio images into time-lapse calcium measurements for individual cells or regions of interest within cells (Fig. 2). To perform this analysis:

1. Use the region of interest (ROI) tool to define the areas of the image in which you want to measure calcium responses (*see* **Notes 12** and **13**).

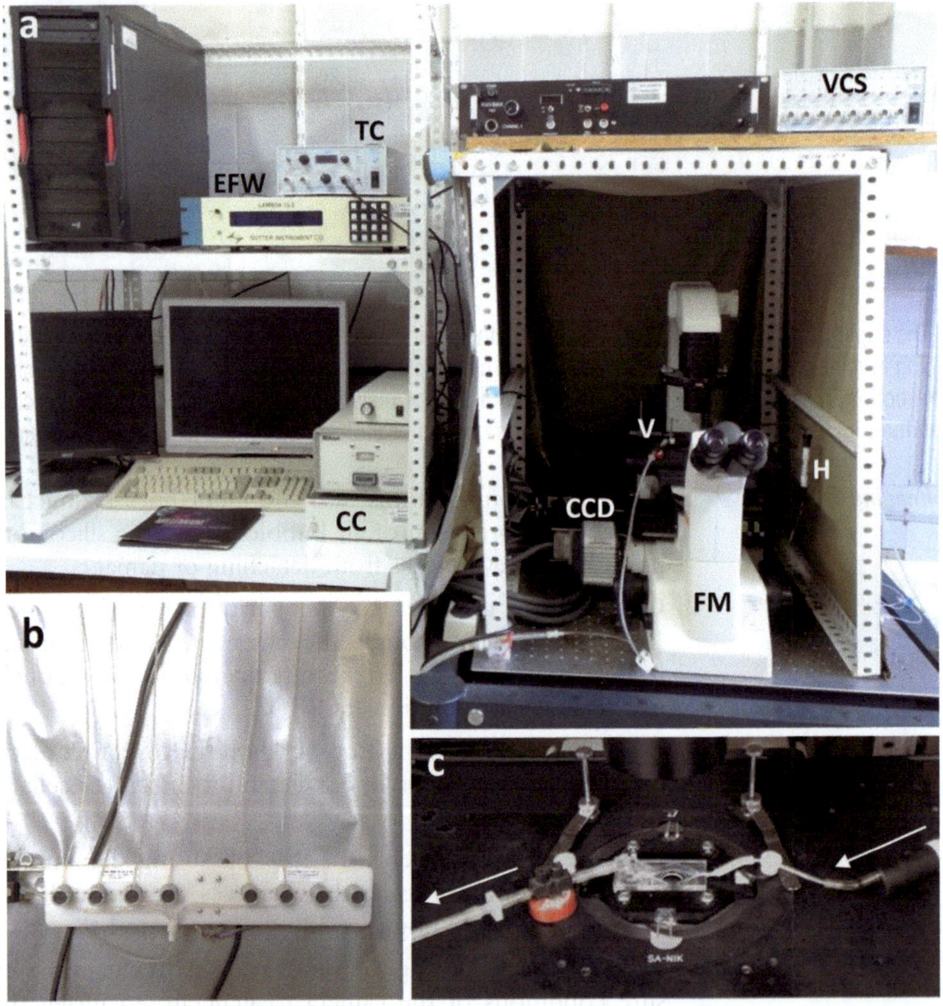

Fig. 2 Video-microscopy equipment for intracellular calcium measurements in single cells. (**a**) The video-microscopy equipment consists of an inverted fluorescence microscope (FM), a valve control system (VCS), an excitation filter wheel (EFW), a CCD camera (CCD) coupled to a camera controller system (CC), an automatic temperature controller (TC) coupled to a solution inline heater (H) and a vacuum trap (V) for solution removal. (**b**) Lateral view of the video-microscopy setup showing a detail of the 8-channel perfusion valve system. (**c**) Detail of the superfusion chamber mounted onto the microscope stage. The solution flow direction is depicted by arrows

2. Use the software to collect time-lapse F_{340}/F_{380} ratio measurements for each ROI in each image.
3. Use the ROI tool to measure the intensity of the background in a variety of locations in the images for each wavelength.
4. The formula given by Grynkiewicz et al. [13] allows for the conversion of Fura-2 ratio values to intracellular calcium concentrations:

$$[Ca^{2+}]_i = K_d * (R - R_{min})/(R_{max} - R) * Q$$

where $[Ca^{2+}]_i$ is the intracellular calcium concentration, K_d is the Ca^{2+} dissociation constant of Fura-2, R is the observed fluorescence intensity ratio at both wavelengths (F_{340}/F_{380}), R_{min} is the F_{340}/F_{380} ratio in the absence of calcium, R_{max} is the F_{340}/F_{380} ratio in the presence of a saturating concentration of calcium, and Q is the ratio F_{380min}/F_{380max}. Since K_d value of Fura-2 is affected by a number of variables (pH, viscosity, temperature, protein concentration, ionic strength, and divalent cations concentration), a calibration curve will be required to calculate the K_d in each particular experimental system [16, 17]. This calibration must be routinely accomplished at the end of the experiment (*see* **Note 14**).

5. Import the ratio measurements into an analysis program and subtract the background value. Moreover, you will need to view the calcium traces for specific ROIs, to average multiple ROIs and convert the ratio measurements to intracellular free calcium values. Accessible software like Excel can be used to perform all these analyses.

4 Notes

1. Calcium imaging experiments can be performed using a variety of physiological phosphate buffers, it is important to avoid phenol red in the buffers, because it greatly increases the fluorescent background.
2. We use Locke's solution, which is isotonic with blood plasma similarly to physiological saline and easily prepared. It is frequent to remove Mg^{2+} from Locke's solution to enhance P2X7 receptor activation. In this case, it is necessary to maintain the osmotic balance of the medium increasing glucose concentration from 5.5 to 7.9 mM.
3. In some conditions, it is convenient to remove calcium from Locke's solution to check whether intracellular calcium elevations are due to calcium influx from the extracellular medium. In this case, we supplemented Locke's solution with a 1:1 mixture of 1 M Tris buffer and 1 M EGTA, pH 7.4 (12 mL of mixture in 1 L of Locke's solution). The final concentration of Ca^{2+} in this solution will be similar to the intracellular levels of cytosolic free calcium in resting cells (in the nM range).
4. When testing the calcium imaging equipment, it is frequent to use stimuli like Locke's medium containing either 60 mM KCl for excitable cells or 2 μM ionomycin for nonexcitable cells, both triggering a strong intracellular calcium rise.

5. To perform Fura 2-AM loading buffer, we use Locke's solution supplemented with BSA in order to block the activity of serum esterases present in the culture medium that may degrade Fura-2-AM.
6. We use an inverted fluorescence microscope equipped with a Xenon arc lamp, an excitation filter wheel, a 430 nm dichroic mirror, a 510 nm band-pass filter, and a CCD camera (Fig. 3).

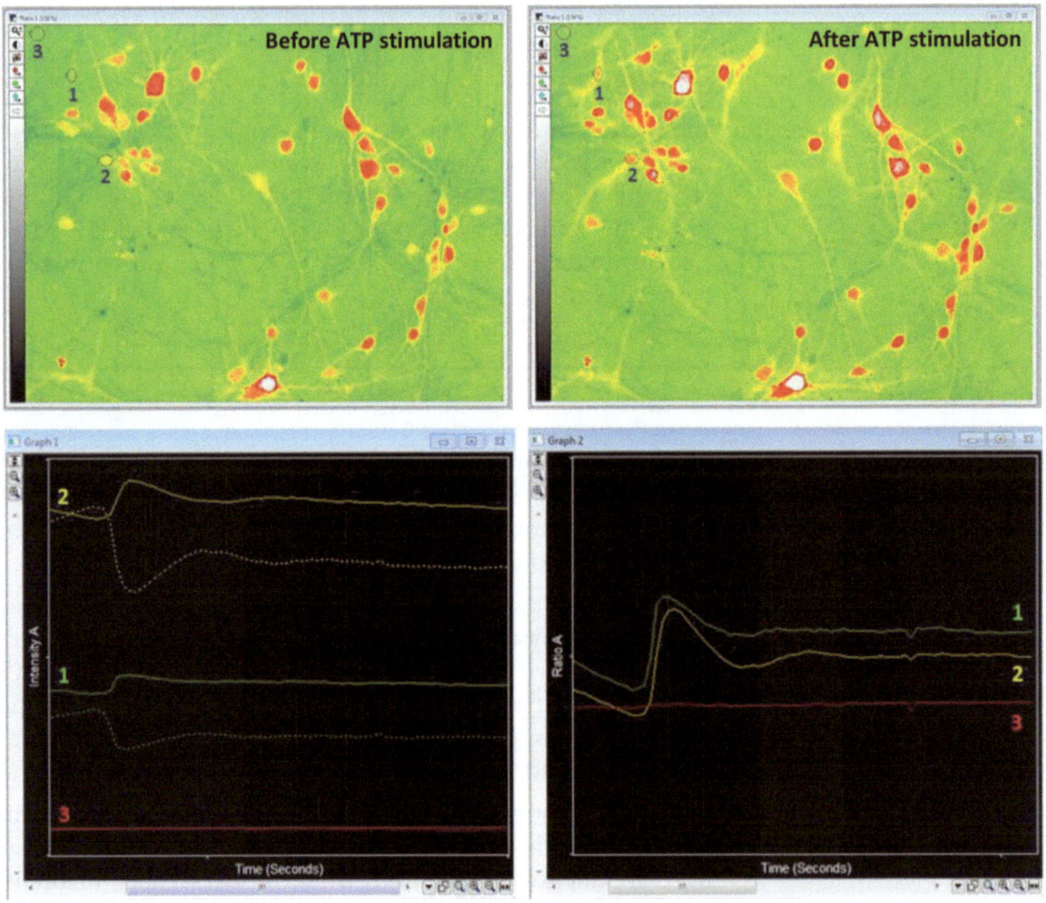

Fig. 3 ATP increases intracellular calcium levels in primary cultures of cerebellar granule neurons. Upper panels: Fura-2 ratiometric images (F_{340}/F_{380}) showing cytosolic Ca^{2+} levels of granule neurons before (left) and after (right) stimulation with 1 mM ATP. In pseudocolor images, red/white colors correspond to high Ca^{2+} concentrations, whereas green/blue colors indicate low Ca^{2+} levels. Lower panels: Representative traces of intracellular calcium responses to 1 mM ATP in Fura-2 AM-loaded granule cells. The numbers correspond to the regions of interest (ROI) depicted in the upper panels: ROI 1 (axodendritic fiber of granule neuron), ROI 2 (cell body of granule neuron) and ROI 3 (background). Left: fluorescence intensity at both registered excitation wavelengths, F_{340} nm (continuous line) and F_{380} nm (dotted lines). Right: registers obtained over the time are divided to determine the F_{340}/F_{380} ratio

A computer running MetaFluor 6.2r & PC software controls the microscope, although several other software packages are available for ratiometric imaging. Depending on the cell subtype and the cellular region of interest, different objectives can be used. Our fluorescence microscope consists of Plan Fluor 20×/0.5 water objective, Plan Apo 60×/1.40 oil immersion objective and S Fluor 100×/0.5–1.3 oil immersion objective, all from Nikon.

7. When doing this procedure it is important to use glass coverslips instead of plastic, since plastic interferes with fluorescent at ultraviolet wavelengths.
8. Depending on the cell culture, it may be appropriate to coat glass coverslips with a cellular adhesive (like poly-L-lysine or laminin) to prevent the cells from detaching or moving during experiments.
9. It may be difficult to load Fura-2 AM into certain types of cells [18]. For this reason, in many studies, Fura-2 AM loading is facilitated by the addition of Pluronic F-127, a nonionic low-toxicity detergent that improves the cell loading of Fura-2 AM by facilitating its solubilization in physiological media.
10. The coverslip must be secured with vacuum grease to the chamber and two tubes at either side of the chamber allow for perfusion of solutions through the chamber.
11. In general, cells should not be illuminated with UV light for more than 10 or 15 s and the intensity of the excitation light should be reduced with a neutral density filter to prevent phototoxicity.
12. It is generally useful to have at least one ROI that covers the cell body of the cell.
13. When defining the ROIs, it is useful to play the movie of the images to verify that the cells do not move during the course of the experiment.
14. The easiest way of measuring Fura-2 ratio values as a function of calcium concentrations is to use an in vitro calibration kit that contains multiple buffered calcium solutions and Fura-2 free acid. These can be obtained from Thermo Fisher Scientific (Invitrogen) and contain precise instructions for use. Nevertheless, many calcium studies avoid performing that calibration and simply express the calcium determination as time-lapse F_{340}/F_{380} ratio. This option is commonly accepted as an accurate measurement of intracellular free calcium levels, especially when comparing calcium transients in different experimental conditions.

Acknowledgments

This work was supported by Ministerio de Economia y Competitividad (MINECO, BFU 2014-53654-P), Red de Excelencia Consolider-Ingenio Spanish Ion Channel Initiative (BFU2015-70067REDC), Comunidad de Madrid (BRADE-CM S2013/ICE-2958), and Fundación Ramón Areces (PR2018/16-02). F. Ortega is the recipient of a Ramón y Cajal contract (RYC-2013-13290).

References

1. Abbracchio MP, Burnstock G (1994) Purinoceptors: are there families of P2X and P2Y purinoceptors? Pharmacol Ther 64 (3):445–475
2. Burnstock G (2007) Purine and pyrimidine receptors. Cell Mol Life Sci 64 (12):1471–1483. https://doi.org/10.1007/s00018-007-6497-0
3. Berridge MJ, Bootman MD, Roderick HL (2003) Calcium signalling: dynamics, homeostasis and remodelling. Nat Rev Mol Cell Biol 4 (7):517–529. https://doi.org/10.1038/nrm1155
4. Petersen OH, Michalak M, Verkhratsky A (2005) Calcium signalling: past, present and future. Cell Calcium 38(3–4):161–169. https://doi.org/10.1016/j.ceca.2005.06.023
5. Mateo J, Garcia-Lecea M, Miras-Portugal MT, Castro E (1998) Ca^{2+} signals mediated by P2X-type purinoceptors in cultured cerebellar Purkinje cells. J Neurosci 18(5):1704–1712
6. Garcia-Lecea M, Delicado EG, Miras-Portugal MT, Castro E (1999) P2X2 characteristics of the ATP receptor coupled to $[Ca^{2+}]_i$ increases in cultured Purkinje neurons from neonatal rat cerebellum. Neuropharmacology 38 (5):699–706
7. Hervas C, Perez-Sen R, Miras-Portugal MT (2003) Coexpression of functional P2X and P2Y nucleotide receptors in single cerebellar granule cells. J Neurosci Res 73(3):384–399. https://doi.org/10.1002/jnr.10676
8. Nobile M, Monaldi I, Alloisio S, Cugnoli C, Ferroni S (2003) ATP-induced, sustained calcium signalling in cultured rat cortical astrocytes: evidence for a non-capacitative, P2X7-like-mediated calcium entry. FEBS Lett 538 (1–3):71–76
9. Carrasquero LM, Delicado EG, Bustillo D, Gutierrez-Martin Y, Artalejo AR, Miras-Portugal MT (2009) P2X7 and P2Y13 purinergic receptors mediate intracellular calcium responses to BzATP in rat cerebellar astrocytes. J Neurochem 110(3):879–889. https://doi.org/10.1111/j.1471-4159.2009.06179.x
10. Gomez-Villafuertes R, del Puerto A, Diaz-Hernandez M, Bustillo D, Diaz-Hernandez JI, Huerta PG, Artalejo AR, Garrido JJ, Miras-Portugal MT (2009) Ca^{2+}/calmodulin-dependent kinase II signalling cascade mediates P2X7 receptor-dependent inhibition of neuritogenesis in neuroblastoma cells. FEBS J 276 (18):5307–5325. https://doi.org/10.1111/j.1742-4658.2009.07228.x
11. Traves PG, Pimentel-Santillana M, Carrasquero LM, Perez-Sen R, Delicado EG, Luque A, Izquierdo M, Martin-Sanz P, Miras-Portugal MT, Bosca L (2013) Selective impairment of P2Y signaling by prostaglandin E2 in macrophages: implications for Ca^{2+}-dependent responses. J Immunol 190 (8):4226–4235. https://doi.org/10.4049/jimmunol.1203029
12. Gomez-Villafuertes R, Rodriguez-Jimenez FJ, Alastrue-Agudo A, Stojkovic M, Miras-Portugal MT, Moreno-Manzano V (2015) Purinergic receptors in spinal cord-derived ependymal stem/progenitor cells and their potential role in cell-based therapy for spinal cord injury. Cell Transplant 24 (8):1493–1509. https://doi.org/10.3727/096368914X682828
13. Grynkiewicz G, Poenie M, Tsien RY (1985) A new generation of Ca^{2+} indicators with greatly improved fluorescence properties. J Biol Chem 260(6):3440–3450
14. Tsien RY, Rink TJ, Poenie M (1985) Measurement of cytosolic free Ca^{2+} in individual small cells using fluorescence microscopy with dual excitation wavelengths. Cell Calcium 6 (1–2):145–157
15. Sanchez-Nogueiro J, Marin-Garcia P, Leon D, Leon-Otegui M, Salas E, Gomez-Villafuertes-R, Gualix J, Miras-Portugal MT (2009) Axodendritic fibres of mouse cerebellar granule neurons exhibit a diversity of functional P2X

receptors. Neurochem Int 55(7):671–682. https://doi.org/10.1016/j.neuint.2009.06.009

16. Lattanzio FA Jr, Bartschat DK (1991) The effect of pH on rate constants, ion selectivity and thermodynamic properties of fluorescent calcium and magnesium indicators. Biochem Biophys Res Commun 177(1):184–191
17. Petr MJ, Wurster RD (1997) Determination of in situ dissociation constant for Fura-2 and quantitation of background fluorescence in astrocyte cell line U373-MG. Cell Calcium 21 (3):233–240
18. Di Garbo A, Alloisio S, Nobile M (2012) P2X7 receptor-mediated calcium dynamics in HEK293 cells: experimental characterization and modelling approach. Phys Biol 9 (2):026001. https://doi.org/10.1088/1478-3975/9/2/026001

Chapter 24

Assays to Measure Purinoceptor Pore Dilation

Ben J. Gu, Pavan Avula, and James S. Wiley

Abstract

The P2X7 receptor is a classic purinoceptor/ion channel. After activated by ATP, it opens a cation selective channel, which dilates to a large pore over tens of seconds, allowing the entry of big molecules. This unique feature is often used to evaluate this receptor's function with DNA-binding dyes (MW 300–400 Da), such as ethidium bromide and Yo-Pro-1. Here we describe two-color flow cytometry based protocols for measuring P2X7 pore dilation. One is ATP-induced ethidium uptake by real-time multicolor flow cytometry for standardized and accurate quantitation, and the other is a quick whole blood assay which is particularly useful for ex vivo study.

Key words Real-time flow cytometry, Pore dilation, Ethidium, FlowJo, WinMDI, Area under curve, Excel formula

1 Introduction

The concept of purinoceptors pore dilation was developed in different laboratories from studies using various techniques (electrophysiology, fluorescent dye uptake, radioisotope influx) to examine the permeability properties of immune cells expressing the purinergic P2X7 receptor. Activation of the P2X7 receptor by its agonist, ATP causes an immediate opening of an ionic conductance to small cations such as Ca^{2+} (within several hundred milliseconds) which is followed by a slower increase in permeability over tens of seconds which allows the uptake of large organic fluorophores such as ethidium (cation 314 Da) or YOPro-1 (cation 375 Da). This permeability behavior is variously described as "channel to pore transition" or "pore dilation" and is considered a unique property of P2X7 [1–3]. Evidence for channel dilation in other purinoceptors such as P2X2 or P2X4 are based only on electrophysiology with controversial interpretation [4] and await study with a different technique. Only one study has used two techniques to show P2X7 channel dilation in fresh human B-lymphocytes either with ATP added with permeant at zero time ("channel") or with ATP

Pablo Pelegrín (ed.), *Purinergic Signaling: Methods and Protocols*, Methods in Molecular Biology, vol. 2041,
https://doi.org/10.1007/978-1-4939-9717-6_24,

added 10 min prior to permeant to fully dilate the native P2X7 ("pore") expressed on this cell type. Permeability was assessed both by isotopic influx of Ba^{2+} (a surrogate for Ca^{2+}) or by the uptake of the ethidium measured by flow cytometry. Both Ba^{2+} influx and ethidium uptake are essentially unidirectional with a magnitude that reflects channel conductance. Both Ba^{2+} and ethidium influx were greater for cells preincubated with ATP for 10 min as compared to simultaneous addition permeant with at zero time [5]. Reducing the temperature from 37 to 24 °C magnified the permeability difference greatly such that permeant flux through the "pore" was at least double that through the "channel" [5]. The dramatic difference was not due to opening of a second channel but rather was associated with the slow dissociation of P2X7 from the underlying nonmuscle myosin of the cytoskeleton [6], suggesting that this dissociation confers a subtle conformational change in the open channel state of P2X7. A model based on channel dilation of the P2X7 receptor has been proposed [3].

Ethidium bromide is a phenanthridinium intercalator which binds both DNA and RNA and is generally excluded from viable cells because of the relatively large molecule weight of $ethidium^+$. Once bound to nucleic acids, the fluorescence is enhanced 20–30 fold. The excitation maximum is shifted to 512 nm and the emission maximum is shifted to 605 nm. Other cell markers can then be detected together with ethidium, such as FITC-conjugated anti-CD14 monoclonal antibody (mAb) and/or APC-conjugated anti-CD16 mAb. Time-resolved flow cytometry generates the mean fluorescence intensity of analyzed cells over a certain time period. This technique allows a sensitive measurement of the initial rate of ethidium uptake, which is essentially unidirectional because of binding of this permeant cation to nucleic acids. The multicolor real time flow cytometry method described here allow quantitative assessment of the pore dilation of functional P2X7 receptors on the surface of different subtypes of leukocytes (both native and transfected) as well as excluding dead cells from analysis [7–9].

2 Materials

Prepare all solutions using ultrapure water (Milli-Q water with a BioPak adapter, to attain a sensitivity of 18 MΩ-cm at 25 °C) and analytical grade reagents. Prepare and store reagents as indicated. Diligently follow all waste disposal regulations when disposing of waste materials. We do not add sodium azide to reagents.

2.1 Reagents

1. ATP.
2. Ethidium bromide.
3. Sodium chloride.

4. Potassium chloride.
5. D-Glucose.
6. Bovine serum albumin (BSA).
7. Tetramethylammonium hydroxide (TMA).
8. 4-(2-Hydroxyethyl)-1-piperazineethanesulfonic acid (HEPES).
9. Ficoll-Paque™ PLUS.
10. Mouse anti-human CD14 and CD16 monoclonal antibodies.
11. Flow cytometry calibration beads.
12. Quantum R-PE MESF beads.
13. Right Reference Standard Phycoerythrin High Level beads.

2.2 Buffers

1. KCl buffer: 145 mM KCl, 5 mM KOH, 10 mM HEPES, pH 7.5.
2. K media: KCl buffer with 5 mM D-glucose and 1 mg/mL BSA (*see* **Note 1**).
3. NaCl buffer: 145 mM NaCl, 5 mM KOH, 10 mM HEPES, pH 7.5.
4. Na media: NaCl buffer plus 0.1 mM $CaCl_2$, 5 mM D-glucose, and 1 mg/mL BSA (*see* **Note 1**).
5. Mg^{2+} buffer: 20 mM $MgCl_2$ in NaCl buffer.
6. 100 mM ATP stock: weight ATP powder for make a 20 mL stock of 100 mM (*see* **Note 2**). Dissolve ATP powder into 17 mL KCl buffer, keep stirring, slowly add 2 mL of 18% (w/v) TMA into the solution, measure pH, add more KCl buffer or TMA to bring pH to 6.8–7.0 with a total volume of 20 mL. Do not over raise pH >7.0. Aliquot and store at −80 °C.
7. Sterile phosphate buffered saline (PBS) solution.

2.3 Equipment

1. Flow cytometer, we use a FACSCalibur (BD Bioscience) with Time Zero Module (Cytek, https://www.cytebio.com) and a circulating water bath (Fig. 1, *see* **Note 3**).
2. Desktop centrifuge and adaptors for 5 mL FACS tubes and 15 and 50 mL conical tubes.
3. Class II biosafety cabinet.
4. 37 °C water bath.
5. Small tube vortex.
6. 5 mL FACS tubes, 15 and 50 mL conical tubes.
7. A set of micropipettes including P10, P20, P100, and P1000.

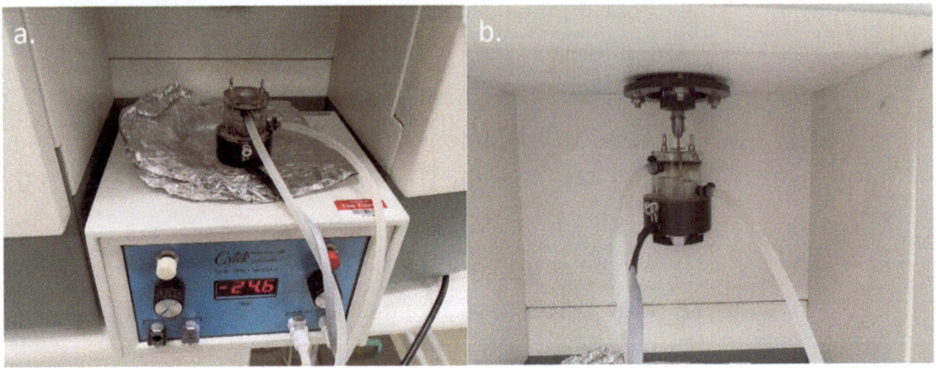

Fig. 1 (**a**, **b**) A picture of Time Zero module integrated with a FACSCalibur flow cytometer

2.4 Cells

1. Heparin anticoagulant human peripheral blood from healthy individuals.

2.5 Software

1. WinMDI software (*see* **Note 4**).
2. FlowJo software (*see* **Note 5**).
3. Microsoft Excel.

3 Methods

3.1 Flow Cytometry Analysis

1. Mix 15 mL heparin anticoagulant blood with equal volume of PBS.
2. Carefully load on top of 15 mL Ficoll-Paque in a 50 mL centrifuge tube.
3. Spin at ~400 × *g* for 30 min with brake off.
4. Collect the visible layer of peripheral blood mononuclear cells (PBMCs) at the interface between Ficoll and blood. Add these to PBS to make 50 mL. Stopper the top and gently mix by inverting tube several times. Count the cell concentration with a hematology cell counter using microscopy or using an automated cell counting machine.
5. Wash PBMCs once with 30–40 mL PBS spun at ~350 × *g* for 8 min (*see* **Note 6**) and resuspend in Na media at a concentration of ~2 × 10^7/mL.
6. Incubate cells with fluorescent conjugated anti-CD14 and/or anti-CD16 monoclonal antibody for 20 min at 4 °C with gentle rock (*see* **Note 7**).
7. Add 3–4 mL of Na media, mix gently by inverting, then centrifuge at 200 × *g* for 5 min, resuspend the cells in Na media at about 2 × 10^6/mL (*see* **Note 8**).

8. Wash cells with 3–4 mL of K media and centrifuge at 200 × *g* for 5 min again.
9. Resuspend the cells in K media at ~2 × 10^6/mL, aliquot to 1 mL/tube. Keep at 4 °C.
10. Turn on the FACSCalibur flow cytometer and the computer at the same time, and leave the circulating water bath at 39 °C (*see* **Note 9**).
11. Setup the FACSCalibur acquisition parameters: FL2 (for ethidium detection) voltage was set at around 595 V with a gain of 5.0, at which the linear channel fluorescence intensity for Quantum PE standard beads with MESF 300747 was 48 ± 1 (256 linear scale) and the peak channel for right reference standard PE high level beads (MESF ~560,000) was 100 ± 1. The compensation of FL1–FL2 and FL2–FL1 was 7% and 8% respectively. No compensation of FL4 and FL2 is needed. Both FL1 and FL4 are in log scale. The time interval is 5 s. Acquisition will include parameters of FSC, SSC, FL1, FL2, FL4, and time.
12. Warm up the cells at 37 °C for at least 5 min.
13. Add 5 μL of 5 mM ethidium bromide and a small stir bar to the tube.
14. Insert the tube into the water-jacketed adapter of Time Zero Module and turn on stirring with lowest speed.
15. Insert the adapter to the sample acquisition platform.
16. Acquire PBMCs at approximately 1000 events per second.
17. Ten microliters of 100 mM ATP is added 40 s later (*see* **Note 10**).
18. Continue acquisition for another 5 min.

3.2 Analysis with WinMDI

1. Run WinMDI software.
2. Choose "Display|Density Plot" on the menu, select the file, in the popup window "Format 2D Display," choose "256 × 256" for "Display Array Resolution" which gives a better resolution.
3. Left-click on the density plot, choose "Regions" to create a polygon gate R1 based on forward scatter and side scatter, to gate lymphocytes or monocytes population (Fig. 2a).
4. Choose "Display|Dot Plot" to draw another dot plot of CD14 (or CD16) versus ethidium. On "Format Dotplot" window, choose "All" for "Plot number of events." Left-click on the dot plot, choose "Regions" to create a "SortRect" gate R2 based on the CD14/CD16 positivity. Make sure this gate does not include events over 254 channels, which are mainly dead cells (Fig. 2b).

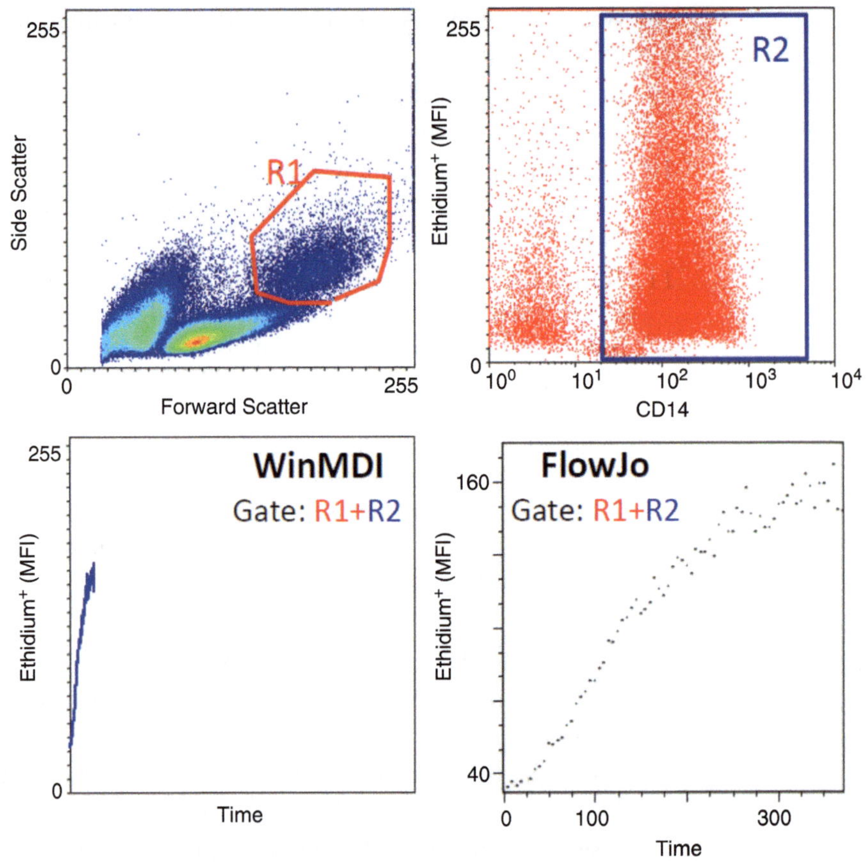

Fig. 2 Typical flow cytometry diagrams. (**a**) Forward and Side Scatters showing monocytes gated by R1; (**b**) dot plots showing $CD14^+$ monocytes gated by R2. The straight upright distribution of $CD14^+$ events indicates the correct compensation. Note R2 gate excludes the top events (ethidium intensity >254 channel) which are considered as dead/leaky cells. (**c**) Kinetic curve drawn by WinMDI. (**d**) Kinetic curve drawn by FlowJo

5. After the two gates are set, draw another dot plot of time versus main fluorescence. On "Format Dotplot" window, choose "All" for "Plot number of events" and check "Kinetics mode," "Overlay kinetics line," and "Draw kinetics only" to draw a kinetics line in the dot plot (KinPlot) (Fig. 2c). Left-click on the plot, choose "Gates," select "And" for both R1 and R2. Choose the KinPlot window, then choose "File|Save as" to save the date to a tabled text file (*see* **Note 11**). The file can then be imported to Microsoft Excel.

3.3 Analysis with FlowJo

1. Open FlowJo software and import the file.
2. Gate on lymphocytes or monocytes populations by forward and side scatter first.
3. Select the gate, double-click to open a new window and draw a rectangular gate on CD14 (or CD16) positive cells, make sure do not include the top events.

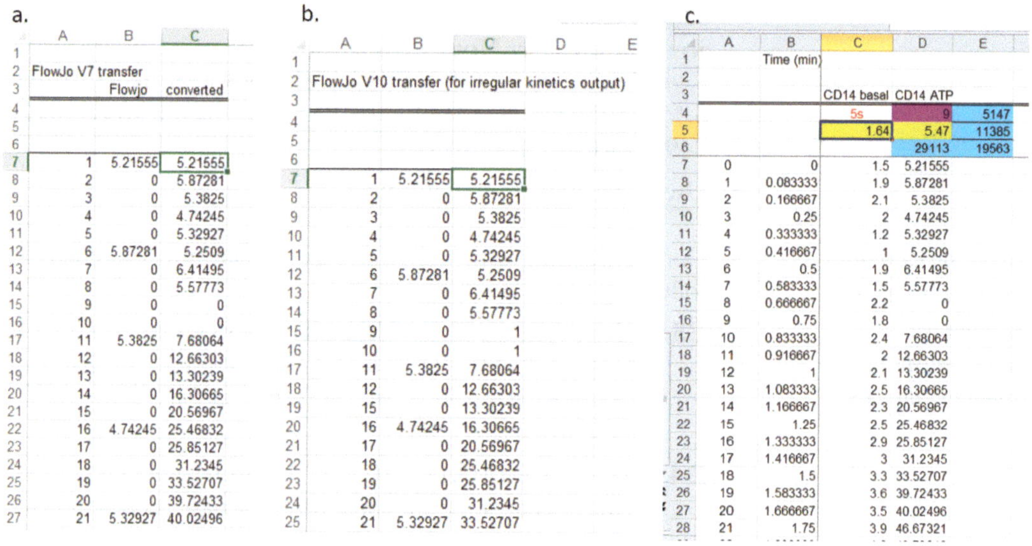

Fig. 3 Examples of Excel worksheet containing formulas to (**a**) convert regular FlowJo time series to WinMDI-like format; (**b**) convert irregular FlowJo time series to WinMDI-like format; (**c**) calculate area under ethidium uptake curve

4. Choose this gate, then press "Ctrl + F8" to open a new window with kinetic curve, change *Y*-axis to FL2 (ethidium) (Fig. 2d).
5. In this window, choose "Edit" then "Copy Time Series" to copy the kinetic data and paste to Microsoft Excel.

3.4 Import Time Series to Excel

1. WinMDI tabled text file contains four columns: **x** is the number of intervals, **y** is the mean fluorescence intensity, **sd** is the standard variation of events, and **npts** is the events count in each interval. The row with **npts** being 0 indicates when the ATP is added, usually in this row **y** is also 0. The column **y** can be imported to Excel directly.
2. FlowJo time series values can be directly pasted to Excel. However, there are many 0's in the column which need to be removed. This can be done by apply a formula as below (Fig. 3a):

$$C7 = INDIRECT(ADDRESS(7+5 * (ROW()-7),COLUMN()-1,3,1)).$$

3. Use "Ctrl + D" to get all values.
4. The above formula only works when the time series data are in a regular format (e.g., there are four 0's between values). If there are not enough events in some intervals or FlowJo version 10 is used, the time series may become irregular; an array formula is therefore needed to extract all values. First, find where ATP is

added, change the value "0" to "1," then apply the following array formula (*see* **Note 12**, Fig. 3b):

C7 = {IF(COUNTIF(B$7:B$493,">0")<ROW(B7)-6,"", INDEX(B:B, SMALL(IF(B$7:B$493<>0,ROW(B$7:B$493)),ROWS(B$7:B7))))}.

5. Use "Ctrl + D" to get all values.

3.5 Calculate Area Under Curve as a Quantitative Measurement of Ethidium$^+$ Uptake

1. Converted time series can then be copied and pasted to the following formulas (Fig. 2c):

C5 = AVERAGE(C$7:INDIRECT(ADDRESS(D$4+5,COLUMN(),4,1),1)) (*to calculate the average linear mean channel of basal line in the same period before ATP is added into the test tube*).

D4 = IFERROR(MATCH(0,D$7:D$30,0),MATCH(1,D$7:D$30,0)) (*to locate in which interval that ATP is added*).

D5 = AVERAGE(D$7:INDIRECT(ADDRESS(D$4+5,COLUMN(),4,1),1)) (*to calculate the average linear mean channel of basal line before ATP is added*).

D6 = 2.5*(SUM(INDIRECT(ADDRESS(D$4+8,COLUMN(),4,1),1):INDIRECT(ADDRESS(D$4+68,COLUMN(),4,1),1))+SUM(INDIRECT(ADDRESS(D$4+9,COLUMN(),4,1),1):INDIRECT(ADDRESS(D$4+67,COLUMN(),4,1),1))-SUM(INDIRECT(ADDRESS(D$4+8,COLUMN()-1,4,1),1):INDIRECT(ADDRESS(D$4+68,COLUMN()-1,4,1),1))-SUM(INDIRECT(ADDRESS(D$4+9,COLUMN()-1,4,1),1):INDIRECT(ADDRESS(D$4+67,COLUMN()-1,4,1),1))-(D$5-C$5)*120) (*to calculate the first 5 min area under ethidium uptake curve after addition of ATP*).

E4 = 2.5*(SUM(INDIRECT(ADDRESS(D$4+8,COLUMN()-1,4,1),1):INDIRECT(ADDRESS(D$4+32,COLUMN()-1,4,1),1))+SUM(INDIRECT(ADDRESS(D$4+9,COLUMN()-1,4,1),1):INDIRECT(ADDRESS(D$4+31,COLUMN()-1,4,1),1))-SUM(INDIRECT(ADDRESS(D$4+8,COLUMN()-2,4,1),1):INDIRECT(ADDRESS(D$4+32,COLUMN()-2,4,1),1))-SUM(INDIRECT(ADDRESS(D$4+9,COLUMN()-2,4,1),1):INDIRECT(ADDRESS(D$4+31,COLUMN()-2,4,1),1))-(D$5-C$5)*48) (*optional: to calculate the first 2 min area under ethidium uptake curve after addition of ATP*).

E5 = 2.5*(SUM(INDIRECT(ADDRESS(D$4+8,COLUMN()-1,4,1),1):INDIRECT(ADDRESS(D$4+44,COLUMN()-1,4,1),1))+SUM(INDIRECT(ADDRESS(D$4+9,COLUMN()-1,4,1),1):INDIRECT(ADDRESS(D$4+43,COLUMN()-1,4,1),1))-SUM(INDIRECT

(ADDRESS(D$4+8,COLUMN()-2,4,1),1):INDIRECT (ADDRESS(D$4+44,COLUMN()-2,4,1),1))-SUM (INDIRECT(ADDRESS(D$4+9,COLUMN()-2,4,1),1): INDIRECT(ADDRESS(D$4+43,COLUMN()-2,4,1),1))-(D$5-C$5)*72) (*optional: to calculate the first 3 min area under ethidium uptake curve after addition of ATP*).

E6 = 2.5*(SUM(INDIRECT(ADDRESS(D$4+8,COLUMN()-1,4,1),1):INDIRECT(ADDRESS(D$4+56,COLUMN()-1,4,1),1))+SUM(INDIRECT(ADDRESS(D$4+9,COLUMN()-1,4,1),1):INDIRECT(ADDRESS(D$4+55,COLUMN()-1,4,1),1))-SUM(INDIRECT(ADDRESS(D$4+8,COLUMN()-2,4,1),1):INDIRECT(ADDRESS(D$4+56,COLUMN()-2,4,1),1))-SUM(INDIRECT (ADDRESS(D$4+9,COLUMN()-2,4,1),1):INDIRECT (ADDRESS(D$4+55,COLUMN()-2,4,1),1))-(D$5-C$5)*96) (*optional: to calculate the first 4 min area under ethidium uptake curve after addition of ATP*).

3.6 ATP Induced Ethidium+ Uptake by Whole Blood

1. 200–500 μL of heparin-anticoagulant whole blood is incubated shaking with FITC-conjugated anti-human CD14 monoclonal antibody (*see* **Note 7**) at 4 °C for 20 min with gentle shaking.
2. Whole blood is then diluted to 1.0 mL in NaCl buffer prewarmed at 37 °C.
3. Add 5 μL of 5 mM ethidium bromide and place the tube in a 37 °C water bath.
4. Add 10 μL of 100 mM ATP stock solution and incubate the tube at 37 °C for 10 min (*see* **Note 13**).
5. Add 1 mL of prechilled Mg^{2+} buffer to the tube.
6. Mix and add another 2 mL of cold NaCl buffer.
7. Allow red cells to sediment down to the bottom for 5–10 min, collect 3 mL of supernatant (*see* **Note 14**).
8. Spin down the supernatant at 200 × *g* for 3 min.
9. Resuspend the pellet in 0.5 mL NaCl buffer.
10. Analyze with a flow cytometer (*see* **Note 15**).

4 Notes

1. All buffers are filtered with 0.22 μm filter, aliquoted into 50 mL tubes and stored at −80 °C.
2. The free molecule weight of anhydrous ATP is 551.14, does not include the MW of disodium and $\times H_2O$ which is different from batch to batch. This part need to be added for calculation.

3. This device allows for precise temperature control and stirring of cells suspension to which a stimulus is delivered within one second, allowing uninterrupted measurement of cellular response. This Time Zero system is compatible with BD FACS-Calibur and FACSFortessa. It consists of two modules, the Time Zero module with water-jacketed tube holder and the Air Supply module. An additional circulating water bath is also needed for temperature control. Because subsecond cell response is not crucial for pore dilation assay, the Air Supply module can be ignored. Just unscrew the sample platform and fit the water-jacketed tube holder on the sample bar of a BD flow cytometer (Fig. 1). However, the tube has to be physically removed and replaced after the addition of ATP, which incurs a delay of 2–5 s before resuming recording. A tiny stir bar (1 × 3 mm) has to be placed in the bottom of tube in order to mix cells. A major advantage of the Time Zero system is the device for magnetic stirring of the reaction cuvette, which maximizes the number of cell–cell interactions as well as rapidly mixing agonist or probe into the suspension. It is also a good idea to leave a small amount of water inside the water-jacket tube holder to ensure good thermal conductivity to the tube.
4. WinMDI is a freeware developed by Joseph Trotter of the Scripps Institute; the last version is 2.9, which works under Windows XP. You will need Windows Virtual PC with Windows XP model to be installed if you are using Windows 7. It does not work with Windows 10 or Apple computers. The software only recognizes 8 digits in filename. Nevertheless, this is the only freeware which can analyze the kinetic response.
5. FlowJo is a widely used flow cytometry software, which can be used to analyze kinetic response as well, although its kinetic function is not as good as WinMDI. The latest version (Version 10) has serious bugs in kinetic function therefore it is recommended that version 7.6.5 to be used for kinetic analysis.
6. After Ficoll density centrifugation, the middle layer of PBMC aspirated from the tube will inevitably contain some of the high density Ficoll-Hypaque. Therefore, a higher g-force (350 × *g*) and a longer duration (8 min) are needed to spin down the cells.
7. The optimum amount of antibody should be previously established by titration of varying dilutions of antibody with cells to find the lowest concentration which gives saturation of binding. This is usually less than the recommended dilution by the manufacturers.
8. If the FACSCalibur is not available in a short time, the cells can be left on ice and in Na media for up to 2 h. Cells in K media will become leaky after 1 h or so, resulting in increased ethidium baseline.

9. Because of the long tubing connecting the circulating water bath and the tube adaptor of Time Zero module.
10. Adding ATP into the tube requires a bit practice. This needs to be done quickly, preferably within 10 s. To do this, one has to release the arm level, take the tube and the adaptor out, add ATP, then reinsert the tube back, as well as the adaptor. When tube and the adaptor are in place, pull the tube down to the bottom, to ensure there is enough space for the small stirring bar.
11. Sometime you may get an error message saying "Out of Memory" when you save the date to a table text file. Just quit WinMDI and restart the program.
12. The array formula is different with normal formula. The {} cannot be added manually, you will need to press "Ctrl + Shift+Enter" instead of "Enter" to make it an array formula after you finished editing.
13. In a situation where the cells cannot be washed (e.g., an ex vivo study on drug effect), a whole blood assay is needed to maintain the drug effect during ATP activating P2X7 pore formation.
14. It is not recommended to use hypotonic red cell lysing buffer to lysis red cells in this method, as it greatly increases the baseline of ethidium$^+$ uptake in all leukocytes.
15. Ethidium fluorescence intensity is collected in log scale. To determine the voltage setting for FL2, we can use digitonin (0.1%) or saponin to completely permeabilized cells to achieve maximum fluorescent intensity of ethidium$^+$ (Fig. 4).

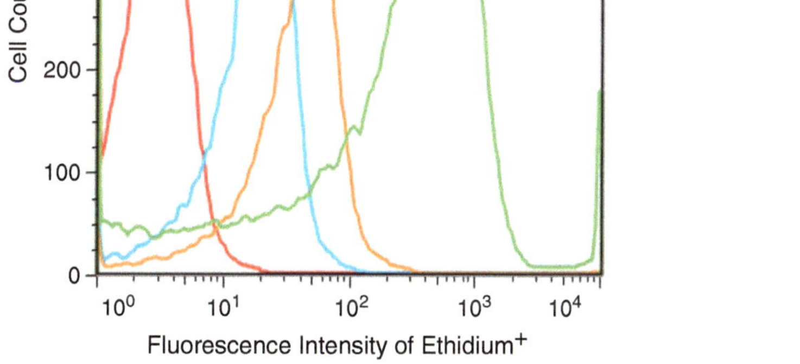

Fig. 4 Example histograms of whole blood ethidium uptake assay

References

1. North RA (2002) Molecular physiology of P2X receptors. Physiol Rev 82:1013–1067
2. Ferrari D, Pizzirani C, Adinolfi E, Lemoli RM, Curti A, Idzko M, Panther E, Di Virgilio F (2006) The P2X7 receptor: a key player in IL-1 processing and release. J Immunol 176:3877–3883
3. Pelegrin P (2011) Many ways to dilate the P2X7 receptor pore. Br J Pharmacol 163:908–911
4. Li M, Toombes GE, Silberberg SD, Swartz KJ (2015) Physical basis of apparent pore dilation of ATP-activated P2X receptor channels. Nat Neurosci 18:1577–1583
5. Wiley JS, Gargett CE, Zhang W, Snook MB, Jamieson GA (1998) Partial agonists and antagonists reveal a second permeability state of human lymphocyte P2Z/P2X_7 channel. Am J Phys Cell Phys 44:C1224–C1231
6. Gu BJ, Rathsam C, Stokes L, McGeachie AB, Wiley JS (2009) Extracellular ATP dissociates nonmuscle myosin from P2X(7) complex: this dissociation regulates P2X_7 pore formation. Am J Physiol Cell Physiol 297:C430–C439
7. Gu BJ, Zhang WY, Bendall LJ, Chessell IP, Buell GN, Wiley JS (2000) Expression of P2X7 purinoceptors on human lymphocytes and monocytes: evidence for nonfunctional P2X7 receptors. Am J Phys Cell Phys 279: C1189–C1197
8. Jursik C, Sluyter R, Georgiou JG, Fuller SJ, Wiley JS, Gu BJ (2007) A quantitative method for routine measurement of cell surface P2X7 receptor function in leucocyte subsets by two-colour time-resolved flow cytometry. J Immunol Methods 325:67–77
9. Gu BJ, Wiley JS (2012) Broad applications of multi-colour time-resolved flow cytometry. In: Schmid I (ed) Flow cytometry: recent perspectives. InTech, Croatia, pp 185–202

Chapter 25

Detection of Inflammasome Activation by P2X7 Purinoceptor Activation by Determining ASC Oligomerization

Helios Martínez-Banaclocha and Pablo Pelegrín

Abstract

The purinoceptor P2X7 is highly expressed in cells of the innate immune system, including monocytes and macrophages. Its activation is a potent signal to activate the NLRP3 inflammasome and induce the release of proinflammatory cytokines of the IL-1 family. In this chapter, we present a method to monitor NLRP3 inflammasome activation in human monocytes upon P2X7 receptor stimulation by detecting intracellular oligomers of ASC by flow cytometry. This method could be used to evaluate the degree of inflammasome activation in blood samples from patients suffering different chronic inflammatory diseases.

Key words NLRP3, Flow cytometry, ASC, Monocyte, Inflammasome, ATP, P2X7

1 Introduction

Extracellular ATP is considered a intercellular messenger molecule that signals upon activation of a wide range of purinoceptors in target cells [1, 2]. In steady state, extracellular concentrations of ATP are usually low due to the action of ectonucleotidases enzymes that will degrade ATP; however, in conditions of tissue damage or inflammation, extracellular ATP concentrations increase and then it signals as a danger molecule to alert the immune system of the host [3, 4]. High concentrations of extracellular ATP (in the range of mM) activates the P2X7 receptor that induces the activation of the nucleotide-binding domain and leucine-rich repeat containing receptor with a pyrin domain 3 (NLRP3) inflammasome [5, 6]. Upon activation NLRP3 oligomerizes, forming scaffolds to nucleate the apoptosis-associated speck-like protein containing a CARD (ASC) in large filaments that will finally result in large ASC oligomers that are commonly named as ASC specks [7, 8]. ASC interacts with the CARD domain of pro-caspase-1, inducing its activation within the inflammasome [9]. Caspase-1 will now

Pablo Pelegrín (ed.), *Purinergic Signaling: Methods and Protocols*, Methods in Molecular Biology, vol. 2041,
https://doi.org/10.1007/978-1-4939-9717-6_25, © Springer Science+Business Media, LLC, part of Springer Nature 2020

execute a specific type of cell death called pyroptosis upon gasdermin D cleavage and plasma membrane pore formation [10]. Caspase-1 also cleaves the inactive precursors of the cytokines interleukin (IL)-1β and IL-18 into their mature forms [11]. Also casapse-1 is able to process other cytosolic substrates and induces the release of different intracellular proteins [12]. The detection of IL-1β processing by Western blot or ELISA is one of the hallmarks of NLRP3 inflammasome activation upon P2X7 receptor activation. Here, we describe the use of a flow cytometry technique to determine intracellular ASC speck formation [13] in human monocytes after ATP stimulation.

2 Materials

Prepare all solutions using ultrapure water and analytical grade reagents. Prepare and store all reagents at room temperature (unless indicated otherwise). Diligently follow all waste disposal regulations when disposing of waste materials.

2.1 Solutions and Stimulations

1. Sterile Ficoll solution (*see* **Notes 1** and **2**).
2. Sterile phosphate buffered saline (*see* **Note 3**).
3. Culture media: Roswell Park Memorial Institute (RPMI) 1640 medium with 10% of heat-inactivated fetal calf serum (FCS).
4. ATP dilution buffer: 147 mM NaCl, 10 mM HEPES, 13 mM D-(+)-Glucose, 2 mM KCl, 2 mM $CaCl_2$, and 1 mM $MgCl_2$, pH 7.4. After preparation sterilize the buffer through 0.22 μm filters (*see* **Note 4**).
5. 1 μg/ml ultrapure lipopolysaccharide from Escherichia coli 0111:B4 (LPS) solution in culture media (*see* **Notes 5** and **6**).
6. 100 mM ATP solution in ATP dilution buffer pH 7.4 (*see* **Note 7**).
7. Flow cytometry staining buffer: phosphate buffered saline with1% of FCS and 0.1% of sodium azide (*see* **Note 8**).
8. Cell permeabilization buffer: phosphate buffered saline with 3% of FCS, 0.1% of sodium azide and 0.1% of saponin (*see* **Note 8**).
9. Cell fixation buffer: 2% paraformaldehyde in phosphate buffered saline (*see* **Note 9**).

2.2 Materials and Equipment

1. Cell culture CO_2 incubator set at 5% CO_2 and 37 °C.
2. Biological Safety Cabinets, Class II.
3. Flow cytometer equipped with argon laser and filter settings for the employed fluorophores (*see* **Note 10**).
4. Flow cytometry data analysis software, such as FCS express from De Novo Software, but other similar software is also suitable.

5. Vortex.
6. Centrifuge.
7. Micropipettes and pipet tips for 10, 100, 200, and 1000 μl.
8. 15 and 50 ml sterile conical centrifuge tubes.
9. Plastic sterile Pasteur pipettes.
10. Centrifuge for conical 15 and 50 ml tubes.
11. 10 ml sterile glass pipettes and pipette vacuum cleaner.
12. Sterile cell-culture treated 24-well plates.
13. Flow cytometry polypropylene tubes.
14. Specific antibodies for the detection of ASC (*see* **Note 11**).
15. Goat or donkey anti-rabbit Alexa Fluor 488 secondary antibody (*see* **Note 12**).
16. Mouse anti-human CD14 fluorophore conjugated antibodies for the detection of monocytes within peripheral blood mononuclear cells (PBMCs) (*see* **Note 13**).
17. Peripheral blood human sample (*see* **Notes 14** and **15**).
18. Ice.

3 Methods

3.1 Peripheral Blood Mononuclear Cell Isolation

1. Add 3 ml of anticoagulated blood to a conical centrifuge tube and 3 ml of sterile phosphate buffered saline.
2. Mix the blood and buffer by inverting the tube several times or by gently pipetting the mixture in and out of a pipette.
3. Agitate by inversion the sterile Ficoll solution bottle several times to ensure thorough mixing.
4. Add 3 ml of sterile Ficoll media to a conical centrifuge tube.
5. Carefully layer the diluted blood sample on the top of Ficoll media solution (*see* **Note 16**).
6. Centrifuge at 400 × *g* for 40 min at room temperature with centrifuge brake turned off.
7. Carefully recover the layer of mononuclear cells using a sterile pipette and transfer to a sterile conical centrifuge tube.
8. Add at least 3 volumes of phosphate buffered saline to the mononuclear cells in the centrifuge tube.
9. Suspend the cells by gently pipetting repeatedly in and out of a pipette.
10. Centrifuge at 400 × *g* for 10 min at 18 °C (*see* **Note 17**).
11. Remove the supernatant.
12. Resuspend the mononuclear cells in 8–10 ml of phosphate buffered saline.

13. Centrifuge at 400 × *g* for 10 min at 18 °C (*see* **Note 17**).
14. Remove the supernatant.
15. Resuspend the mononuclear cells in 1 ml of culture media.
16. Count the cells (*see* **Note 18**).

3.2 Cell Stimulation

1. Seed 5 × 10^5 cells in 500 μl of culture media per duplicate and per condition to test in sterile flow cytometry tubes or into 24-well culture plates.
2. Incubate the cells with LPS final concentration for 2 h at 37 °C and 5% CO_2. For negative control use two replicates incubated with culture media without LPS.
3. Add 3 mM of ATP (final concentration) to the cultured cells in the presence of LPS and incubate for 30 min at 37 °C and 5% CO_2 to activate the P2X7 receptor and subsequently the NLRP3 inflammasome (*see* **Notes 19** and **20**) (Fig. 1).

3.3 Assessing Inflammasome Activation by Flow Cytometry

1. After inflammasome activation, centrifuge tubes or cell-culture plate at 350 × *g* for 5 min to pellet floating cells and carefully remove the supernatants (*see* **Note 21**).
2. Resuspend the cells in 500 μl of cell fixation buffer.
3. Incubate cells during 10 min on ice (*see* **Note 22**).
4. Wash the cells two times with 2 ml of staining buffer and centrifuge cells at 400 × *g* during 5 min at room temperature.

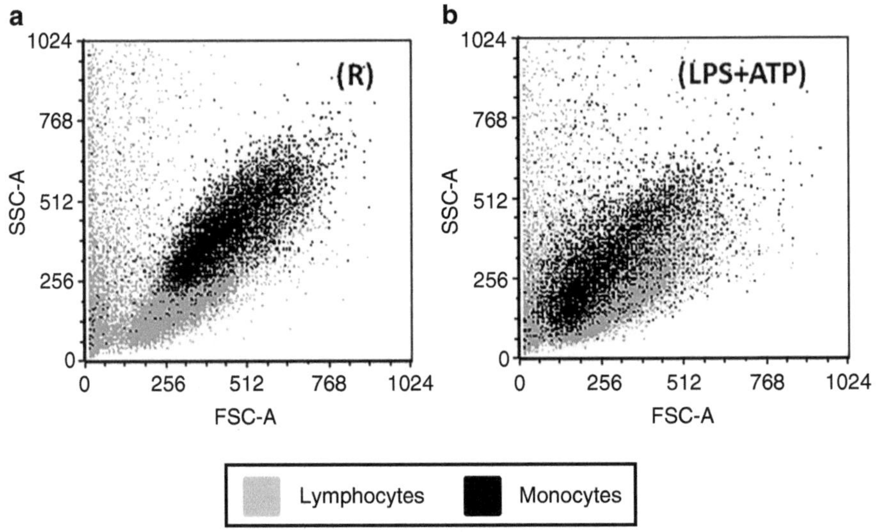

Fig. 1 Size and granularity of human monocytes before and after NLRP3 inflammasome activation triggered by P2X7 receptor activation. Flow cytometry plots showing the granularity (SSC) and size (FSC) of (**a**) untreated resting (R) or (**b**) LPS + ATP treated human lymphocytes and monocytes

5. Incubate the cells with 250 μl of permeabilization buffer and 250 μl of a 1:500 dilution of anti-ASC antibody in staining buffer to achieve a final dilution of 1:1000 in the desired tubes (*see* **Note 23**).
6. Incubate 1 h at room temperature.
7. Wash cells with 2 ml of staining buffer at 400 × *g* during 5 min at room temperature.
8. Add 100 μl of a 1:1000 dilution of donkey anti-rabbit IgG Alexa Fluor 488 antibody in cell permeabilization buffer per tube and incubate tubes for 45 min at room temperature in the darkness (*see* **Note 23**).
9. Wash cells with 2 ml of staining buffer at 400 × *g* during 5 min at room temperature.
10. Resuspend the cells in 500 μl of staining buffer.
11. Acquire the cells in a flow cytometer and analyze the staining of ASC in monocytes (Fig. 2), as well as the distribution of ASC FITC width vs ASC FITC area in $CD14^+$ monocytes to determine and quantify the percentage of ASC specking monocytes (Fig. 3).

4 Notes

1. If you have Ficoll solution stored at 4 °C, it must be allowed to warm at room temperature prior to its use.
2. Ficoll® is a neutral, highly branched, high-mass, hydrophilic polysaccharide, patented by GE Healthcare. It can be found commercialized by different companies as Ficoll-Paque®.
3. For peripheral blood mononuclear cell isolation a sterile 0.9% NaCl solution could replace the use of phosphate buffered saline during the isolation steps.
4. For long storage periods keep the buffer at 4 °C.
5. Prepare 1 mg/ml stock of LPS solution in sterile DMEM or RPMI media without serum, aliquot and store at −20 °C.
6. 100 ng/ml of LPS also primes the cells.
7. This is the stock ATP solution, aliquot and store at −20 °C, do not refreeze aliquots.
8. Store the buffer at 4 °C until use. Flow cytometry staining buffer and Cell permeabilization buffer can be stored at 4 °C for up to 6 months.
9. To prepare this buffer, we normally use a commercial 32% solution of paraformaldehyde from Electron Microscopy Sciences, although paraformaldehyde from other companies is also suitable. Paraformaldehyde is toxic and carcinogenic. Wear

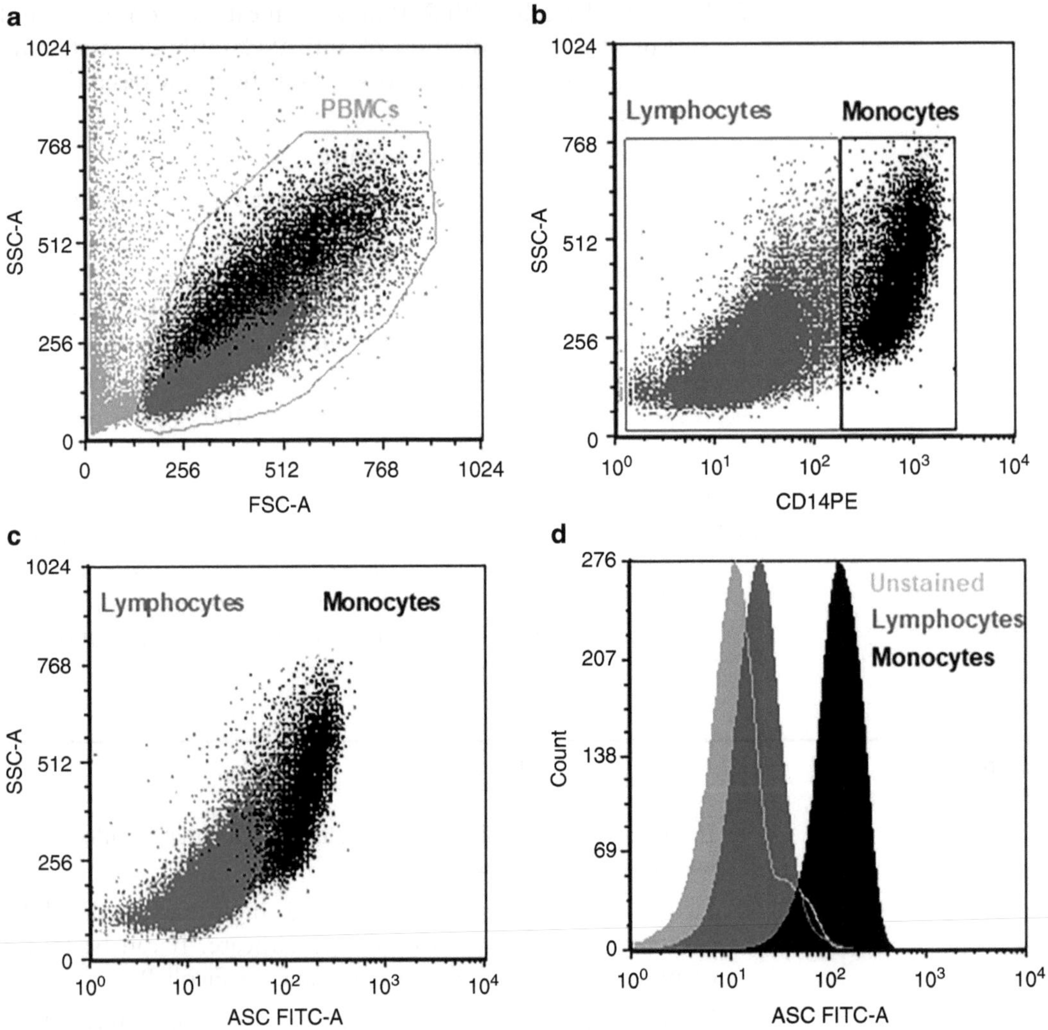

Fig. 2 Staining of intracellular ASC in human monocytes by flow cytometry. (**a**) Peripheral blood mononuclear cells (PBMCs) were gated based on SSC and FCS dot plot. (**b**) Monocytes were gated based on SSC and CD14-PE dot plot. (**c**) Dot plot showing intracellular ASC FITC staining of lymphocytes (grey) and monocytes (black). (**d**) Histogram showing the mean fluorescent intensity of ASC-FITC in lymphocytes (grey) and monocytes (black). Cells were stained with rabbit anti-ASC antibody clone N-15-R from Santa Cruz followed by the staining with secondary antibody directly conjugated with Alexa 488

protective equipment according to the local regulations. Paraformaldehyde solutions need to be made up fresh every time. In most methods of immunostaining paraformaldehyde is used at 4%; here we use 2% paraformaldehyde to facilitate surface staining after fixation.

10. We normally use a flow cytometer equipped with a 488 nm laser with 505/45 nm emission optics.

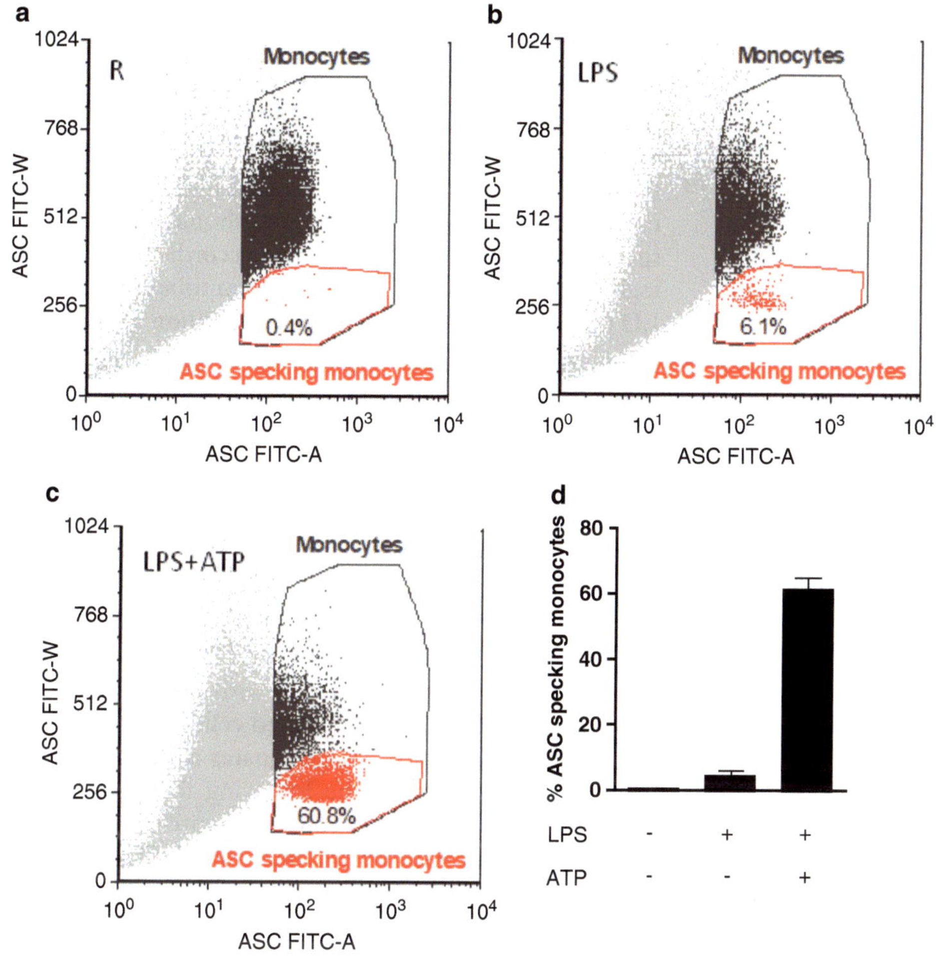

Fig. 3 Staining and quantification of intracellular ASC specks in human monocytes by flow cytometry after P2X7 receptor stimulation. Dot plots showing intracellular ASC specks in (**a**) resting, (**b**) LPS, and (**c**) LPS + ATP treated monocytes; black dots represent monocytes not forming intracellular ASC specks; red dots represent monocytes forming intracellular ASC specks. Cells were fixed and stained with rabbit anti-ASC antibody clone N-15-R from Santa Cruz followed by the staining with secondary antibody directly conjugated with Alexa 488. (**d**) Quantification of the percentage of monocytes with intracellular ASC specks from $n = 5$ healthy donors

11. There are different anti-human ASC antibodies suitable to detect intracellular ASC specks by flow cytometry; here we use the rabbit polyclonal anti-ASC from Santa Cruz (clone N-15-R), although other antibodies or clones might be also suitable.
12. These conjugated antibodies are exclusive of Life Technologies, although other fluorophore-conjugated antibodies from other companies are also suitable.
13. Alternatively to CD14, monocytes can be also identified in PBMCs with anti-CD33 staining. Here we use PE conjugated mouse anti-human CD14 (clone 61D3) from Tonbo

Biosciences and APC-Vio770 conjugated mouse anti-human CD33 (clone AC104.3E3) from Miltenyi Biotech. These antibodies clones work properly when are used after fixation with 2% paraformaldehyde. If other antibodies clones are used, the staining in fixed cells should be optimized.

14. Fresh blood should be used to ensure high viability of isolated mononuclear cells. Human samples should be obtained under specific consent from donors and according to specific local legislation. Blood could be collected in tubes containing either EDTA or sodium heparin as anticoagulation agents.
15. Other macrophages or monocytes types could be also used, that is, THP-1 cells, mouse peritoneal macrophages, or mouse bone marrow derived macrophages.
16. In this step try not to mix the Ficoll with the blood sample, so lay blood sample slowly, with the tube inclined.
17. A centrifugation at high speed (400–500 × *g*) increases the mononuclear cell recovery. To remove platelets lower centrifugation speed is recommended (60–100 × *g*). In both cases temperature could range from 18–20 °C.
18. Make a dilution of obtained cells with trypan blue to exclude death cells and put 10 μl of diluted cells in a Neubauer chamber. Count at least three large squares and calculate the concentration of cells using the following formula: (Number of cells × 1000)/(Number of squares counted × dilution). For a dilution of 1/10 with trypan blue the dilution factor will be 0.1.
19. Inflammasome activation in cells usually results in disintegration of the plasma membrane due to gasdermin D pore formation and cell morphology changes, resulting in a loss of cell size (FSC) and granularity (SSC). This must be taken into account when analyzing inflammasome activated cells by flow cytometry.
20. This method is also suitable to study and quantify ASC specking macrophages upon different stimuli activating the NLRP3 inflammasome (as nigericin), Pyrin inflammasome (as Clostridium difficile toxin B), and other inflammasomes (as AIM2 or NLRC4) in human PBMCs and other cell types.
21. The cell supernatants could be centrifuged to remove detached cells, and stored at −80 °C for posterior analysis of other inflammasome activation parameters as the release of IL-1β or IL-18 by ELISA or the presence of extracellular lactate dehydrogenase as marker of pyroptotic cell death.
22. If cell stimulation is performed in flow cytometry tubes, fixation can be done directly in tubes with 2% paraformaldehyde solution during 10 min on ice. If cell stimulation is performed

in cell-culture plates remove the supernatant after centrifugation, put the plate on ice, add 2% paraformaldehyde to wells, gently pipette up and down several times to detach the cells, transfer the cells to flow cytometry tubes and fix them on ice during 10 min.

23. To optimize staining of ASC specks, it is important to titrate the antibodies to get the best signal. We found that the primary anti-ASC antibody from Santa Cruz works well at a final dilution of 1:1000. In the case of the fluorescently conjugated secondary antibody from Life Technologies, the optimal signal was obtained with a 1:1000 dilution. Unless stated otherwise, this concentration was used throughout in the experiments described here.

Acknowledgments

This work was supported by the European Research Council (ERC-2013-CoG 614578), EU COST Program (BM1406), Fundación Séneca (20859/PI/18) and Ministerio de Economia, Industria y Competitividad–Fondo Europeo de Desarrollo Regional (project no. SAF2017-88276-R).

References

1. Baroja-Mazo A, Barberà-Cremades M, Pelegrin P (2013) The participation of plasma membrane hemichannels to purinergic signaling. Biochim Biophys Acta 1828:79–93
2. Surprenant A, North RA (2009) Signaling at purinergic P2X receptors. Annu Rev Physiol 71:333–359
3. Amores-Iniesta J, Barberà-Cremades M, Martínez CM et al (2017) Extracellular ATP activates the NLRP3 inflammasome and is an early danger signal of skin allograft rejection. Cell Rep 21(12):3414–3426
4. Barberà-Cremades M, Baroja-Mazo A, Gomez AI et al (2012) P2X7 receptor-stimulation causes fever via PGE2 and IL-1β release. FASEB J 26:2951–2962
5. Pelegrin P (2011) Many ways to dilate the P2X7 receptor pore. Br J Pharmacol 163:908–911
6. Young MT, Pelegrin P, Pelegrin P et al (2006) Identification of Thr283 as a key determinant of P2X7 receptor function. Br J Pharmacol 149:261–268
7. de Torre-Minguela C, Mesa del Castillo P, Pelegrín P (2017) The NLRP3 and pyrin inflammasomes: implications in the pathophysiology of autoinflammatory diseases. Front Immunol 8:43
8. Schmidt FI, Lu A, Chen JW et al (2016) A single domain antibody fragment that recognizes the adaptor ASC defines the role of ASC domains in inflammasome assembly. J Exp Med 213:771–790
9. Boucher D, Monteleone M, Coll RC et al (2018) Caspase-1 self-cleavage is an intrinsic mechanism to terminate inflammasome activity. J Exp Med 215(3):827–840
10. Shi J, Gao W, Shao F (2017) Pyroptosis: gasdermin-mediated programmed necrotic cell death. Trends Biochem Sci 42:245–254
11. Compan V, Baroja-Mazo A, Bragg L et al (2012) A genetically encoded IL-1β bioluminescence resonance energy transfer sensor to monitor inflammasome activity. J Immunol 189:2131–2137
12. de Torre-Minguela C, Barberà-Cremades M, Gómez AI et al (2016) Macrophage activation and polarization modify P2X7 receptor secretome influencing the inflammatory process. Sci Rep 6:22586
13. Sester DP, Sester DP, Thygesen SJ et al (2014) A novel flow cytometric method to assess inflammasome formation. J Immunol 194:455–462

Chapter 26

Measuring Leukocyte Migration to Nucleotides

Taylor J. Moon and Michael R. Elliott

Abstract

Extracellular nucleotides are potent damage-associated molecular patterns that shape the immune response to cell stress and tissue damage. These nucleotides are sensed by purinergic receptors and mediate a wide range of cellular effects. Among the best characterized of these effects is cellular migration. While the motility responses of leukocytes to nucleotides can be achieved by microscopic live-cell imaging approaches, such systems are time-consuming and require costly equipment and analysis tools not readily available to all researchers. Transwell migration chambers are a widely used alternative to microscopy due to their relatively low cost and moderate through-put capacity. However, extracellular nucleotides are labile and rapidly degraded in serum-containing cell cultures due to the presence of phosphohydrolases. Thus, evaluating leukocyte migration to nucleotides presents a number of challenges not seen with more stable classes of chemoattractants like proteins and lipids. Here we describe a method to measure leukocyte migration to nucleotides that is cost-effective, rapid and produces robust and reproducible migration of leukocytes using transwell migration chambers.

Key words Cell migration, Chemotaxis, Nucleotides, ATP, UTP, Transwell, Boyden chamber, Leukocytes, Purinergic signaling

1 Introduction

Transwell chambers, also known as modified Boyden chambers, are one of the most commonly used systems to study the signaling processes that control cell migration to chemoattractants [1]. These assays are relatively simple to set up and, due to the availability of disposable transwell plates from a number of commercial vendors, are one of the first-line techniques used to study cell migration under very controlled experimental conditions. Conceptually, this is a fairly straightforward experimental approach. Chemoattractant-containing media is placed in the lower well of a tissue culture plate followed by addition of a removable upper chamber containing the cells interest in medium without chemoattractant. The lower portion, or "floor," of the upper chamber is comprised of a synthetic (typically polycarbonate) mesh or "filter" manufactured to contain a discrete number of regularly spaced

Pablo Pelegrín (ed.), *Purinergic Signaling: Methods and Protocols*, Methods in Molecular Biology, vol. 2041,
https://doi.org/10.1007/978-1-4939-9717-6_26,

circular pores of defined diameter. The presence of chemoattractant in the lower but not upper chamber establishes a chemoattractant gradient at the interface of the upper and lower chambers that acts to stimulate chemotaxis. The assembled chamber is then incubated for a specific time and under controlled environmental conditions determined by the user necessary for sufficient chemotaxis to take place. Cells that traverse from the upper chamber through the filter and into the lower chamber are quantified using a variety of basic cell counting techniques, and the data are typically expressed as the fraction of cells originally placed in the upper chamber that have successfully migrated to the lower chamber (i.e., "percent migration").

As migration through the transwell pores is dependent on establishment of a chemoattractant gradient, the stability of such chemoattractants is a major, if often underappreciated, factor in these assays. This is particularly true for migration induced by ATP and UTP. These nucleotides can have potent and rapid effects on leukocyte motility and have been shown to be important endogenous signals that enable swift response of immune cells to cellular damage in many contexts, including apoptotic cell clearance, anti-tumor immunity, and neuronal cell death [2–6]. However, extracellular triphosphate and diphosphate nucleotides are rapidly hydrolyzed to relatively inert monophosphates through the actions of phosphohydrolases present in serum-containing media as well as ectohydrolases expressed on the surface of migrating cells used in the assay. Chief among these ectoenzymes is CD39, whose expression has been shown to modulate cell motility and immune responses of many different leukocyte subsets [7]. However, as discussed below, we and others have successfully modified classical transwell chamber assay methods that account for nucleotide degradation in order to assess leukocyte migration to these important signaling molecules.

2 Materials

All materials should be prepared using sterile, tissue culture grade reagents and aseptic techniques throughout the assay to prevent microbial contamination of cultures.

2.1 Cell Culture

1. Here we describe transwell assays using the human monocytic cell line THP-1 as our exemplar cell type. For other cell types of interest, there are a number of important considerations and potential preliminary experiments required in choosing a cell type that will provide robust transwell migration results (*see* **Note 1**).

2. THP-1 growth medium: RPMI 1640 supplemented with 1× final penicillin/streptomycin/L-glutamine, 10% heat-inactivated fetal bovine serum, 10 mM HEPES.

2.2 Transwell Migration Materials

1. Migration medium: RPMI 1640 supplemented with 1× final penicillin/streptomycin/L-glutamine, 0.5% bovine serum albumin, 10 mM HEPES. Once components are mixed and dissolved, filter through 0.2 μm filter top bottle and stored at 4 °C (*see* **Note 2**).
2. Transwell chambers: choose the appropriate plate size and filter pore size for your cell type (*see* **Note 3**).
3. Nucleotides: 1–1000 μM purified nucleotides dissolved in H_2O, stored at −20 °C.

2.3 Materials for Quantifying Migration

1. Flow cytometry quantitation beads: AccuCount Blank Particles (5.0 μm diameter).
2. Flow cytometer.

3 Methods

3.1 Cell Culture

1. THP-1 cells should be routinely maintained at a density of 0.2–0.5 × 10^6/mL.
2. Approximately 16–24 h prior to the transwell experiment, pellet the cells by centrifugation at 350 × *g* for 5 min.
3. Resuspend cells in fresh THP-1 growth medium at 0.3 × 10^6 cells/mL.

3.2 Transwell Chamber Setup

1. Thaw nucleotides on ice.
2. Dilute nucleotides in migration medium at desired concentration, typically from 0.1 to 1 μM.
3. With upper chamber in the well, load chemoattractant into the lower chamber for each condition (*see* **Notes 4–6**).
4. Load duplicate input control wells with 300 μL migration medium and no transwell insert.
5. Place plate in environmentally controlled incubator to warm and equilibrate for at least 30 min (*see* **Note 7**).
6. Collect cells by centrifugation, wash once in migration medium and resuspend in migration medium at 1 × 10^6 cells/mL.
7. Load cells into the upper chambers of duplicate or triplicate transwells and also into input control wells containing same volume of medium as in lower chambers of transwells less the volume of input cells (*see* **Note 8**).
8. Place the plates in the incubator at 37 °C and 5% CO_2.

9. Incubate for 3 h.
10. Remove plates and carefully remove the inserts (*see* **Note 9**)—discard inserts.
11. Add 50 μL of flow cytometry counting beads per well.
12. Pipette medium in the well gently and transfer to flow cytometry tubes (*see* **Note 10**).
13. Record events on the flow cytometer by gating on beads and collecting a fixed number of bead events per sample (typically 2000–5000 bead events per sample). The number of cell gate events will allow you to calculate a relative migration amount compared to input controls.
14. Calculate the percent migration (*see* **Note 11**).

4 Notes

1. Due to the instability of nucleotides, it is important to evaluate your cell type of interest expresses the main ectophosphohydrolase, CD39, on the surface. This can be done by flow cytometry using commercially available anti-CD39 antibodies. If CD39 is expressed at high levels on your cell type of interest, cells may be pretreated with ENTPD1 inhibitors (e.g., POM1) to block this endogenous phosphohydrolase activity prior to starting the assay.
2. Most FBS contains detectable phosphohydrolase activity that is not eliminated by heat-inactivation at 56 °C. Therefore, it is recommended that these migration assays are carried out in serum-free, defined media. The simplest approach for this is to replace FBS with 0.5% BSA. However, the performance and/or survival of some cell types can be compromised when incubated in BSA-containing medium. In these cases, media containing defined growth factors is a suitable alternative.
3. The appropriate pore diameter for a particular cell type should be determined empirically, as there are a number of factors that can affect the ability of a cell to move through a pore. In general, most leukocyte cell lines have a diameter of 10–15 μm, and will require a 5 or 8 μm pore diameter. However, smaller cell types, such as primary lymphocytes, with a diameter of 7 μm may require a smaller pore size of 3 μm in order to prevent high levels of cell migration in the absence of chemoattractant.
4. The volume of medium/chemoattractant to add to lower chamber varied depending on the well size. For 24 well plates, we have seen best results loading 500 μL per well.
5. It is important to load the lower chamber while the transwell insert is in place. This is best accomplished by dispensing the

medium through the openings in the upper chamber to prevent trapping bubbles under the filter.

6. When loading the lower chamber, take care not to introduce bubbles from pipet tip into them chamber, which can then be trapped under the filter and disrupt migration of cells through the filter.
7. The timing for prewarming medium in lower chamber should not exceed 1 h.
8. For 24-well plate assays, 200 μL (2×10^5 cells) in the upper chamber (and also input wells) is recommended.
9. Use fine-tip tweezers to remove inserts gently and prevent spillage of upper chamber contents into the lower chamber.
10. Insufficient pipetting in the step will lead to low and/or inconsistent cell count results. Pipet up and down at least 15 times and swirl as you do so to ensure you get all the cells from the well.
11. To calculate percent migration use the input control cell numbers as 100% migration, as these wells represent the cells you would record if all of them could migrate through the membrane. Each migration condition is calculated a part of this whole and thus becomes a percentage for reading out this assay.

$$\frac{100 \times \frac{(\#\text{Cells in lower chamber for experimental condition})}{(\#\text{Beads in lower chamber for experimental condition})}}{\frac{(\#\text{Cells in input chamber})}{(\#\text{Beads in input chamber})}}$$

Acknowledgments

This work was supported by NIH grants R01 AI114554, P30 AI027767, and T32 AI049815.

References

1. Zigmond SH, Foxman EF, Segall JE (2001) Chemotaxis assays for eukaryotic cells. Curr Protoc Cell Biol. Chapter 12:Unit 12.1–12.1.29
2. Cekic C, Linden J (2016) Purinergic regulation of the immune system. Nat Rev Immunol 16:177–192
3. Ferrari D, McNamee EN, Idzko M, Gambari R, Eltzschig HK (2016) Purinergic signaling during immune cell trafficking. Trends Immunol 37:399–411
4. Burnstock G, Boeynaems J-M (2014) Purinergic signalling and immune cells. Purinergic Signal 10:529–564
5. Idzko M, Ferrari D, Eltzschig HK (2014) Nucleotide signalling during inflammation. Nature 509:310–317
6. Medina CB, Ravichandran KS (2016) Do not let death do us part: "find-me" signals in communication between dying cells and the phagocytes. Cell Death Differ 23:979–989
7. Antonioli L, Pacher P, Vizi ES, Haskó G (2013) CD39 and CD73 in immunity and inflammation. Trends Mol Med 19:355–367

Chapter 27

Assessment of Cell Adhesion After Purinoceptor Activation

Juan Jose Martínez-García and Pablo Pelegrín

Abstract

Cell adhesion is a characteristic feature of phagocytic myeloid cells, which is important for several inflammatory processes, such as migration, invasion, and proliferation. Purinergic signaling in macrophages plays an important role in cell adhesion of this cell type to different extracellular substrates. This protocol describes the use of two different detection methods to quantify cell adhesion upon P2X7 receptor activation by extracellular ATP.

Key words Cell adhesion, P2X7 receptor, Macrophage, Extracellular matrix, Extracellular ATP, MTT, Yo-Pro-1

1 Introduction

Cell adhesion is an important feature of macrophages, and is involved in different processes of myeloid cells during the inflammatory response such as cell migration, invasion, and proliferation. These processes are possible due to the ability of macrophages to adhere to different extracellular substrates, including the binding to activated endothelial cells that allow monocyte extravasation into areas of infection and injury [1]. Macrophages have also the ability to adhere to different components of the extracellular matrix (ECM), such as collagen, laminin, or fibronectin, via integral plasma membrane proteins termed integrins, that link the ECM with the intracellular actin cytoskeleton [2, 3].

P2X7 receptor plays an important role in the innate immune response [4]. P2X7 receptor is a trimeric ion channel located in the plasma membrane, that presents three binding sites for ATP in its extracellular loop [5]. P2X7 receptor activation by extracellular ATP allows cations transit through its pore, inducing potassium efflux, and calcium and sodium influx. Potassium efflux is associated to the activation of the macrophage by inducing the release of proinflammatory cytokines [6, 7]. Moreover, it has been shown that P2X7 receptor activation by extracellular ATP also induces a

Pablo Pelegrín (ed.), *Purinergic Signaling: Methods and Protocols*, Methods in Molecular Biology, vol. 2041,
https://doi.org/10.1007/978-1-4939-9717-6_27,

shedding of different adhesion molecules from the plasma membrane of macrophages, which would facilitate cell migration [8–10]. So far, there are no reported protocols to quantification cell adhesion after purinoceptor activation.

In this chapter we present a method to quantify cell–substrate adhesion before and after P2X7 receptor activation in murine macrophages using two different detection techniques for the adherent cells. First, we use a colorimetric detection method using the yellow-colored 3-(4, 5-dimethylthiazol-2-yl)-2, 5-diphenyl tetrazolium bromide (MTT) reduction to the purple-colored formazan, by the action of succinate dehydrogenase enzyme of the mitochondria, and result in a signal proportional to the amount of attached cells [11]. Second, we present a fluorescent method to quantify the number of adherent cells by the use of the cell-impermeant and negative charged molecule Yo-Pro-1. Yo-Pro-1 has 629 Da that upon binding to nucleic acids emits fluorescence that is proportional to the amount of DNA presented in the adherent cells [12]. Therefore, in this chapter we show how to quantify the number of adherent cells after ATP stimulation by staining the cells with the nuclear dye Yo-Pro-1 or by a colorimetric assay using the MTT dye.

2 Materials

Prepare all solutions using ultrapure water and analytical grade reagents. Use sterile solutions and material manipulated in Biological Safety Cabinets Class II. Diligently follow all waste disposal regulations when disposing of waste materials. We do not add sodium azide to the reagents.

2.1 Cell Culture

1. Bone marrow differentiated macrophages (BMDMs) (*see* **Notes 1** and **2**).
2. Dulbecco's Modified Eagle's medium (DMEM).
3. Complete cell medium: high glucose DMEM, supplemented with 20% fetal calf serum and 1% penicillin and streptomycin.
4. Cell culture CO_2 incubator set at 5% CO_2 and 37 °C.
5. Biological safety cabinets, Class II.
6. 96-Well plate, clear, flat and micro-clear bottom, cell culture treated, sterile with lids for MTT measurements (luminescence) (*see* **Note 3**).
7. 96-Well plate, black, flat and micro-clear bottom, cell culture treated, sterile with lids for Yo-Pro detection (fluorescence) (*see* **Note 3**).
8. Inverted phase contrast microscope for cell-state checking during cell plating and stimulations.

9. Sterile pipettes.
10. Sterile 0.2, 0.5, and 1.5 mL tubes.
11. Sterile conical 15 mL tubes.
12. Water bath set at 37 °C.
13. Centrifuge for conical 15 mL tubes.
14. 10 μg/mL collagen I solution in PBS media (*see* **Note 4**).
15. 4 μg/mL laminin solution in PBS media (*see* **Note 4**).
16. 6 μg/mL fibronectin solution in PBS media (*see* **Note 4**).

2.2 Solutions and Stimulation of Cells

1. Orbital shaker.
2. Ice.
3. Stimulation buffer: 10 mM HEPES, 13 mM D-Glucose, 2 mM $CaCl_2$, 2 mM $MgCl_2$, 2 mM KCl, and 147 mM NaCl (*see* **Notes 5** and **6**).
4. 1 μg/mL solution of lipopolysaccharide from Escherichia coli 055:B5 (LPS) in complete cell medium (*see* **Note** 7).
5. 5 μM adenosine 5′-triphosphate (ATP) solution in stimulation buffer (*see* **Notes 8** and **9**).
6. Sterile phosphate-buffered saline (PBS).
7. 20 μM solution of the P2X7 receptor specific inhibitor 3-[[5-(2,3-Dichlorophenyl)-1H-tetrazol-1-yl]methyl]pyridine hydrochloride (A438079) in stimulation buffer (*see* **Note 10**).
8. 20 μM solution of the P2X7 specific inhibitor *N*-(1-{[(cyanoimino)(5-quinolinylamino)methyl]amino}-2,2-dimethylpropyl)-2-(3,4-dimethoxyphenyl)acetamide (A740003) in stimulation buffer (*see* **Note 10**).

2.3 Detection of Cell Adhesion

1. Aluminium foil.
2. Centrifuge with adapter for 96-well plates.
3. Dimethyl sulfoxide (DMSO).
4. 0.75 mg/mL working solution of 3-(4,5-dimethylthiazol-2-yl)-2,5-diphenyl tetrazolium bromide (MTT) in PBS (*see* **Note 11**).
5. 2 μM solution of 4-[(3-methyl-1,3-benzoxazol-2(3H)-ylidene)methyl]-1-[3-(trimethylammonio)propyl]quinolinium diiodide (Yo-Pro-1) with 2% Triton X-100 in stimulation buffer (*see* **Notes 12** and **13**).
6. Plate reader able to detect absorbance at 590 nm with a reference filter of 620 nm (for MTT assay) or able to detect fluorescence at 485 nm excitation and 515 nm emission wavelengths (for Yo-Pro-1 assay).

3 Methods

3.1 Cell Stimulation

1. Warm complete medium and stimulation buffer to 37 °C in a water bath.
2. Seed cells from a differentiated BMDM culture in a 96-well plate, previously coated or not with ECMs (*see* **Note 4**), at a confluence of 10^5 cells per well in a final volume of 100 μL of complete cell medium (*see* **Note 14**). Plate at least three or four replicate wells for each condition to be tested.
3. Allow cells to adhere to the bottom of the wells by incubating the plate for 2 h at 37 °C and 5% CO_2.
4. Once cells are adhered into the wells, carefully discard the medium with a sterile pipette.
5. At this point, BMDM could be stimulated with LPS. For that, add 1 μg/mL of LPS in a final volume of 200 μL of complete cell medium (*see* **Note 15**). If no LPS stimulation is required, please go to **step 8**.
6. Incubate the plate during 4 h at 37 °C and 5% CO_2 to prime macrophages with the LPS.
7. After LPS incubation, discard the complete cell medium with LPS using a sterile pipette.
8. Wash the cells using 200 μL per well of stimulation buffer.
9. Then add 20 μM of P2X7 receptor specific antagonists A438079 or A740003 diluted in 100 μL of stimulation buffer per well, and leave some wells without inhibitor just with 100 μL of stimulation buffer (*see* **Note 16**). If no antagonist is required, please go to **step 11**.
10. Incubate the plate during 10 min at 37 °C and 5% CO_2.
11. Add ATP to a final concentration of 5 mM to wells treated and untreated with the different P2X7 receptor antagonists. Leave some wells without ATP stimulation as controls (*see* **Note 17**).
12. Incubate the plate during 30 min at 37 °C and 5% CO_2.
13. After ATP stimulation, shake the plate vigorously using an orbital shaker at 400 rpm during 30 s to eliminate detached cells from the bottom of each well.
14. Wash the cells twice using 200 μL per well of stimulation buffer. Then continue the determination of remaining attached cells by the preferred method (*see* Subheading 3.2 or 3.3) (*see* **Note 18**).

3.2 Assessing Cell Adhesion with Colorimetry

1. Add 100 μL of DMEM medium to each well (*see* **Note 19**).
2. Add 0.25 mg/mL MTT into the wells with the DMEM (*see* **Note 20**).

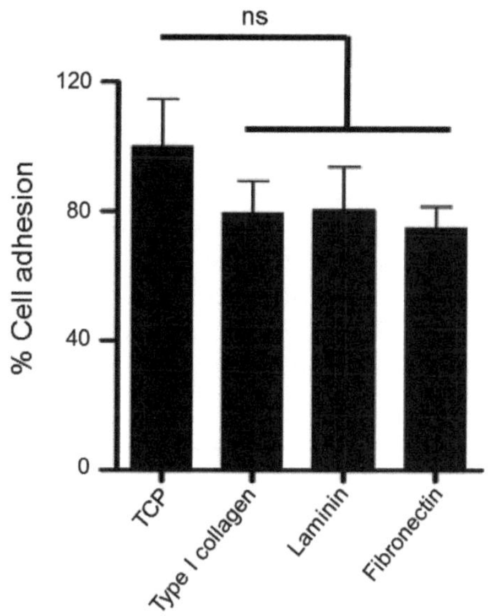

Fig. 1 Detection of adherent macrophages to different ECMs. Percentage of BMDM adhesion measured by MTT to uncoated wells (*TCP* tissue culture plate) or to wells coated with type I collagen, laminin or fibronectin. Each value represents the average of $n = 3$ independent experiments; *ns* no significant difference

3. Incubate the MTT for 4 h at 37 °C and 5% CO_2 to allow the reduction of tetrazolium salt to formazan.
4. After MTT incubation, centrifuge the plate at 433 × *g* during 5 min. Then, formazan crystals deposits on the bottom of wells.
5. Discard the supernatants (*see* **Note 21**).
6. Add 200 μL per well of DMSO.
7. Shake the plate covered with aluminium foil during 30 min at room temperature to solubilize formazan crystals homogeneously on the wells (*see* **Note 22**).
8. Read absorbance at 590 nm with a reference wavelength of 620 nm.
9. Figure 1 presents an example of cell adhesion to different ECMs. 620 nm absorbance values were subtracted from 590 nm values. In this indirect method the corrected absorbance value is directly proportional with the number of adherent cells. Calculate the proportion of adherent cells taking as reference the wells without ATP treatment.

3.3 Assessing Cell Adhesion with Fluorescence

1. Add 100 μL to each well of Yo-Pro-1 with Triton X-100 solution (*see* **Note 23**).
2. Incubate the plate on ice during 5 min.

3. Using an orbital shaker, shake vigorously the plate during 30 min in dark conditions, for that aluminium foil could be used to cover the plate.
4. Read fluorescence using a plate reader at 485 ± 9 nm excitation and 515 ± 9 nm emission wavelengths.
5. Yo-Pro-1 fluorescence is proportional to the amount of the DNA of the adherent cells and therefore to the total number of adherent cells present on each well. Calculate the proportion of adherent cells taking as reference the wells without ATP treatment.

4 Notes

1. To differentiate BMDMs, mouse bone marrow is extracted from tibia and femur and cultured during 6 days at 37 °C with a 5% CO_2 with 100 ng/mL of macrophage-colony stimulating factor (M-CSF) supplemented into complete culture medium.
2. Alternatively to BMDMs, other adherent cells expressing P2X7 receptor could be also used. Adherent cells expressing different purinoceptors could be also used to investigate the role of these receptors on cell adhesion.
3. The same 96-well plate is used during cell culture and during the detection method; depending on the selected method to determine cell adhesion the plate should be different from the beginning.
4. To coat the wells of the 96-well plate with different ECMs, add 120 μL of ECM sterile solution into each well and incubate overnight at 4 °C. Then remove ECM solution and wash the wells twice with PBS before cell seeding.
5. To store stimulation buffer for long periods, filtrate through 0.22 μm filter and store aliquots of 50 mL at 4 °C.
6. This solution is used as the diluent buffer for ATP stimulations and contains the physiological amount of extracellular cations. This solution could have different variants with the aim to block some cationic transit. For example, a higher potassium solution (with 100 mM KCl) blocks potassium efflux from the cells during P2X7 receptor activation.
7. LPS stock solution is prepared at 1 mg/mL in sterile DMEM media without serum, aliquot and store at −20 °C.
8. ATP is prepared at 100 mM stock solution in sterile water and pH should be adjusted to 7.4, aliquot the solution and store at −20 °C. Do not refreeze ATP stock solution.

9. For some experiments, ATP is incubated at different concentrations, including 3 and 5 mM to trigger P2X7 receptor activation. However, other concentrations of working concentrations for ATP or other nucleotides (as ADP or UTP) are suitable if different purinoceptors are going to be studied.
10. P2X7-specific antagonists are prepared at 100 mM stock solution, aliquoted and stored at −20 °C. Other specific P2X7 receptor antagonists are also suitable to use. If other purinoceptors are going to be studied, the use of P2X7 receptor selective antagonists should be replaced with specific antagonists for the different purinoceptors.
11. MTT stock solution is prepared at 5 mg/mL in PBS and stored at −20 °C. MTT working solution is stable at 4 °C for few weeks.
12. Yo-Pro-1 stock solution is prepared at 1 mM in DMSO and stored at −20 °C.
13. Triton X-100 is added together 2 μM Yo-Pro-1 in the same solution, to permeabilize the cells and allow Yo-Pro-1 to enter the cells and bind to the DNA.
14. Optimal confluence for the cells has been assessed by measuring the signal of MTT or Yo-Pro-1 from cells plated at different confluences (between 1.7×10^4 and 1.5×10^5 cells per well of a 96-well plate).
15. LPS will increase adherence of the macrophages. If nonimmune cells are used, there is no need to stimulate with LPS.
16. The use of P2X7-specific antagonists will determine if the effect of ATP on cell adhesion is regulated via the P2X7 receptor. The use of different antagonists for other purinoceptors in this step will indicate their participation in this process.
17. It is recommended that P2X7-specific antagonists are also present during ATP stimulation.
18. Both methods presented in this protocol are incompatible in the same plate, so the researcher should choose the method that better adapts to their cells.
19. Medium used during MTT incubation must not contain bovine serum albumin (BSA), or serum, because BSA will precipitate with formazan.
20. Before MTT addition, check the adherent cells with an inverse microscope under phase contrast. For the right MTT solution, 50 μL of the 0.75 mg/mL solution should be added to each well containing 100 μL of DMEM to reach a final volume of 150 μL.
21. Be gentle and make sure the adherent cells are not loosed when discard the supernatant after centrifugation.

22. After DMSO addition and shaking step, alternatively to the instant measurement in a plate reader, DMSO could be discarded and the plate could be stored in a freezer at −20 °C for a month.
23. Perform this step in the Biological Safety Cabinet, Class II without light, as light can quench Yo-Pro-1 fluorescence. From this point onward, protect the plate from direct light until measurement.

Acknowledgments

This work was supported by the European Research Council (ERC-2013-CoG 614578), EU COST Program (BM1406), and Ministerio de Economia, Industria y Competitividad–Fondo Europeo de Desarrollo Regional (project no. SAF2017-88276-R).

References

1. Dowling C, Kiely P (2015) Targeting protein kinase C downstream of growth factor and adhesion signalling. Cancers 7:1271–1291
2. Hynes RO (2002) Integrins: bidirectional, allosteric signaling machines. Cell 110:673–687
3. Taherian A, Li X, Liu Y, Haas TA (2011) Differences in integrin expression and signaling within human breast cancer cells. BMC Cancer 11:293
4. Roger S, Gillet L, Baroja-Mazo A, Surprenant A, Pelegrin P (2010) C-terminal calmodulin-binding motif differentially controls human and rat P2X7 receptor current facilitation. J Biol Chem 285:17514–17524
5. Yan Z, Khadra A, Li S, Tomic M, Sherman A, Stojilkovic SS (2010) Experimental characterization and mathematical modeling of P2X7 receptor channel gating. J Neurosci 30:14213–14224
6. Pelegrín P (2011) Many ways to dilate the P2X7 receptor pore. Br J Pharmacol 163:908–911
7. Roger S, Pelegrin P, Surprenant A (2008) Facilitation of P2X7 receptor currents and membrane blebbing via constitutive and dynamic calmodulin binding. J Neurosci 28:6393–6401
8. Furlan-freguia C, Marchese P, Gruber A, Ruggeri ZM, Ruf W (2011) P2X7 receptor signaling contributes to tissue factor-dependent thrombosis in mice. J Clin Invest 121:2932–2944
9. Sengstake S, Boneberg E-M, Illges H (2006) CD21 and CD62L shedding are both inducible via P2X7Rs. Int Immunol 18:1171–1178
10. Mishra A, Guo Y, Zhang L, More S, Weng T, Chintagari NR et al (2016) A critical role for P2X $_7$ receptor-induced VCAM-1 shedding and neutrophil infiltration during acute lung injury. J Immunol 197:2828–2837
11. Price P, Mcmillan TJ (1990) Use of the tetrazolium assay in measuring the response of human tumor cells to ionizing radiation. Cancer Res 50:1392–1396
12. Idziorek T, Estaquier J, De Bels F, Ameisen JC (1995) YOPRO-1 permits cytofluorometric analysis of programmed cell death (apoptosis) without interfering with cell viability. J Immunol Methods 185:249–258

INDEX

A

Alkaline phosphatase activity 107–115
Amperometric, enzyme-based microelectrode
biosensors 197–206
Annexin A5 staining, *see* Cell death determination

B

Bimolecular fluorescence complementation
(BiFC) 156, 158–161
Bioluminescence resonance energy transfer (BRET)
NanoBRET..................................164–171, 175, 178
Biotin labelling 138, 140

C

Calcium imaging, *see* Intracellular calcium determination
Calcium influx, *see* Intracellular calcium determination
Cell culture
ACN
DNA transfection 187
B16F10D
DNA transfection 185
bone marrow differentiated macrophages
(BMDM)... 352, 354
HEK293T
DNA transfection 125, 126, 150
human peripheral blood mononuclear cells
(PBMCs)... 129, 193,
278, 326, 327, 332, 337, 340–342
monocyte-derived macrophages 276,
278–279, 282
mouse primary blood, spleen/lymph nodes
lymphocytes 119, 122, 126
mouse primary peritoneal macrophages 127,
281, 342
neurons 5, 29, 80,
81, 210, 346
3T3 .... 228, 229
1321N1
DNA transfection 287
THP-1 .... 342, 346, 347
WEHI-3B
DNA transfection 185
Yac-1
DNA transfection 119
Cell death determination
colorimetric (*see* MTT assay)
fluorescence (plasma membrane pore
dilatation)... 315
Cell migration 345, 348, 351, 352
Confocal microscopy 167, 177,
210, 212, 215–217

D

Derivation of macrophages from mouse bone
marrow 342
3-(4, 5- dimethylthiazol-2-yl)-2, 5-diphenyl
tetrazolium bromide (MTT) 352, 353

E

Ecto-domain shedding of cell surface
proteins 121, 127
Ecto-nucleotidase activity 108, 109
Electrophysiology, *see* Patch-clamp recording
Enzyme histochemistry............................. 108, 109, 114
Ethidium staining, *see* Cell death determination

F

Flow cytometry
real-type flow cytometry 125, 126, 324
Fluorescence microscopy 165, 271,
313, 316, 318, 319
Fluorescent antagonist 166, 169, 175, 178
Förster resonance energy transfer (FRET) 155,
156, 164, 183, 223–231, 244

H

Homology modelling 46, 55, 65–74

I

Immunoblots..254
Immunofluorescence 108, 113, 114, 193
In situ hybridization
in *Xenopus* embryos 87–105
Intracellular calcium determination 316

Pablo Pelegrín (ed.), *Purinergic Signaling: Methods and Protocols*, Methods in Molecular Biology, vol. 2041,
https://doi.org/10.1007/978-1-4939-9717-6, © Springer Science+Business Media, LLC, part of Springer Nature 2020

L

Luminescence
  ATP bioluminescence assay 197
  resonance energy transfer (*see* Bioluminescence resonance energy transfer)
  whole animal in vivo luminescence 183

O

Optogenetics .... v, 301

P

Patch-clamp recording
  automated planar patch-clamp recording..... 285–299
  HEK293T cells..262
  macrophages and monocytes.................................282
  Xenopus oocytes 243, 262
Polyacrylamide gel electrophoresis (PAGE)..............139, 141–143, 147
Polymerase chain reaction (PCR) 83, 95, 271
Propidium iodide staining, *see* Cell death determination
Protein biotinylation, *see* Biotin labelling
Protein cross-linking 151
Protein expression.. 117, 142, 210, 251, 255, 262
Protein extraction 245, 252, 256
Protein structure prediction, *see* Homology modelling

R

Real-time live-cell imaging 216
RNA interference (RNAi) 77–83

T

Transwell migration, *see* Cell migration

V

Video-microscopy, *see* Real-time live-cell imaging

W

Western blot, *see* Immunoblot

Y

Yo-Pro-1 staining, *see* Cell death determination

MIX
Papier aus verantwortungsvollen Quellen
Paper from responsible sources
FSC® C105338

If you have any concerns about our products,
you can contact us on
ProductSafety@springernature.com

In case Publisher is established outside the EU,
the EU authorized representative is:
Springer Nature Customer Service Center GmbH
Europaplatz 3, 69115 Heidelberg, Germany

Printed by Libri Plureos GmbH
in Hamburg, Germany